THE BRAINS OF MEN AND MACHINES

ERNEST W. KENT

University of Illinois
Chicago, Illinois

BYTE/McGraw Hill
70 Main St
Peterborough, N.H. 03458

THE BRAINS OF MEN AND MACHINES

Library of Congress Cataloging in Publication Data

Kent, Ernest.
The Brains of Men and Machines.

Bibliography: p.
1. Brain. 2. Artifical Intelligence. 3. Electronic Digital Computers. I. Title.
QP376.K39 1980 001.53'5 80-18115
ISBN 0-07-034123-0

McGraw-Hill Publications Co.
1221 Avenue of the Americas
New York, New York 10020

TO LAURA

TABLE OF CONTENTS

I Basic Principles

I. WHAT THIS BOOK WILL TRY TO DO

This is a book on the organization of the brain, and how its unique structure and functional characteristics may be applied to the development of intelligent robotic systems. Obviously, this approach is not that of the usual neuroscience textbook. I will not assume that you have much of a background in biology, psychology, or other disciplines central to modern brain research, such as neurochemistry. Although I will omit and simplify for the sake of presenting conceptually exciting paradigms and observations, this investigation of the hardware and software of natural intelligence will be thorough enough to provide ideas that you could use in the creation of artificial intelligence. If you are familiar with computer electronics and programming, you will find that the book is essentially self-contained; I will discuss the functions of the brain in as much depth as is necessary to support the particular lines of reasoning that will be developed, and if you are particularly interested in some area for its own sake, a bibliography of selected readings is provided to guide you to the relevant literature.

A great deal has already been written about the similarities and differences between brains and computers. It is a subject which fascinates a great many people, and I have found that my friends who work (or play) in the computer world are no exception. Most of what has been written on this subject is of a general nature and explores topics such as the possibility of artificial consciousness. Although I express my views on such subjects, the bulk of the present work is intended instead as a more detailed description of brain hardware for the person who is seriously interested in exploring new directions in electronic computer and control systems. In particular, it is addressed to those interested in designing machinery to handle tasks which are currently handled only by brains. This area is perhaps best described as being part of the emerging science of robotics.

There are several reasons for writing such a book now. One is that computer hardware advances, particularly in the development of the microprocessor chip, have freed us from some of the constraints which originally dictated computer design, and the brain is a potentially useful source of ideas on what may be done with this new ability. The computer, as traditionally designed, and the brain are both powerful machines, which excel at very different sorts of jobs. Traditional computers are strong at high-speed mathematics and logic; on the other hand, machines are very poor at "common sense" kinds of operations at which man excells. To be convinced of this, it is only necessary to note the difficulties which have plagued the development of machinery aimed at general-purpose robots or to consider the limited results of software attempts at artificial intelligence.

The book is divided into ten chapters which are grouped into several topical sections. In this initial chapter, I will examine the primary building blocks of the brain, and gain an overview of its major structural and functional features. Following this, the general plan of the book will be to proceed from the periphery of the brain inward to the most central levels of its functional apparatus—the goal-directing systems. I begin this process in Chapters 2 through 4 with an examination of those parts of the nervous system that correspond roughly to the input and output subprocessors of a large computer system. These subsystems handle not only the first, rather invariant aspects of input analysis, but also accomplish the final, detailed organization of muscular activity into behavioral output. Chapters 5 and 6 discuss the brain's central machinery in which the higher levels of processing convert the preprocessed input into statements of output under the directions of the goal-defining systems. This general functional overview is concluded in Chapter 7 with an examination of a goal-directing mechanism which has no clear analogy in standard computer systems, but which seems to have much the same relation to brain functions that the programmer's goals have to the function of a computer executing a program. I then examine two interesting details of brain operation, the specialization of the two cerebral hemispheres for different types of processing, in Chapter 8, and the operation of the brain's data storage system, in Chapter 9. Finally, Chapter 10 contains speculations on the relationships between brain, machine, and mind.

I will attempt to show that the factors in current computer designs which render them extremely inefficient for robotics and artificial intelligence applications result from a design approach that is very different from that of the brain. The "new electronics," initiated by large-scale integration (LSI), has the potential of permitting designs which can employ brain-like methods profitably. Cognitive psychology has already had some influence on computer programs that display some aspect of intelligence, and computer science has probably had an even greater impact on theories of cognitive psychology; however, considerations of brain hardware have had virtually no influence on either artificial intelligence programming or computer design. This is in part because of the belief that the function of the program is independent of the machine hardware, as in a universal Turing machine. This is certainly true, but it is also true that some machine designs may be much more efficient at some kinds of problems than are others. When we attempt to write programs which will display "common sense" in a generalized environment, we are stopped at a very trivial level because the time required to process responses rises at an astronomical rate with only minor increases in the complexity of the task. At the heart of this problem is the fact that because of hardware constraints, serial operations are being applied to problems that are best handled the way the brain does them—with massive parallel processing. The development of inexpensive single-chip processors and related LSI circuits may give us the ability to design machines which will be efficient at this sort of operation.

I am not going to suggest that we try to build a hardware replica of the brain, because we have neither the knowledge nor the hardware. However, I would like to suggest that if we are interested in approaches to a science of robotics, it will be very instructive to examine the only example of a thinking machine that is available to us and try to identify principles of operation that can be put to good use with the kinds of hardware that we do have. I am not going to tell you how to build a machine that can do what a brain can do because I don't know how. I do propose to tell you, in terms of computer concepts, how a brain works. From there, you're on your own. (I may offer some speculations from time to time on possible paths of approach, but that is not the main point of the book. In fact, because I am not an engineer, you should probably take such speculations very lightly!)

Although it is generally very difficult to explain the functions of the brain to someone with no background in neuroscience, I have found that this is not the case with computer-oriented people, because brain operation involves, or can be easily translated into, concepts and operations with which they are already familiar. If you can understand digital and analog electronics and computer software, you can understand in principle (if not in detail) how your brain works. You don't have to have a background in physiological psychology, neurophysiology, neuroanatomy, neuropharmacology, and biochemistry. This investigation will not produce an exhaustive treatment of brain theory, because I assume that you will be more interested in potentially useful ideas about brain architecture than in the fine points of neurophysiology. For that reason, I will usually omit discussions of the merits of different theories and simply present those positions which seem to be best supported or most useful at the present time. Some of these positions, particularly in the areas of the higher functions, of necessity will be speculative. At the risk of driving some of my colleagues into livid rages by slighting their favorite positions, or by selecting theoretical stances in some areas beyond what may be firmly supported by the existing data, I will try to present a conceptually unified model of brain operation. Generally I will not directly reference detailed sources of support for individual facts or ideas but leave it to the reader to consult the bibliography. In this way I will be able to attempt a more or less coherent picture of the brain's overall operation without getting bogged down in too many divergent views. A disadvantage is that some of what I present will be simplified or preliminary, and some of it will ultimately be proven quite wrong. This would be a bad approach for a text on the brain, but not necessarily for a source of ideas for machine design. After all, an idea which has occurred to someone schooled in brain studies as a plausible explanation of how some function is accomplished in the organic brain may well be a clue to a reasonable way to approach that function in an artificial system, even if it is later proven to be done otherwise in the brain. I can't give you ideas that don't exist yet, but you may be able to put to use what is available now.

This also explains why I will omit, or only touch on, some topics which are central to brain function. They are relevant to an organically based machine, but not to one based on silicon and copper. (Unfortunately, this forces me to bypass some of my own favorite topics, such as the pharmacology of reward!)

The core of the book will be a consideration of the brain's input, output, logical, and motivational mechanisms. The first three of these are common items in computer function, although the brain handles them differently. The fourth is not, and constitutes a real addition to the potential functions of a computer system. We shall see that the motivational system essentially takes the place of the programmer in a traditional computer. Some of these topics were covered in a series of articles in BYTE magazine entitled "The Brains of Men and Machines," parts 1, 2 and 3, January, February and March 1978. Material from those articles is heavily represented in the early chapters discussing the lower levels of input/output (I/O) processing. The later chapters extend these operations into the higher reaches of brain processing and take up additional topics.

II. BRAINS AND COMPUTERS

Why don't we build computers like brains? Everyone has heard the old adage to the effect that a computer equal to the human brain would require a machine the size of the Empire State Building, with the electrical output of Niagara Falls to power it, but that was always just an excuse. Aside from the fact that whoever coined that notion didn't have any idea of what a brain-like computer would require, we would have built it anyway, if we'd known how. Moreover, when that guess was made, we were making computers out of 12AX7 vacuum tubes, and it's a long way from the 12AX7 to a Z80 very large-scale integration (VLSI) microprocessor. We would certainly build an artificial brain today, if we knew how.

I assure you that the problem has nothing to do with any mysterious properties of the brain. If we have learned anything about the brain, it is that it is a machine, a complex machine to be sure, but nonetheless a machine. This notion frequently upsets people; they are bothered by the suggestion that they might be "only a machine." This is the wrong interpretation. The statement that the brain is a machine does not mean that it is "only" a machine in the sense of the simple machines of limited ability that we have produced. Rather, it is a statement that extends our concept of what a machine can do. It does not denigrate what the brain can do. Of course, it raises the possibility that we *can* design a machine that thinks like a man, with all of the attendant problems of philosophy and theology which that raises. One could take refuge in the notion that brain and mind may somehow be different, but the evidence is that, when the brain is manipulated experimentally, all of our mental processes, our sensations and perceptions, our feelings and emotions, our intellectual processes, memories, and even our states of consciousness are manipulated, too.

Why then are we having so much trouble making our computers behave in brain-like fashions? Both are machines, and both are information-processing machines at that. Moreover, our computers are supposed to be universal Turing machines that theoretically can do anything. However, although our computing machines may theoretically be capable of imitating all others, Turing didn't say they would handle all problems with equal ease. In point of fact, the brain's architecture is quite different from that of a computer, and these differences are very instructive with regard to the task of building a "thinking machine."

The brain and the computer have both developed in an evolutionary manner, with "survival of the fittest" determining what features were retained and what were discarded. Their designs differ because nature and computer engineers have different notions of what constitutes fitness. There are two aspects of this difference: the problems the machine is required to solve and the hardware available to build the machine. The successful brains, whose genes contributed to the next generation, were those that were able to solve problems such as recognizing and avoiding dinosaurs, and recognizing and catching frogs. Ability at higher mathematics was never a very important criterion in determining successful brain design, and our poor brains get quickly strained when they are required to do much of it. Primarily, computer design was judged in terms of its ability to perform mathematical functions. As a result, computers are successful at mathematics, but are failures at catching frogs.

The hardware available to computer engineers and to organic evolution was also different and this has determined in part the differences in the architectures of successful brains and successful computers. Although, as we shall see, the logic gate and the neuron have a great deal in common, some of their differences have turned out to have far-reaching consequences. The brain does not have speed on its side—neurons operate in milliseconds not nanoseconds—but it never lacks for quantity. (You want million-bit bytes and ten thousand-legged gates? Sure, how many trillion?) On the other hand, computer engineers were limited in quantity because of the expense and difficulty of assembly of components, which dictated designs that were hardware-conservative. This was compensated by using speed to substitute for quantity. Thus, our computers have emphasized small bytes and few registers, but achieved high data throughout with iterative reuse of these components at great speeds. Bus-oriented design and other hardware-conservative adaptations arise from these same considerations.

In contrast, parallel, multiprocessor designs with hierarchical organization, which the brain uses with wild abandon, are seen in comparatively primitive form in our current computers and computer networks. The brain has no compunctions about freely mixing digital and analog computing elements and uses each to its best advantage where needed. Thus, the brain and the computer have each found a design best suited to the problems they are required to solve and to the hardware available, albeit large brains can perform some computer-like functions poorly, and large computers can perform some brain-like functions poorly.

Curiously enough, both the brain and the computer seem to have settled on a single basic organization whether the device is large or small. We all know how a computer works in principle, although different machines differ in detail. The central processing unit (CPU), the bus, the clock, the memory, and the input/output (I/O) interface are all arranged according to the same basic plan in large or small machines. Similarly, the brain of the white rat in my laboratory is basically of the same design as the brain reading this page. All the same parts are included in both and connected in the same manner. The differences are in capacity and relative development of the parts. The brains of different species may even be more similar to one another than various makes of computers.

What I am suggesting is that the computer has developed an architecture optimized for logical and mathematical problems, rather than for the display of basic common sense. Thus, as presently configured, our computing machines are terribly inefficient at the kinds of problems brains solve easily, and prodigious feats of programming, vast amounts of memory, and all the speed that can be mustered give us only the most trivial results. On the other hand, brains have also found an optimal architecture for the kinds of processing problems which they face. This suits them very well to certain activities such as those involving concurrent manipulation of large amounts of data, real-time operations, and finding adequate solutions to complex problems as opposed to exact solutions to simpler ones.

III. THE BRAIN'S COMPONENTS

First, it will be useful to understand the basic unit of brain structure. This is called a neuron. It is a living entity, a cell like all the other cells in your body, but it is specialized for information processing. You can think of it as performing the same function as a logical gate in a digital machine or an operational amplifier in an analog machine. In fact, it is a very versatile device and can do either or both jobs. The brain uses neurons, billions and billions of them, to do everything it does. Let's look at the diagram of a neuron in figure 1.1. The output of the neuron appears on the long thin part labeled "axon." Think of the axon as a wire. The brain uses it for transmit-

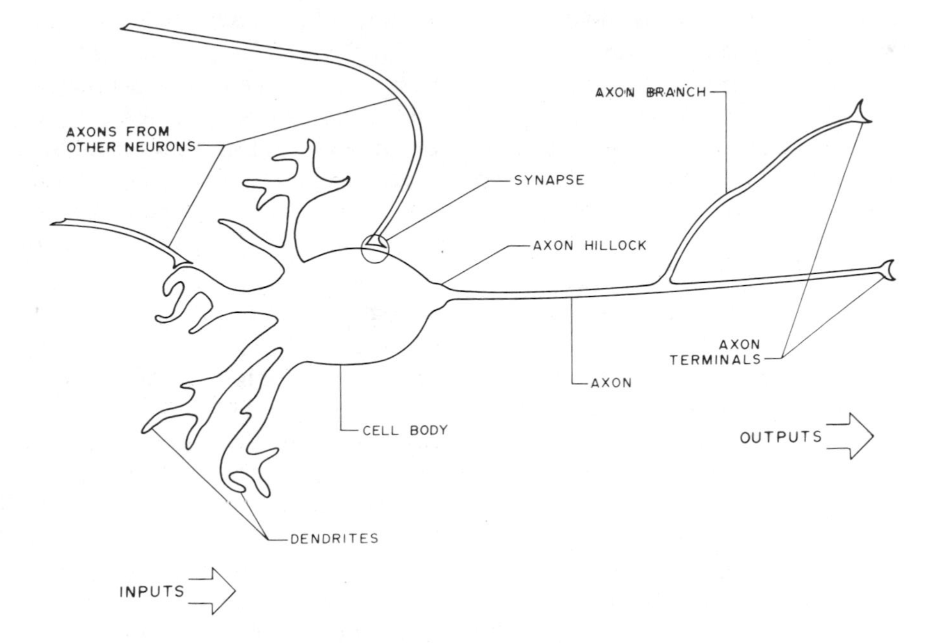

Figure 1.1: *The principal parts of the neuron. The neuron is the basic "gate" of the nervous system. Computations are performed in the cell body and the results are transmitted down the axon and its branches in pulse-coded digital form. The synapses are the inputs to the gate, where the pulse-coded information is reconverted to analog form. The inputs are subjected to weighted summation, and when a threshold is reached, a new output pulse is placed on the axon. In this diagram, information flow is roughly from left to right through the neuron.*

ting information over distance. The difference is that this wire transmits only pulse streams, not direct current levels. It is a digital wire. All the pulses are approximately the same height and duration and can be thought of as binary bits. Only 1s and 0s are allowed. Now look at the part labeled "axon hillock." That's a Schmitt trigger. It puts a pulse on the axon whenever the analog voltage in the part labeled "cell body" crosses a preset threshold value. Whenever this occurs, the voltage in the cell body (ie: the potential difference between the interior and exterior of the cell) is reset to the baseline or initial value. This voltage rises towards the Schmitt trigger's threshold, or falls away from it, by the action of pulses impinging on the cell body from the axons of other neurons. The place where an axon meets a cell body is called a "synapse," and it transmits only in one direction. Think of it as a diode. The effect of activity at a synapse may be positive or negative; that is, a pulse at a given synapse may either add to or subtract from the cell body voltage. This is a function of the synapse which transmits the pulse (just like the small inversion circles on logic gate inputs). Remember that the voltage in the cell body is an analog voltage and represents an algebraic sum of the inputs. Moreover, the inputs may have different weights. Those farthest from the axon hillock have the least effect; those nearest have the greatest effect. Waiting (ie: the more recently arrived pulses have a greater weight in the sum) may be reduced greatly by placing the input far out on an extension of the cell body called a "dendrite." The effect of an input outlasts the pulse that produces it, so that inputs which do not arrive synchronously may still sum with one another (within brief time limits). Think of it as a pulse stretcher at each input, plus a time constant in the cell body. The effect of the finite time constant is to give each input pulse a temporal weighting.

When the summed inputs to the cell body exceed the threshold of the axon hillock, and a pulse is placed on the axon as a result, we say that the neuron has "fired." An equivalent circuit (for our purposes) of the basic neuron is shown in figure 1.2. If you study these figures for a moment, you will see that the neuron has digital inputs, which are converted to analog values, and operated on algebraically in an analog fashion. The result is then converted back to digital form for transmission.

The neuron is a very powerful tool. It can act as an AND gate (coincidence of several equally weighted inputs required to reach firing threshold), an OR gate (any of several inputs can drive the cell past firing threshold), or any of a variety of other functions (NAND and NOR can be achieved by cells with a resting potential above firing threshold, and inhibitory inputs). When you consider that some of the inputs may represent feedback from the neuron's own axon, and that an average neuron may have ten thousand inputs, you begin to appreciate the possibilities. It can integrate and differentiate and do other useful functions by virtue of feedback and analog operation, just like an operational amplifier. (In this case, the analog values are encoded as pulse frequency of the output.)

No one would try to design a very large electronic system out of gates as complex as this unless someone finds a way to make them in quantity on a chip. Making an n-input gate might not be difficult, but gaining access to each input from outside the chip is a problem. They are worth studying, however, because consideration of their operation reveals some important features of brain architecture. First, although neurons can be synchronized by simultaneously driving them with an overriding input, they are normally asynchronous devices. The system can use local "clocks" where they are useful, but it doesn't have a system clock. This permits an interesting development. In synchronous systems, information can be coded only in terms of which lines are active. If you want to indicate the on or off status of some condition, you put a 0 or 1 on the appropriate line. But, if you want to indicate a numerical value greater than 1, you have to do it by coding it as the simultaneous activity, or lack of it, on several lines. For example, we represent numbers by the bit pattern in a byte on the data bus. The brain can, and in several instances does, function in this mode. We refer to it as "place coding." Because the information need not be synchronous, other information may be coded in the temporal relation of the pulses. Examination of the electrical behavior of a neuron reveals that the greater the positive input drive to the cell body, the more quickly the analog voltage

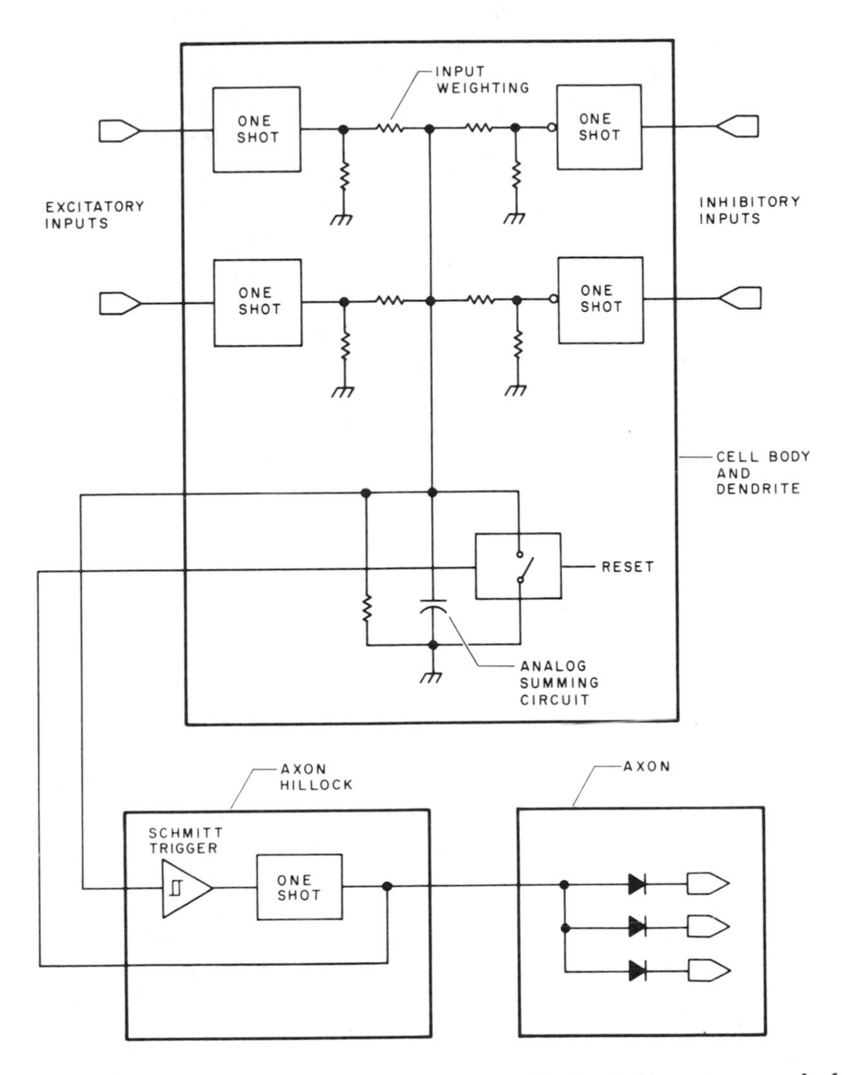

Figure 1.2: *An equivalent circuit for the neuron. This circuit illustrates several of the important features of neural operation with regard to its role as an analog-to-digital device. The various parts of the circuit may be compared to figure 1.1.*

will reach threshold, and the more quickly a new pulse will be placed on the axon after the reset following the preceding pulse. This means that it is very easy to use frequency coding to indicate the magnitude of the summed input activity. Because of the "pulse stretchers" at the inputs to the neuron, and the time constant of the cell body, sequential pulses on the incoming axon can sum with one another to produce a greater analog voltage. This voltage is proportional to the frequency of the incoming pulses and thus transforms frequency-coded input back to the analog mode.

Interaction of inputs from different axons is termed "spatial summation," and interaction of sequential pulses on the same axon is termed "temporal summation." One of the interesting things that the brain can do with this capability is to use both place and frequency coding simultaneously on the same line. Another advantage is the ability to represent very large numerical quantities with what might be called a "temporal byte," or the integration over a brief time period of a single input line. As an example, consider how the brain encodes sensory information from your skin. The type of sensation (ie: heat, cold, pressure, etc) as well as the location of the sensation is place coded. That is, *what* you feel is a function of which axon is active. The magnitude of the sensation, *how much* you feel, is frequency coded on the same line. Thus, the brain structure receiving the information can determine the type and location of the stimulation with a "spatial byte" (place code), which determines the set of active lines, and the intensity of the stimulation with a "temporal byte" of frequency code.

The brain's basic "byte" has both a spatial and a temporal dimension. Two independent sets of information can be encoded in these two dimensions, and they can then interact in the receiving structure (ie: the postsynaptic neuron) in a way determined by physical and chemical properties of the cell's membrane. Notice that the spatial aspect of the byte is essentially digital information, and that the temporal aspect of the byte is essentially analog information, although it is encoded in the frequency of digitial pulses. (See figure 1.3.)

A mathematical treatment of this scheme would be an information theorist's nightmare (although it can be done). However, from a practical standpoint it has some clear advantages. Digital information output from a structure which determines the type of action to be taken and analog information output from another structure which determines intensity of action required may "gate" one another, in a third location, to produce an output stream which simultaneously determines the nature and magnitude of the action to be taken. One could think of this process as specifying the enabling of a set of switches with an appropriately coded digital byte, while presenting a set of analog values to the switched lines. We do see this sort of thing in some I/O applications, but the brain makes use of this, and much more complicated interactions, in its internal processing. It may use the information from one aspect of the byte to determine the nature or extent of the operation to be performed on the information in the other aspect of the byte.

Two additional properties of the neuron need to be mentioned to complete our understanding of the basic gate. The first is a different type of inhibitory input. The "negative synapse" described above causes the voltage across the neuronal membrane to move away from the firing threshold. This action simply antagonizes (with a specified weight) the combined action of *all* the positive inputs. It is also possible to have a "negative synapse" which antagonizes only a specific input: this is called "presynaptic inhibition"; it may be thought of as a disable input to one of the input one-shots.

The final desirable property of the neuron as a computer element is that

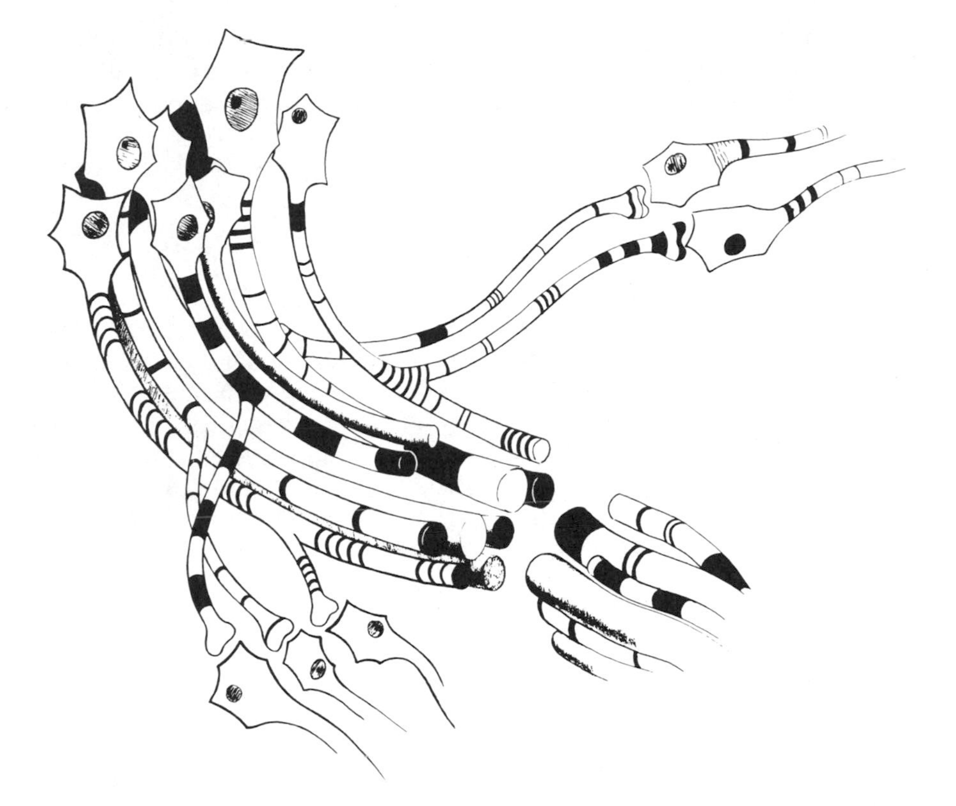

Figure 1.3: *This figure illustrates the "two-dimensional byte" of the neural bus structure. A bundle of axons constituting a "bus" is shown coming from a group of neurons in the upper left. Neural pulses traveling down the axons are shown as dark areas. In a cross section through the "bus," we see that some axons are active and others inactive. These "spatial byte" permutations may be decoded by selecting fibers from the bundle to terminate on particular receiving elements. This is frequently accomplished, as shown here, by taking off collaterals from the desired elements of the bus. In addition, the active axons carry intensity information, which is seen in this figure as the pulse frequencies indicated by the spacing of the dark areas. The origin of the selected axon determines the kind of information it carries, and its relative activity carries information about the intensity associated with that kind of data.*

speed of transmission of pulses down the axon can vary over a wide range (although it is always the same in any given axon). This means that we may use high-speed axons to move data quickly, but low-speed axons may be employed as delay lines. Because axons can have branches coming off at any point, we may have tapped delay lines. We shall see some stunning examples of the utility of this feature.

It should be apparent by now that the basic neuron is an enormously powerful tool. In practice, few situations call for all of the complexity of this device, and it is frequently seen as simply a switch, AND gate, or some other very domestic sort of creature. Indeed, in many situations, neurons take on a variety of specialized shapes and connections which optimize them for one or another function, to the exclusion of others. In all cases, however, their operation may be understood in terms of the basic design I have discussed.

IV. THE BRAIN'S GENERAL PLAN

Now that we have some terms for the basic elements, let us take a leap to the other end of the size spectrum and examine the overall structure of the brain. The exact anatomy is actually of little relevance for our purposes, but it may help to have a visual image of the device as we discuss the features of its parts. Figure 1.4 shows the general appearance of the human brain, together with some of its internal structure. Figure 1.5 shows the general organization of the parts as they would appear if the brain were taken out of the body, unfolded, and flattened out in a neat plane view.

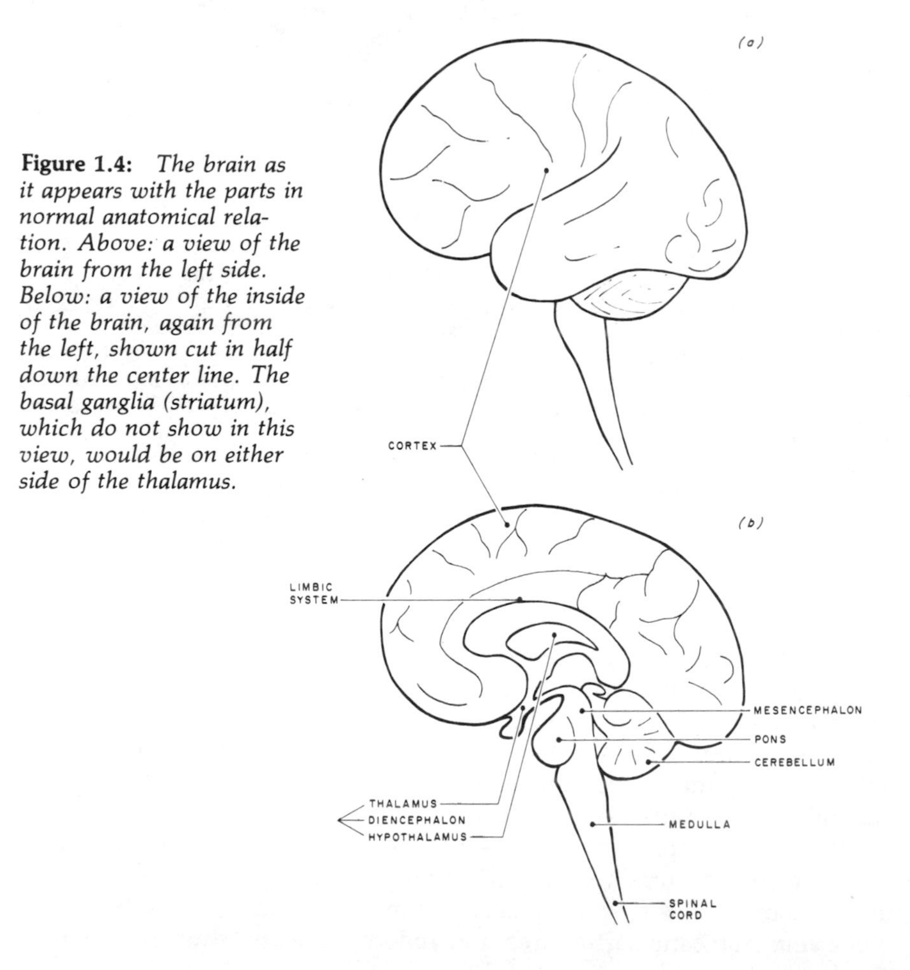

Figure 1.4: *The brain as it appears with the parts in normal anatomical relation. Above: a view of the brain from the left side. Below: a view of the inside of the brain, again from the left, shown cut in half down the center line. The basal ganglia (striatum), which do not show in this view, would be on either side of the thalamus.*

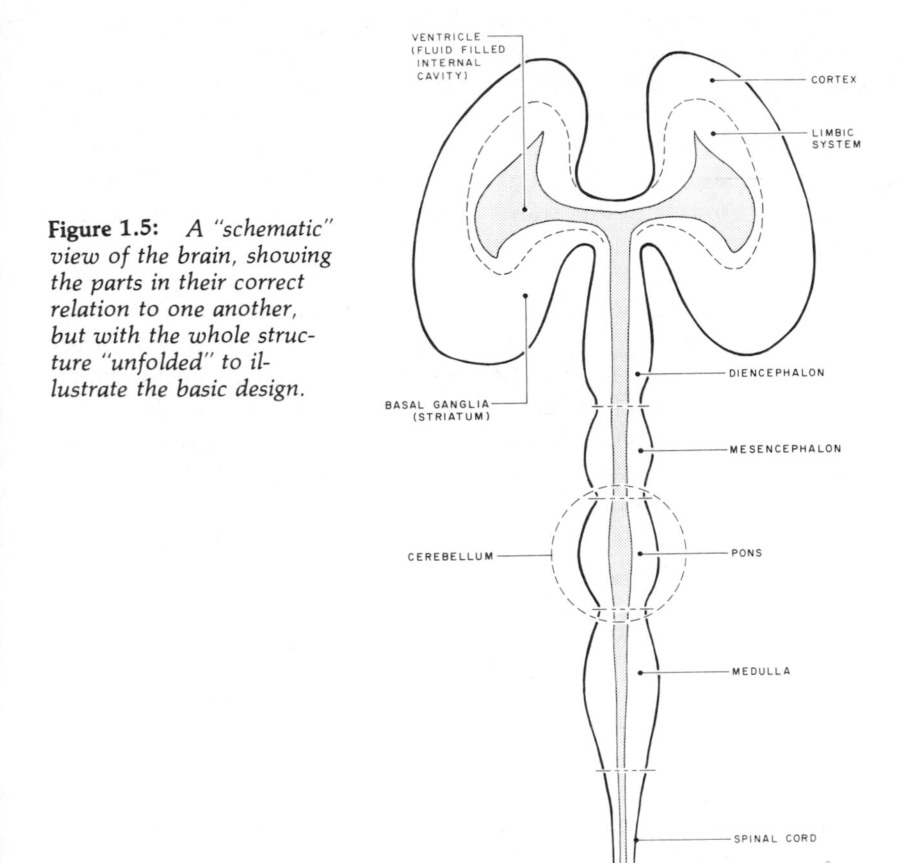

Figure 1.5: *A "schematic" view of the brain, showing the parts in their correct relation to one another, but with the whole structure "unfolded" to illustrate the basic design.*

The functional structures of the brain can be generally divided into two categories, fiber tracts and nuclei. Fiber tracts are simply bundles of axons going from one place to another; they are the cabling and wiring of the brain. The nuclei are groups of cell bodies. Each nucleus may be thought of as analogous to a central processor with a dedicated function and (in most cases) a hard-wired or read-only memory (ROM) program. Most of the nuclei are irregular clusters of cell bodies, but in some, the cell bodies are arranged in orderly layers which form a folded sheet of cells. In this case, the tissue is called a cortex rather than a nucleus, but the idea is the same. (The most familiar of course is the cerebral cortex, of which humans are very proud because it is better developed in man than in any other species.) The cells in the nuclei can be divided into two types: *local neurons*, whose function is in the data processing internal to the nucleus, and whose axons do not leave the nucleus, and *output neurons*, which give rise to the axons that make up the fiber tracts and communicate with other nuclei.

There are many nuclei of all levels of size and sophistication. Unfortunately, their names have arisen historically, rather than systematically and are in Latin and occasionally unpronounceable (eg: *Nucleus Para-*

fascicularis). The names of the fiber tracts are equally difficult (eg: *Tractus Habenulointerpeduncularis)*. The only thing to do with the names of neuroanatomy is to endure them or ignore them. We shall try to ignore them. It will help, however, if you will take the time to learn the names of a few of the major divisions of the brain which are shown in figure 1.5 and to remember their basic relation to one another. The most important items, from bottom to top, are the spinal cord, the medulla, the pons, the cerebellum, the mesencephalon, the diencephalon (and its two major subdivisions, the thalamus and the hypothalamus), the limbic system, the striatum (also called basal ganglia), and the cerebral cortex.

This bottom to top sequence corresponds in a general way to a sequence of increasingly more global levels of control, from most detailed and specific to most general and abstract. It also corresponds roughly to the evolutionary sequence from oldest and most primitive to most recent and advanced. All of the apparatus shown here is present in some form by the time the evolutionary level of the mammals is reached.

The basic architecture of the system is hierarchical. Each of the major functions of the system is partially organized at each level of the system, rather than particular levels being devoted to particular major functions. At the lowest levels, there are a multitude of relatively simple processing elements doing similar jobs, and at the higher levels there are a few very complex and powerful processing elements defining system goals and priorities and organizing the activities of the lower levels to achieve them.

On the input side, the lowest levels gather raw data, which is then progressively abstracted, sorted, and refined at each stage according to general guidelines which may be hard-wired or provided by higher levels. The highest levels then receive abstract symbolic information about the general state of the environment rather than details. ("There is a black cat there" as opposed to "the following points of the visual field are dark.") Similarly, output functions begin at the highest levels, which determine general goals and strategies and transmit these in the form of statements about more limited momentary objectives to lower levels, which in turn send information about desired actions and timing to the lowest levels for execution in detail.

Thus, at each level, are a number of relatively independent processing elements pursuing their own jobs in parallel, while trading information with echelons above and below, and laterally with one another. It follows that it doesn't make sense to ask where in the brain any large-scale function is processed. Different aspects of it will be handled in different portions of functional subsystems which are represented at all major levels of the physical system. It might sound hopeless to try to follow the operation of such a device, but in practice there is order, not chaos. At the lower levels where semi-independent processors are most numerous, there is the least diversity among them. The organization is in many ways like a military command chain, and one doesn't have to study each lieutenant and platoon individually to comprehend the principle. A general plan of this organization is shown in figure 1.6. In the figure the three principal functional

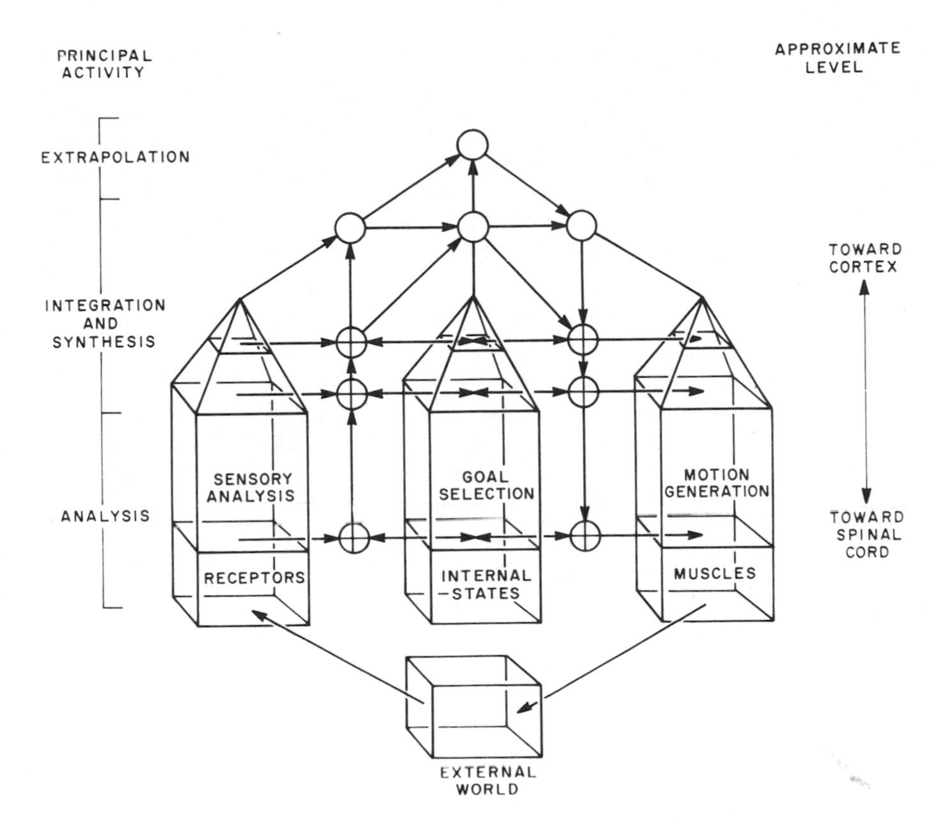

Figure 1.6: *A conceptual representation of the basic organization of the brain's architecture. The organization is both parallel and hierarchical. The three major functional systems (input, goal direction, and output) are shown with their relation to the internal and external worlds. Each of the major systems has representation at high, low, and intermediate levels of processing paths may exist in the system. Square areas indicate data bases in which massive amounts of information are processed simultaneously, in parallel. Circles indicate sites of major interactions between systems.*

systems of the brain — input, output, and goal selection or "motivation" — are shown as vertically organized systems. (The highest levels of these systems and their interactions constitute what we would more commonly call the "logical" aspects of brain function, but we shall see how these develop out of simpler functions.) At the lowest levels, the square cross sections are intended to indicate large assemblages of data which are presented and handled simultaneously, in parallel, by numerous simple processors. At the higher levels, the processors are fewer and more complex and deal with data in smaller but more highly organized formats.

Three features of such an organization are immediately relevant to a machine that must deal with the real-world environment and do so in real time. These are hierarchical decision-making capability, parallel processing of I/O data, and "fail safe" backup function. With regard to the first, the vast majority of the decisions that have to be made are trivial and can be handled at lower levels without taking up time that the higher levels can use to work on more complex problems. We already see the beginnings of this concept in "intelligent terminals" and in the two or three levels of preliminary I/O processing buffering a big machine. This sort of operation need not be limited to big installations. With the appearance of inexpensive microprocessors, it should now be possible to have, for example, a devoted processor for driving each joint of a robot limb by continuously computing required forces, based on limb positions or velocities requested by higher levels, and resistive forces and other local input from below. The brain provides such a processor for each fiber of each muscle, and it will be instructive to examine their approaches to the problems involved.

The second feature of the brain's architecture, parallel I/O processing, provides the speed needed to deal with a complex world in real time, even in the face of slow components. By breaking down the tasks into small parts that can be handled simultaneously by numerous simple processing elements, the time required for the task is not greater than the time required for the execution of one of its components. This is a fairly obvious statement of course, but its real importance is apparent when we examine the differences between the types of problems typically faced by a brain operating on the real world and the types of problems traditional computers are designed to handle.

In the solution of mathematical and logical problems, it frequently does not make sense to attempt simultaneous solutions to parts of the problem. The results of one operation are essential to the beginning of the next. Such problems are inherently serial in nature. The brain, being unable to apply the power of its parallel organization to such problems, becomes terribly slow. On the other hand, the two tasks that require the greatest feats of processing from the brain, analysis of the flood of information impinging on the sense organs and the design and execution of movements in space, lend themselves perfectly to the parallel processing of small subunits. Here the brain can vastly outperform our typical current computers, which have only one, or at most a few, processing units capable of simultaneous operation. For example, all of the data in the visual field is available simultaneously on the surface of the retina. Rather than dealing with it point by point, the brain sucks it all in at once in one enormous byte and sets to work on the analysis of many small parts of the visual field simultaneously. (We shall examine its algorithms in detail later.)

Even with the small inexpensive processors available to us now, we could obviously never afford to match the brain in quantity. However, we don't have to go to the other extreme and try to analyze input data serially, point by point, with a single, very fast processor, as has been typically attempted. The job is just too large for even the fastest machine to do in this way. There are certain advantages in having a basically parallel system in terms of the feature extraction process. On the other hand, we do have a speed advantage and it certainly should be possible to simulate the operation of a number of the brain's processors with only one of ours in the same time frame. (There will be some increase in complexity where the results of neighboring units are interactive.) Just how to optimize this sort of tradeoff is, of course, a matter for some study. A first step which we shall take here will be to examine some of the tricks and shortcuts in the feature extraction process that the brain itself uses to save time.

The third system characteristic which results from the brain's hierarchical organization is high survival value. We will learn nothing about Asimov's first two "laws" of robotics (the protection of human beings) by studying the brain. The third law, ensuring the survival of the robot, has always been a major concern of brain architecture. It is annoying but not usually fatal when the big machine in the computer center develops a fault. When it happens to a brain (or a robot) the entire device may be destroyed. The redundancy inherent in the brain's basic structure is valuable in this regard, but there is more to it than that. Recall two facts about the brain: there is an evolutionary order of development of its structure, and the major functions have representation at all levels. These two facts are related. Whereas our computers have never been expected to incorporate pieces of earlier models, the brain, in its present state, contains most of the components from its earlier forms. The simplest early brains obviously had to be capable, in their own inelegant way, of surviving. During the course of evolution, complex structures capable of more sophisticated handling of the same basic functions became available. Rather than eliminating the older structures and duplicating their functions, the newer structures simply took control of the older ones and used them as subprocessors.

A general principle of organization evolved in which the higher level structures control the lower, not by turning them on when needed, but by inhibiting their actions except as desired. The beauty of this system is that if a higher center is suddenly damaged, the older, more primitive units, which are normally inhibited, are released to function on their own. Thus, damage tends not to eliminate vital functions, but only to downgrade the complexity with which the job can be performed. This is especially true of functions such as defense. The typical result of damage to higher brain centers is a "nasty" animal (ie: one that can fight adequately), but which fails to make fine discriminations about the appropriate stimulus conditions for doing so, and which defaults in the safe direction by attacking any strong stimulus source. Of course, this control paradigm has its limits. This is particularly true of the most highly developed brains where some of this type of organization is sacrificed to give the highest centers direct access to the lowest for feedforward in the control of complex operations.

In the case of damage to lower centers, the multitude of processing elements available allows some of the higher levels to be reprogrammed to take over the functions of lower level systems by simulating their operation. The process takes a time to organize, but it can be quite effective if the organism can survive for a few weeks while reorganization takes place. Of course, the fact that information from intact sensory modalities may be used to compensate for lost information from disabled modalities is amazing in itself. This kind of contingent reorganization, or dynamic configuration, is a key property of integrated neuronal systems and allows for the tremendous adaptive capacity exhibited by most higher organisms.

While it is apparent that it is not possible to give a definite answer to the question of *where* a general function of any complexity is performed in the brain, it may be useful (for orientation to the device) to identify some of the anatomical divisions of the brain shown in figure 1.4 and 1.5 with some of the functions which have important representation at those levels.

The lowest level of the central nervous system, the *spinal cord*, is the major route of input and output for the brain. With the exception of a few special cases, most of the sensory input from the body and most of the output to the muscles passes through this structure. Although it contains immense fiber tracts, the cord should not be considered merely a cable. Numerous nuclei within the spinal cord perform many important functions as intelligent terminals on both the input and output sides. Moreover, some simple actions are processed entirely at the level of the spinal cord from input to output. Everyone has seen an example of such a "spinal reflex" in the knee-jerk produced by their doctor's rubber hammer.

The *medulla* and *pons* are also involved in "intelligent terminal" types of I/O activities, but at the so-called *supra-segmental* level of control. That is, these structures are frequently concerned with directing the activities of the local I/O processors in the cord to coordinate actions which involve the entire body rather than body subunits (such as a single limb). For example, the pattern of motor activity involved in walking requires the coordination of the whole body to maintain balance as the center of gravity shifts. Such things as the *decision* to walk and the choice of direction are the province of higher centers. The medulla and pons also have important relations to a number of the special senses, such as hearing and balance, which are not represented throughout the body. Some of this sensory information is utilized immediately as input to supra-segmental reflexes, and some is processed for output to higher centers. A complex, physiological organism requires a great deal of regulation of its internal environment; the temperature must be exact, the heart rate must be regulated, inhalation must be controlled. These "housekeeping" routines also have representation at this level of the brain.

The *mesencephalon* is in many ways similar to the pons and medulla in its functions. In general, there is a transition from higher to lower degrees of abstraction in these I/O systems as one progresses from the mesencephalon down to the medulla. In addition, the mesencephalon is one end of a system originating in the limbic system of the forebrain, which is important in regulating the type and intensity of the high-level processing performed by the more advanced structures of the forebrain.

There are two systems of the brainstem (composed of the medulla, pons, and mesencephalon) which should be mentioned in this context. The first, a system known as the *reticular formation*, has important functions in the brain which are analogous to the vectored interrupt system of the computer. It continually monitors the input of all of the sensory systems, more for quantity of activity than for detailed analysis, and controls the degree of activation of various portions of the higher centers on this basis. Thus, it can immediately arouse the brain when a novel or intense stimulus is encountered and jump the whole system to a state of attention and analyze the important event. It also seems to exercise similar functions on the output side of operations.

A second "system" of the brainstem may be called the "*amine*" system. It consists of a set of interacting nuclei which use various neurotransmitter compounds of the chemical class known as "amines" in their operation. This system has a great deal to do with the mode of operation of the rest of the brain. Like the reticular formation, with which its activities are integrated, it sends its axons into all parts of the brain to make synaptic contacts with large groups of neurons in the forebrain and the brainstem and cord. Because the axons of these few neurons make millions of synaptic connections with vast numbers of neurons, we might expect that their function was a regulatory bias rather than the transmission of very specific information, and this appears to be the case. These nuclei are involved in such functions as regulation of walking, sleeping and dreaming states, and apparently other "altered states of consciousness." Some hallucinatory drugs such as LSD are thought to exert some of their effects here. Disorders of these amine systems appear to underlie such abnormal operational modes as schizophrenia. Some portions of the amine system also regulate the selection and intensity of the various detailed patterns of activity which are generated by higher structures. Thus, in Parkinson's disease, which involves a disorder of part of this system, the ability to execute voluntary activities is impaired, although the conception of them is not.

The systems of the brainstem thus exert a great influence over the general types of activity in which the higher levels of the device engage. A very interesting development here is the closing of this loop by return projections from the highest level of the brain, the cortex, which allows the machine to gain control over its own status. This loop is fundamental to the phenomenon of consciousness.

Lying above the pons is the structure known as the *cerebellum*. This device is a subprocessor for some types of motor output. It is basically involved in the parallel-to-serial conversion of output that is not going to be continuously modified by feedback control. It can accept a parallel byte which defines an action to be undertaken, modify it to incorporate the current status of many variables of limb position and loading, and convert the instruction to a series of sequenced operations with specified durations. Damage to the cerebellum has no effect on conscious processes but seriously impairs the performance of muscular actions.

Above the mesencephalon, we encounter the *diencephalon*. It's two major subdivisions, *thalamus* and *hypothalamus*, are quite different and are actually groups of many smaller nuclei, each of which has different functions. The hypothalamic nuclei are heavily involved in the sort of "housekeeping" functions mentioned earlier. For example, they control the secretions of the glandular system and are involved in a wide range of functions such as temperature regulation. In its role as chief executive of internal operations, the hypothalamus must continually monitor internal conditions, which in turn are frequently the ultimate sources of the brain's functional orientation.

Thus, if the hypothalamus detects a low level of sugar in the blood, it initiates a state which we experience as hunger. This state represents a reorganization of the brain's systems for the control of goal-directed behavior, which causes the organism to engage in food-seeking behavior. The hypothalamus contains part of the brain's priority interrupt system. On the output side, the hypothalamus is important in changing the state of the internal body functions to correspond to the current "interrupt routine." For example, if the limbic system initiates a "danger encountered" state, which we experience as fear, the hypothalamus must see to it that the body is mobilized for action with regard to blood flow, adrenaline levels, etc. The hypothalamus is also an important link in the limbic system-mesencephalon organization that, in general, specifies goals based on drives and emotional states for use in the selection of overt behavioral activity.

The thalamus, the other major component of the diencephalon, functions as a very high-level I/O processor which prepares information for and, in part, organizes the activity of the cerebral cortex. Arrival of sensory input in the thalamus is sufficient for some rudimentary conscious experience of sensation, at least in some sensory modalities, but any very detailed resolution of the experience requires the enormous digital processing power of the cerebral cortex. In many respects, the various cortical areas act as subprocessors performing detail work for the nuclei of the thalamus which route the information to and from the cortical regions. Some thalamic-cortical systems are concerned with simple feature extraction of the input data. Others deal with extrapolation of current events, and yet others with transmission of these extractions and extrapolations to other parts of the brain which, for example, evaluate them for relevance to current drive states or return sets of similar data from memory.

In this activity, the *limbic system* functions in the analysis of data for information relevant to the organism's needs. Thus, when we are hungry and see food, the limbic system, in conjunction with the other structures mentioned earlier, initiates a state which enables the sensory signal of food to initiate appropriate behavior patterns, which are generated in detail by the cortex-thalamus-striatum apparatus. We experience the operation of this state of the limbic system as pleasure. When the limbic system recognizes unfavorable situations, other states are generated (experienced as fear or anger), which cause other sorts of detailed actions to be generated.

The basal ganglia (*striatum*), like the cerebellum, is involved in the organization of motor output. It generates patterns of movement on the basis of input from analytic cortical areas, which are regulated in intensity by inputs from the amine systems under the direction of the limbic system. It outputs these patterns both to the movement-controlling areas of the cortex and more directly to lower motor mechanisms. Unlike those generated in the cerebellum, the movements generated in the striatum are under continuous control of the cortex, and because the cortex is continually receiving and processing sensory information from the environment, a closed loop system is formed. This system is importantly involved in "voluntary" and learned movements.

I have already mentioned most of the functions of the *cortex* because it is of necessity involved in almost all of the higher functions of the other brain structures. Its operation is essentially a vast decoding and encoding network that gives analytic and synthetic power to the operation of the other systems. Without it, their operation would be reduced in capability, because of the loss of capacity for fine distinctions and discriminations on the one hand and large-scale generalizations and syntheses on the other. The cortex first appears, in other than rudimentary form, with the evolutionary appearance of the mammals. Their behavioral diversity and plasticity as compared with the stereotyped, reflexive, instinctive behavior of the reptiles is probably associated with this structure.

This brings me to the final general point I want to make about the brain before we consider the detailed operation of some of its functions. One of the most important features of the brain is that it learns. Our computers learn in a limited sense during programming, and some programs can learn to improve their performance on the basis of experience. This latter type of learning is characteristic of the way the brain operates, and it applies the process to almost everything it does.

You will notice that when I described some of the general functions of the different regions of the brain, I made no mention of memory. This is because we have no idea where it is. Indeed, the evidence suggests that the brain's memory is incorporated into its structure at whatever point the stored information is to act. Its memory then may be thought of as being distributed throughout its structure. At present, we can offer only speculation as to the physical nature and detailed processes of the brain's memory storage. Fortunately, we know a great deal about the operation of the brain's memory as a "black box" so that we can understand how it enters into the brain's algorithms, and we do not really need to understand its detailed physical nature to effectively use its principles of operation. Our current semiconductor memory chips are inferior to the brain's memory in terms of capacity, but they are superior in speed and accuracy. Some programmable memory and read-only memory chips associated with each of a robot's processing elements would do nicely, especially if supplemented by a peripheral device for slow, mass storage.

There are at least two types of learning that the brain permits and both are pragmatic. It has found that things that have occurred sequentially several times are likely to do so again, so it learns to associate them and act in anticipation. Thus, if the reflexive response of the nervous system to a painful stimulus applied to the foot is to quickly flex the leg, and if such a painful stimulus is repeatedly preceded by a "neutral" stimulus such as a bell sound, the brain will quickly learn to flex the leg whenever the bell sounds. This is the so-called Pavlovian Conditioned Reflex. The potential utility of this scheme is obvious. Assuming the natural reaction to the painful stimulus is of use to the organism, then performing the reaction in response to the antecedent neutral stimulus allows the organism to get a jump on the world and perform more efficiently. This type of action, although not the capacity to learn it, is employed in some computer memory systems which anticipate the next memory.

Two limitations of this type of learning should be noted. The first is that the only thing that can be learned is the early performance of the natural response to the second stimulus. Thus, the organism's behavioral repertoire is not expanded, just made more efficient. The other is that all that is necessary for this type of learning is temporal contiguity of the events; it does not matter whether or not the anticipation is successful in improving the results. If you tape an electrode to the foot so that a shock following a bell occurs regardless of whether the leg flexes or not, the flexion of the leg still becomes conditioned to the bell.

The second type of learning in which the brain engages is called "operant conditioning." This is the type of learning that permits us to expand our behavioral repertoire and base such expansions on the quality of the results. Simply stated, this type of learning is based on the principle that behaviors immediately preceding a reward are increased in future probability of occurrence. "Reward" here refers either to some pleasing event occurring or to some unpleasant state being terminated. Thus, behaviors that lead to good results will tend to recur. If we now add to this a second principle, that the behaviors immediately preceding the reward are the most strongly affected, it follows that more efficient behavioral routes to the reward are more strongly affected than less efficient ones. In this fashion, whole new behavioral patterns are built up out of successful components of more or less random exploratory behavior, and these quickly become welded together into tight and effective behavioral sequences.

The more developed brains rely very heavily on learning to produce most of their behavioral patterns. Less developed ones rely most heavily on prewired inflexible behaviors. In respect to evolution, primitive brains, such as those of fish, amphibians and reptiles, while capable of limited learning, generally rely on wired-in behavior patterns that are available at birth. The advantage is early ability of the immature organism to fend for itself. The disadvantage is inflexible behavior that cannot easily adapt to an environment which differs from that in which the species evolved. On the other hand, advanced mammals, particularly man, are characterized by heavy reliance on learned behavior, which results in a protracted state of infantile helplessness followed by enormous behavioral flexibility and adaptiveness.

The parallel with our current attempts to construct robots is obvious. The major hurdle is designing a system that can operate in a generalized environment instead of being restricted to a specialized one for which it has been pre-programmed. The answer is also obvious. A successful robot must be capable of operant conditioning, including the ability to be rewarded for successful attempts and to feel punished otherwise. This device is carried to such an extreme in advanced brains that even our basic ability to see, our perceptual structure, is learned in infancy. A newborn child has only the most rudimentary ability to interpret its visual environment the relation between movement of the limbs and the result in visual space, the relation between certain output commands and the result on auditory input. In advanced brains, all this and much more is painfully learned by trial and error. The result is the ability to modify behavior towards desired ends rather than to simply react to stimuli in a preprogrammed fashion. The apparently random play of infants is in fact a serious matter. Its emulation in our machines will be of the utmost importance to their handling of the generalized environment.

In addition to the kind of learning described above, which is based on repeated trials and which slowly alters behavior, the brain also has the ability to store "memories" of facts and events. This is the function we most commonly mean when we refer to "memory," although obviously some sense of memory is also involved in learning. Even memory in the usual sense, however, has several forms. There is short-term memory, which is the kind you use to recall what somebody said a moment ago, but which you won't recall next year. Long-term memory, which is more nearly related to learning as described above, is what you use to recall the multiplication table or the names of your relatives. "Memory" is, in fact, not one function, but many.

From this overview of the brain and its functional organization, it is apparent that we must now select a limited set of brain functions to discuss in detail. I think that those of most practical interest to robot designers at the present time are:

- the brain's mechanisms of output control, including coordination and timing, and the use of feedback in the design and execution of movements in space
- the brain's mechanisms of sensory perception, including its principles of feature extraction, and the tricks and shortcuts that it employs in pattern recognition
- the brain's mechanisms for achieving goal-directed behavior, including mechanisms of emotion and motivation and their control of behavior
- the brain's mechanisms of intelligence and learning

I will attempt to cover each of these subjects in some detail in the following chapters.

2 The Output Controllers of the Brain

I. OVERVIEW

It seems likely that any robotics system will require some kind of output controller which organizes the generation and execution of movements in space and time. The required control systems may be expected to range from very simple to very complex. The brain has solved this problem with control organization and effector organs that are as complex as any we will encounter for a long time to come. Observation of terrestrial animals suggests that the jointed-limb scheme which they employ as their chief means of locomotion and manipulation requires a very complex control system. On the other hand, freed from such restrictions as maintenance of uninterrupted blood flow, limited mechanical strength, and narrow temperature range, a robot has other output options such as wheels, treads, or air cushions. Although these later devices may allow simpler control systems, I would suggest that the jointed-limb system may be superior for operation in a generalized environment. Try to imagine a wheeled or treaded robot scaling a cliff, climbing a tree, or even using a stool to dust bookshelves. Because the complex motion-control system which handles the jointed-limb scheme could also handle simpler systems, it may be most appropriate to use this higher degree of organization so that future robot designs may operate in as wide a range of environments as do animals.

With regard to the actual physiological mechanisms which are controlled by the brain, it is interesting to note there are only two types: muscles and glands. These are the only effector organs to which the brain is connected. Thus you may only contract a muscle and release glandular secretions. The whole range of physical behavior is only some combination of these two. In the present discussion we will concern ourselves exclusively with the muscles and those neuronal structures, referred to as the "motor system," which control them.

There are two fundamental control techniques apparently employed by the brain's motor system. The first is to equip each level of decision-making with subprocessors which accept the commands from higher levels; taking into account inputs containing local feedback and environmental information, these subsystems organize appropriate patterns of activity to execute the high-level commands. A whole series of such steps is employed: the output of each stage defines the increasingly discrete objectives for the subsequent stages. In this fashion, a descending "pyramid" of processors is defined which can accept very general directives and execute them, in a reflex fashion, in the face of varying loads, stresses and other perturbations. By itself this system organization is appropriate to controlling multimodal processes, such as maintaining balance while one walks on uneven terrain. However, this approach by itself is not sufficient for achieving the higher level goals of the organism.

The second principle applies to the operation of higher level systems which *do* generate output strategies in relation to behavioral goals. This involves the categorization of output tasks according to their use of input information rather than to the type of behavior which they generate. We shall examine some specific examples which illustrate these ideas.

II. FEEDBACK INPUTS

The operation of the motor command chain depends upon certain sensory inputs which provide feedback and status information for moment-to-moment operations. To illustrate the influence of these inputs on motor output, we may perform the following experiment:

Close your eyes and put one hand somewhere out in front of you, then touch it with your other hand. Most people have no difficulty doing this quite accurately. With your eyes closed, how could you guide your hands to the right spatial locations? The answer is that special sensory systems allow this kind of motor control even though we are not aware of their internal operation. The kinesthetic and vestibular senses inform the brain's output control processors of things such as the relative positions of the limbs, the muscle tensions, and the accelerations of the body in different directions. These senses do not enter our awareness with the impact of the information provided by the other senses (ie: sight, sound, smell, etc). Nonetheless, they are among the most extensive and intricate sensory systems of the brain. When they are damaged the results are immediately apparent. For example, with damage to the systems which report limb position, people are unable to carry out even the small experiment which you just performed. In fact, they are generally unable to execute any muscular action correctly without constantly watching what they are doing.

The kinesthetic sense, or kinesthesis, is the sensory modality which reports on the status of the limbs with respect to joint angle, muscle loading, and muscle extension or stretch. (These three types of input information are used at many levels of the motor output system to control sequencing and to provide feedback information.) The kinesthetic organs which translate these quantities into neural impulses may be understood in terms of mechanical analogues such as strain gauges or pressure transducers. (This is another instance where *place coding* specifies the particular unit and type of quantity in question, and frequency coding carries the intensity information.)

The vestibular organs also have a major influence upon the brain's output control functions. These inner ear mechanisms are responsible for the "sense of balance" and also provide a continuous indication of the head's position and acceleration with respect to gravity and space. All transduction of vestibular data occurs in a localized region in each inner ear rather than being distributed through the body as is the case with the kinesthetic sense; the position of the head with regard to all other parts of the body can be computed by combining the information derived from the kinesthetic and vestibular senses. Thus, the combined output of the kinesthetic and vestibular modalities is made available as input to much of the high- and low-level motor processing in the nervous system.

III. THE FINAL OUTPUT PROCESSORS

In most cases, muscles work in opposing pairs: one muscle opens or extends a joint and the other closes or flexes it. This configuration is necessitated by the fact that muscles exert force in one direction only (ie: contraction). Figure 2.1 demonstrates this arrangement for a typical joint. This diagram also shows some of the neural elements which control the contraction of these muscles. The principal neuron of this system, which provides input to most muscle fibers, is called a lower motor neuron and is labeled L in figure 2.1. This type of neuron and the other neurons associated with it are located in the spinal cord, where they function as the final processing stage before output to the muscle. We shall refer to the lower motor neuron and its associated elements as an "LMN system."

This system is a good place to observe some of the principles of the brain's motor organization. There are a great many LMN systems in the spinal cord. Every muscle is composed of thousands to millions of fibers and in the case of muscles used for precise operations, there may be an LMN system for each individual fiber. In other cases, a single LMN system may control many fibers of a muscle. Basically, an LMN system must accept and reconcile commands from a multitude of other systems which desire control of the muscle in question. It must attend to these commands according to their priority, modify them on the basis of inputs from both the kinesthetic and vestibular systems and on the basis of status information from related LMN systems, provide an appropriate output to the muscle, and make its own status information available to other systems. In a

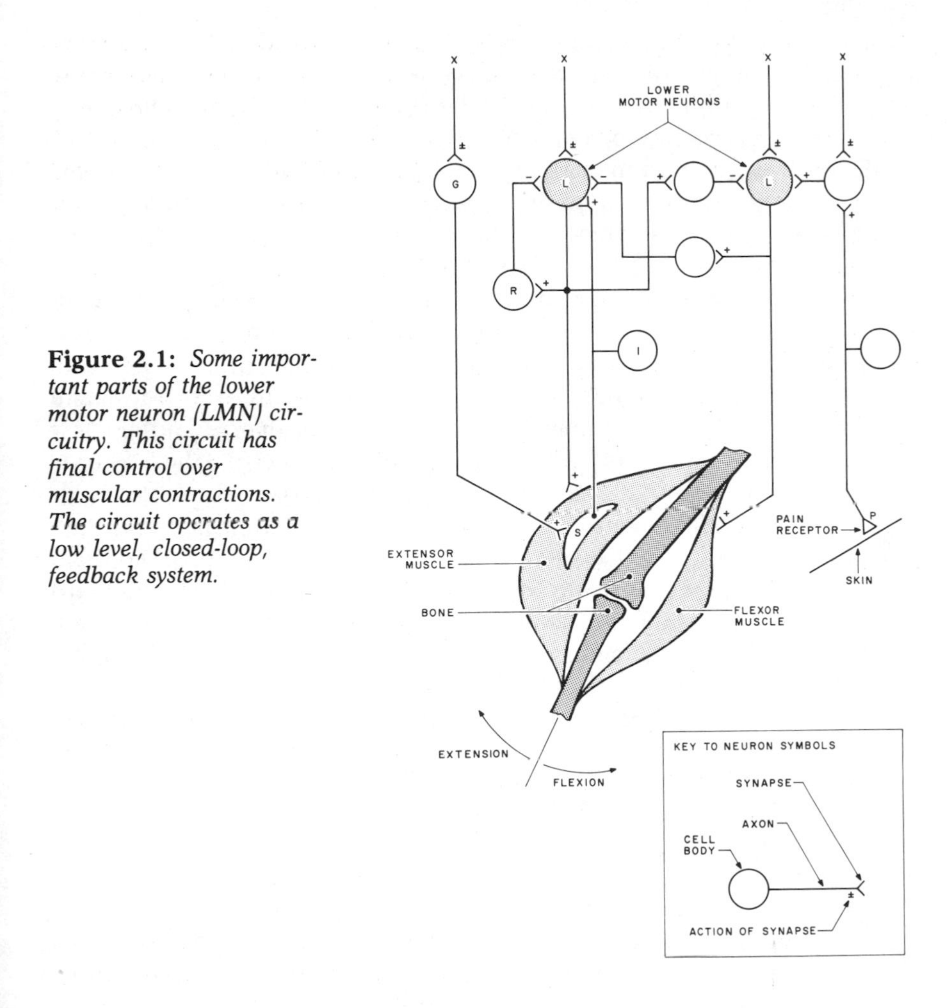

Figure 2.1: *Some important parts of the lower motor neuron (LMN) circuitry. This circuit has final control over muscular contractions. The circuit operates as a low level, closed-loop, feedback system.*

practical robotics application, I see no reason why a single servo actuator and "LMN" processor for each joint would not suffice. There are reasons why a single processor for many joints is less practical, but before addressing this issue, let us examine the LMN system to see what it accomplishes.

In figure 2.1, for clarity, I show only a single LMN driving each muscle. The degree of contraction of the muscle is proportional to the output pulse frequency of the LMN; the higher the frequency, the stronger the contraction. The circuit shown on the right illustrates the simplest type of protective spinal reflex: a pain receptor (nocioceptor) in the skin (P) fires a neuron in the LMN system, which in turn fires the LMN driving the flexor muscle. This simple high-priority operation quickly removes the limb from danger. Inhibitory cross connections between the LMNs driving the two muscles insure that they do not act antagonistically; one muscle relaxes as the other contracts. This reciprocal, synergistic circuitry is generally active in all LMN operations, unless specifically overridden. There are additional outputs, not shown, which ascend the spinal cord to inform higher centers of reflex action in order to allow the motor system to take corrective action via other muscles and limbs which must take up the redistribution of weight, and counteract shifts in the center of gravity, etc.

Higher level inputs to the LMN system may request a variety of actions, such as holding a particular position, moving to a specified position, or moving with a particular velocity. These requests are integrated into the LMN system as follows: the LMN attached to the extensor muscle on the left (figure 2.1) is shown with some of the associated neurons which are involved in the process of carrying out high-level instructions while the muscle tension compensates for an external load. Note that there is a special muscle fiber S (for "spindle") which receives its inputs from the small motor neuron G, (for "Gamma motor neuron") rather than from the LMN which drives adjacent fibers in the extensor muscle. Because the S fiber is mechanically attached to the rest of the muscle, it is passively stretched or relaxed as a result of the neuronal inputs or external forces which extend or contract the main muscle. However, the S fiber's bias, or ambient contractility, is set by its own "private line" input signals from neuron G. This special muscle fiber is the mechanical basis of the kinesthetic transduction organs which allow the monitoring of muscle stretch. The neural component of these stretch receptors (I) is attached to the S fiber, and when this is stretched, the neuron fires at a rate proportional to the degree of stretch. The axon of the I neuron makes an excitatory synapse on the LMN, thus increasing the LMN input drive when the S fiber is stretched. The increased output by the LMN tends to contract the main muscle, and thus relieve the stress of fiber S; the overall system control therefore is via a negative feedback loop.

Suppose that the higher centers in the nervous system wish the LMN system to maintain a joint at a particular angle. This command is specified by a set of constant inputs from above (X) to the LMN and to neuron G. Now suppose that a stress, such as an increased load in the hand, is suddenly applied to the joint. This will tend to flex the joint further, causing the extensor muscle to be stretched beyond the specified degree of contraction. This in turn stretches the S fiber and increases the output of neuron I which in turn increases the output of the LMN. The resulting increase in contractile force of the muscle compensates for the increased load. This kind of local feedback control allows the higher system to remain ignorant of the loading conditions and fluctuations in contraction required to maintain a certain joint extension.

On the other hand, a new input to the "bias" neuron G can cause the S fiber to contract independently of the main extensor muscle, and thus increase the output of the I fiber for a given main muscle length. This bias input defines a new "set point" for the system. It would be difficult for higher motor systems to determine joint extension by monitoring neuron G and fiber I; fortunately, the independent kinesthetic receptor system provides direct indication of joint position.

It is clear that the normal considerations of feedback in control theory are applicable to this situation, and it does not matter for purposes of abstract description if the system is neural or electronic. For example, in this system the mechanical response time of the muscle and joint, which are in the feedback loop, may be slow compared to the response time of the neural elements. As in any other feedback-controlled system, this means that instability and oscillation may result if the feedback system gain does not roll off at higher frequencies. This roll-off is accomplished by the small neuron R (for "Renshaw cell") which produces an inhibitory action on the LMN immediately after each LMN output pulse.

At low input pulse rates from higher systems, the inhibitory effect of the R cell has decreased substantially by the time the next excitatory impulse arrives; the summation of excitatory and inhibitory influences on the LMN cell body (see Chapter 1) exceeds the firing threshold, and the cell will be triggered by its excitatory inputs. However, at higher input frequencies the excitatory input pulse will encounter increasingly greater antagonism from the recurrent inhibitory input produced via R cells (which are activated by the immediately preceding output pulse); a given excitatory input will thus be less effective in bringing the axon hillock above threshold. The inhibitory effect of the R cells progressively reduces the gain of the system with higher input frequencies.

If one looks at the LMN system in the context of the whole hierarchical motor output system, it is apparent that the brain is using a "temporal byte" of frequency-coded, analog information derived from the stretch receptors to specify the degree or quantity of an action. In addition, the set of all of the descending input lines to the numerous LMN systems constitutes a "spatial byte," or place code, which is essentially digital in character, and in which the selected lines (bits) select the set of LMN systems which are addressed and thereby determine the nature of the movement to be performed, but not its speed, force, etc.

IV. MODELING THE LMN IN ROBOTICS SYSTEMS

At first glance, it would seem reasonable to model the behavior of the LMN system with an analog device such as an op-amp controlled by a feedback loop. In practice, this analog model might be quite complex, because the LMN system must integrate about 10,000 synaptic inputs from sources which have different priorities. There is also the problem of how to encode this analog synaptic information from other systems. These difficulties suggest that because our electronic model will contain only a few LMN-type "units" to worry about, it may be more practical to use digital techniques to locate and transfer information. A simple digital processor could emulate the LMN unit, instead of an op-amp; and in the long run this may be the easiest manner of dealing in a controlled way with the interactions of the various inputs to the system.

The next question which arises is why not use one high-speed processor to run all the joints? There are several reasons why this approach might not work. First is reliability: if one LMN system is lost, the others can take compensatory measures automatically. Second, because the output of each LMN system is a factor in the output of each of the others, and because each LMN system is a part of several otherwise distinct feedback loops, a single central-processor system would have to be quite complex. Essentially, it would compute the solution of a number of simultaneous differential equations, or else it would have to deal with each component motion in sequence. Such sequential operations would produce a slow, jerky "movie robot," because each action request must be completed to obtain its results as input data from computing the corrections. A processor with sufficient speed, sophistication and memory to handle the differential equations may be more complex, costly and certainly slower than the multiple simple parallel-processor approach alluded to earlier.

In this latter scheme which the brain has apparently found to be the best approach, programming of the multiple parallel "LMN" processors would be a very simple test-operate-test-exit sequence; the outputs of other simultaneously active units are entered as data each time around the loop. The moment we break out of this sequence to handle several "simultaneous" operations with a serial set of such sequences, things become more complex. However, at processor speeds it should certainly be possible to do this without requiring more than adding some scratch pad programmable memory to the simplest system's read-only memory. The best compromise for a robot remains to be demonstrated.

Finally, a hierarchical system architecture, with interactive parallel units on the lower levels, frees the upper levels of the system to coordinate the actions of the lower parts into the complex behavior of the entire organism or device. By itself this function may require substantial processing power and time.

V. THE HIGHER PROCESSORS

The organization of the motor-control hierarchy is quite conventional and similar to a military command chain. The higher level processing elements, which have the responsibility for coordinating the movements of different limbs, control output commands to the LMN units at the local level, rather than to the muscles directly, and free the LMN units to handle the details. These higher level elements in turn receive orders from, and report to, processing units which are concerned with the coordination of whole-body actions such as the maintenance of posture and balance. The major departure from a "command chain" model is the existence of elaborate lateral information transfer between processing elements at the same level in the hierarchy. The operational principles at each level are quite similar to those we have examined in detail in the LMN units which form the lowest rank in the system. This hierarchical reflex system is capable of receiving and executing commands to perform such high-level reflex actions as running, without further attention. This is because the organization of LMN units and their "supervisors" is a reflex machine quite capable of elaborate control and generation of motion. This reflex system initiates motion only in response to high-level commands, or as a predetermined (reflex) response to specified sensory inputs. It is essentially a very complex automaton.

There are several specialized systems which may issue commands to this "motor automaton." These systems may access the LMN systems directly or may enter the automaton at any level. To understand the division of labor among these systems, we need to focus on the way in which the execution of motion is related to the controls which direct it. There are basically two systems both of which have been used in robot systems. The first is the "dead-reckoning" approach, in which the details of the required action are computed in advance and then executed without regard to their results. (An interesting example of this in a robot system is described in "Newt: A Mobile Cognitive Robot" by Ralph Hollis, June 1977 BYTE page 30.) The other approach is to continually monitor the results of the movement and apply corrections as required. Both of these systems have their uses, advantages, and weaknesses; the brain employs both systems synergistically to effect a given action, although "pure" examples of each approach can be found.

VI. THE CEREBELLUM

One of the two systems mentioned above is associated with the part of the brain called the cerebellum. The cerebellum is not an instigator of action, nor is any conscious experience associated with its activities. However it plays an important role in the expression of both reflexive and voluntary actions which are initiated elsewhere. Among the functions which the cerebellum performs are the translation of parallel to serial output and the control of feedforward correction in open loop control circuits.

Before describing these functions further, it will help to examine the anatomy of the cerebellum. This structure, which lies above the pons, consists of an overlying cortex and a set of nuclei. The neurons of the cerebellar cortex are arranged in a distinctive regular pattern which is endlessly repeated over the surface of the structure. A few elements from this pattern are shown in figure 2.2. Simplified to the bare essentials, this scheme consists of an input neuron "G" (for "Granule cell") which has an axon that runs for some distance parallel to the axons of all of the other input elements, and which in the course of its passage makes contact with a row of output elements "P" (for "Purkinje cell"). Firing an input element thus "selects" a particular set of outputs. Because neuronal impulses may travel

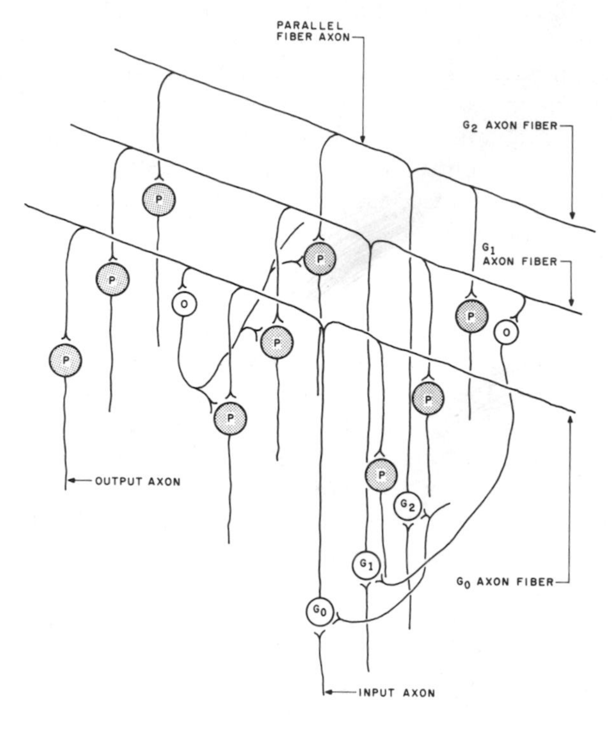

Figure 2.2: *Principle elements of the cerebellar cortex. The output cells (P, for Purkinje cell) are fired in sequence by pulses traveling down the parallel fibers, which are axons of the input cells (G, for granule cells). Each input axon selects a set of output cells for activation and time delays in the parallel fibers help establish sequences of outputs. Other cells of the cerebellar cortex (0) establish interactions between elements of these "main-line" processes, and may act to sharpen tuning of the system and to mediate feedforward interactions between various systems.*

rather slowly in the small diameter axons of the input elements, the arrival of the pulse may sequentially select successive output elements. Thus the cerebellar cortex may act as a tapped delay line, as well as a decoder. If the final output elements are switched to other input elements, elemental sequences may be serially cascaded to form larger patterns.

There are a number of auxiliary elements associated with the G and P neurons, and these are classified as O (for "Other") elements in our diagram. They are capable of performing functions such as selectively inhibiting individual output elements and controlling interactions between adjacent parallel row systems. Thus, these elements may impose modifications on output sequences or call on adjacent systems (which control similar muscle functions) for assistance. Some of these functions have actually been simulated on large digital machines in experimental motion control systems. A schematic of a circuit modeling the essential features is shown in figure 2.3.

The outputs of the cerebellar cortex fall on the neurons of the cerebellar nuclei, which relay them throughout the brain. Likewise, inputs to the cerebellum originate in many portions of the system. In fact there is evidence that different motor system functions may time-share the device! However, a major function of the cerebellum is to allow for interaction between different command systems.

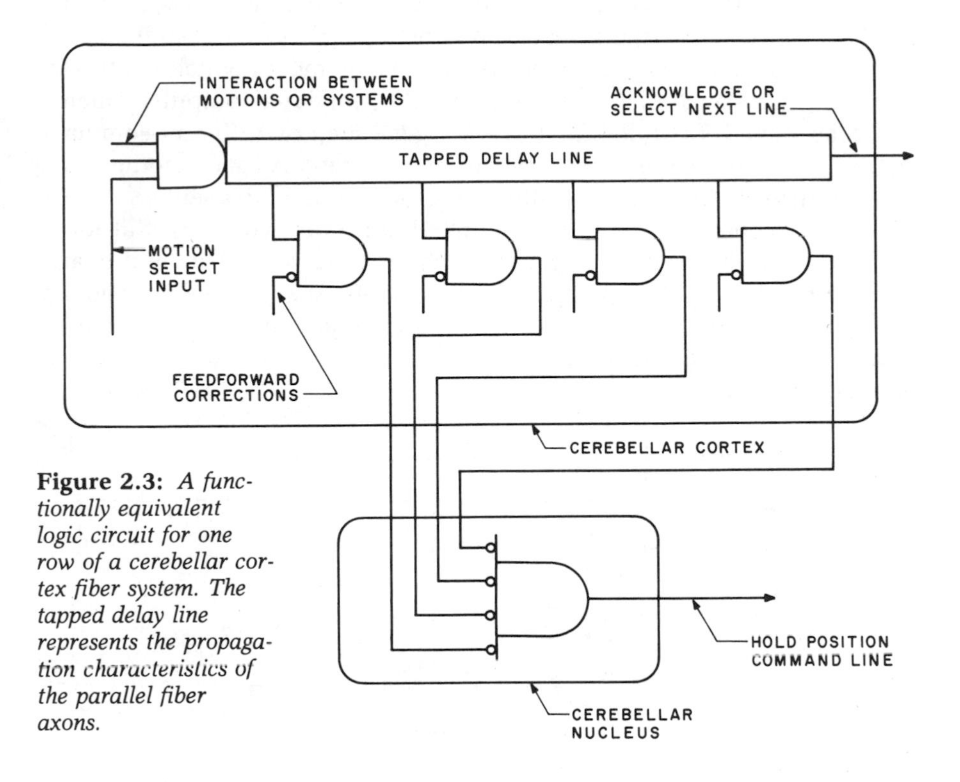

Figure 2.3: *A functionally equivalent logic circuit for one row of a cerebellar cortex fiber system. The tapped delay line represents the propagation characteristics of the parallel fiber axons.*

To illustrate this point, let us see how it is applied to feedforward modification of output. In any system which is not amenable to feedback control, such as one involving actions that are more rapid than the loop time that would be required to control them, or ones that would require very extensive processing of feedback data, it is nonetheless possible to achieve considerable correction for moment-to-moment conditions. This is accomplished by passing the basic output command to both the next level of the output system and to a controller. This latter unit computes deviations from the basic command on the basis of other outputs in progress, or other state data, and forwards these corrections to the lower echelons of the output system. The concept is diagrammed in figure 2.4. Thus, the basic pattern of a reflex motor loop, which performs a function such as walking, may need to be modified by information about head tilt from the vestibular system. At the same time, the reflex vestibular motor systems which can keep the head level may require information about what the stepping generator is about to do, so that it can allow for impending body tilt. If we wish to move swiftly and still avoid a fall, the whole sequence of correcting action needs to take place *before* any muscle action could occur which would generate *feedback* information.

This process is popularly called "coordination," and its quality is dependent on the excellence of the cerebellum. What happens in this process is as follows: sequences of motor actions generated at any level of the hierarchical reflex "automaton" system, or at any high-level system which inputs to it, are sent to the cerebellum either as inputs to the parallel-fiber-decoding systems or as inputs to the "other" elements which control interactions across parallel systems and gate individual output elements. Thus, the waves of parallel-fiber activity generated by different command systems can interact in the cerebellum and modify one another in a predetermined fashion. The resulting modified command is sent forward as a set of corrections to the basic command, and the two interact at lower echelons to produce a corrected action. (Yes, they can get there at the same time. Remember, we have control of transmission speed.) One clear advantage to this approach is the provision of a common site of interaction for systems which are functionally related, but do not possess physical elements in common.

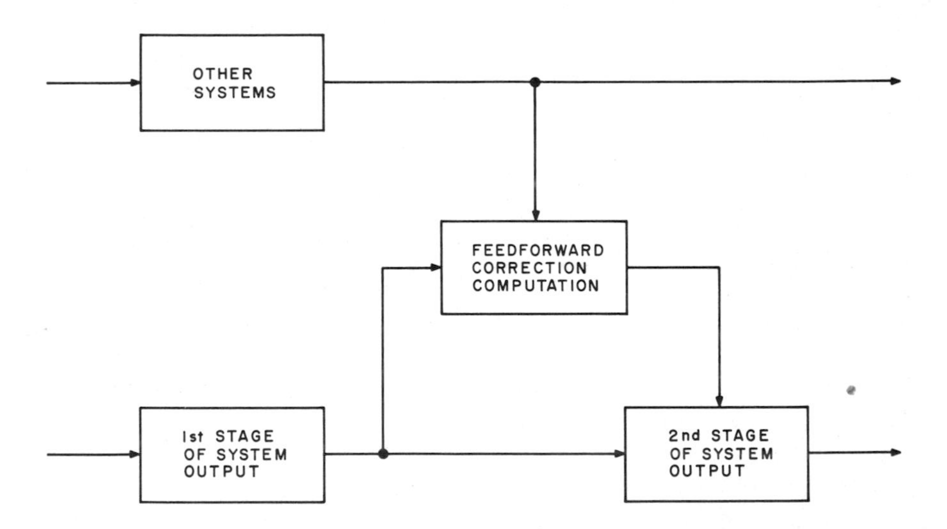

Figure 2.4: *Feedforward as a means of pre-correction in systems where feedback is too slow or unavailable. This is an open loop correction based on the state of other systems or inputs.*

Now that we see how it takes care of interactions and corrections, how does it manage "dead-reckoning" of movement parameters? This process relies on a parallel-to-serial conversion which uses time as an analog of position. A basic function of the cerebellar nuclei is to hold or maintain positions by appropriate outputs to biasing elements in the LMN system.

However, the output elements of the cerebellar cortex, act to inhibit the cerebellar nuclei. Thus, while damage to the cerebellar nuclei results in tremor, oscillation, and similar signs of excess activity, damage to the cerebellar cortex results in deficits related to underactivity; motions may fall short of the target or never materialize at all.

A good example of the pure "dead-reckoning" type of motion is the constant velocity motion of the eyes to a new point of focus, known as a "saccadic" eye movement. (More accurately, the eye muscles are driven by a constant input, although the eye itself can't accelerate and decelerate instantaneously.) It follows that the extent of the motion is determined solely by the duration of the driving signal. Because the motion-generating "automaton" circuits are held in check by activity of the cerebellar nuclei, any action of the cerebellar cortex which inhibits the cerebellar nuclei *disinhibits* the motion generators and the movement begins. If a group of cerebellar cortical output elements is fired in sequence by the same parallel input fiber, the output many inhibit a group of neurons in a cerebellar nucleus; they will keep that group inhibited, and the motion associated with these neurons will progress for as long as the sequential firing of output elements in cerebellar cortex is maintained. The duration of cortical activity will determine the extent of any motion. The cerebellar nuclear cells acts as an OR gate for the timed output sequence of the cerebellar cortex.

It follows, then, that if some high-level command system computes the type, direction and extent of a required motion, it can pass this information to the cerebellum in parallel form as a select request for a particular set of input elements, and perhaps for a set of gating and switching elements as well. This request will set in motion a time sequence of activity in the cerebellum, which will be appropriately modified by interaction with other current activity in the cerebellum and will produce a motion in space with a particular duration of action and spatial extent. Meanwhile, the requesting device is free to go about its business.

There are many movements which rely heavily upon cerebellar regulation of motor activity, and many of them are learned motor skills. For example, playing the piano is clearly a learned sequence of movements, but once learned, the action is too rapid for guidance by feedback from ear or eye. It has been suggested that the learning of such motor sequences may proceed through the formation of new functional connections in the cerebellum, so that the end elements of one sequence become select inputs for the next sequence.

VII. MODELING THE CEREBELLUM

Computation in the cerebellum is primarily an analog process, although these processes have been frequently modeled with a fast processor and an array of digital words to represent the states of output elements. An electronic phenomenon which functions in a manner more analogous to that of the cerebellum is the surface acoustic wave (SAW). SAW devices transform electrical signals into surface waves on a piezoelectric medium, manipulate them in unique ways related to the waves' travel time, and regenerate electrical signals at outputs. A similar result can be achieved with charge-transfer devices. Tapped delay lines are easily made from these devices, and many such lines in parallel have been used for such tasks as electronic focusing of imaging systems. Combining these, this technology would seem to offer a splendid opportunity for developing a "cerebellar chip." An excellent review of these devices can be found in Brodersen and White's article in the March 18, 1977 issue of *Science*.

VIII. THE MOTOR CORTEX

Turning to the other areas of the brain involved in motor control, we find two which are particularly important: the basal ganglia and the motor cortex. These structures operate in an interactive fashion in a supersystem which not only involves some of the thalamic nuclei but also has important connections to systems whose principal functions are best described as cognitive or emotional rather than motor. At this higher level of motor organization, the distinction between concept, desire, and action begins to blur, and these "motor" systems may also be involved in certain motor-oriented aspects of other functions. Thus it is somewhat misleading, but probably necessary, to discuss separate functions for the higher motor systems. The fact that they are parts of a functional supersystem should be borne in mind when we examine the process of generating rational behavior in later chapters.

The somato-motor portion of the cortex was once thought to be the highest level of motor integration in the brain because of the evolutionarily recent development of the cortex. However, it now appears that the motor cortex is more properly regarded as a specialized parallel processor system which has been developed to refine and increase the resolution and processing speed of functions directed from older structures. A notable feature of the somato-motor cortex is a massive projection of large rapid axons which run all the way down the spinal cord and synapse directly on the LMNs. Along the way, these axons give off many branches to higher level motor centers of the medulla, pons, cerebellum, etc. It appears that this direct communication from higher to lower levels of the system allows high-level command systems to "reach around" the motor automaton hierarchy for direct intervention at the LMN level. It is obvious that this type of control must be available for a system to have a flexible behavioral repertoire. This is particularly true if the system is to have the capability of constructing novel behavioral patterns, either to meet a particular problem, or to serve as a basis for learning new behavioral repertoire items.

Although systems such as the cerebellum and the basal ganglia have direct output to the hierarchical motor system to control sterotyped motor activity, they also connect with the somato-motor cortex and apparently provide most of its direction and control. Thus, the somato-motor cortex may be viewed as an extensive decoder for cerebellum and basal ganglia-initiated actions. There is one situation, however, in which the somato-motor cortex itself originates motor function: the control of action based on feedback information from the sense of touch. The reason we refer to it as the "somato-motor" cortex is that this region not only contains the neurons which give rise to the axons controlling the lower motor systems, but also includes the neurons which receive the input from the touch receptors in the skin.

Information derived from these receptors is the basis of the sense of touch, or "somatothesis." The special relation of the sense of touch to the motor system is related to the fact that a great deal of fine motor control is controlled by feedback from the various pressure and sensation transducers which comprise the sense of touch. This relationship is especially significant in man, who places strong emphasis upon the control of precise manipulative movements. Although many movements under feedback control may initially use visual guidance in approaching a goal, final positioning is usually under the control of feedback from touch receptors. When you pick up an object with your hand, feedback is primarily tactile, rather than visual. (Have you ever used a keyboard that didn't provide tactile feedback?) The somato-sensory function of the somato-motor cortex involves elaborate input encoding schemes which are similar to those which we will consider later (in other cortical sensory systems). For now, suffice it to say that this encoded input information may act directly on the motor output of this cortical region and initiate motor activity in those cases where touch information is the appropriate controlling input. In other cases, this information may be used to correct somato-motor region activities which are being initiated and controlled from other structures.

The somato-motor cortex receives its principal control inputs from a group of nuclei in the thalamus, which in turn receive the major share of their input from the cerebellum and the basal ganglia. These thalamic nuclei serve as preprocessors which synthesize directives for the sensorimotor cortex out of requests from several systems.

IX. THE BASAL GANGLIA

Just as the cerebellum is concerned with the operation of feedforward and dead-reckoning kinds of control, the basal ganglia are involved in graded, feedback-controlled movements, particularly those of a learned nature, or those under direct conscious control. It can probably be regarded as the highest level in the command system which has a primarily motor-output function, as opposed to the overall synthesis of action. We will include in the basal ganglia system those mesencephalic and diencephalic structures which function in close conjunction with the basal ganglia nuclei.

At the neuronal level the structure of the basal ganglia is entirely different from that of the cerebellum. There is no obvious pattern of spatial arrangement to its neurons, although both local and output elements can be identified. The local elements are much more numerous than the output elements and form an extensively branched system within the basal ganglia. It appears that most of these local elements have an inhibitory action, so that neighboring elements are quickly turned off by any activity. Some of these connections are recurrent, so that even input-driven elements tend not to remain active beyond an initial response. This is in sharp contrast to the situation in the cerebellum where the entire principle of operation is based on a propagated response in a neuronal network which was initiated by a single input. The response of the basal ganglia to input is called *self-quenching*. That is, an input will initiate a burst of activity, but unless the input is maintained, or augmented by another input, it will rapidly inhibit itself. This is true not only because of the local recurrent inhibitory neurons of the basal ganglia, but also because of global negative feedback loops from the basal ganglia to its inputs which tend to dampen their initial activity.

Notice the similarity of the active basal ganglia to that of a differentiator. If one could consider the spatially-coded byte of active input elements to the basal ganglia as encoding some static scheme of output for motor behavior, the temporal output byte of the basal ganglia might be thought of as having properties similar to the first time derivative of the specified behavior. This output would then be decoded into commands to the motor cortex, cerebellum, and reflex motor system. By outputting commands which decay over time, behavior will not continue unless (1) the command is sustained by some other means, or (2) a new command set is tried which will produce a new set of self-quenching output pulses. This feature is essential if the continuation of a behavior is to be made contingent on its immediate consequences. In fact, the basal ganglia have sets of inputs which are precisely configured to handle this contingency.

The principal outputs of the basal ganglia run through the thalamic nuclei to the motor cortex, through the motor nuclei of the mesencephalon and thence to the subsystems of the reflex motor apparatus, and to the motor nuclei of the pons and from there to the cerebellum. The basal ganglia are in a position to transmit information and commands to all aspects of the motor system. Most of the functions of this system may be understood in terms of principles we have already established.

On the other hand, the inputs to the basal ganglia are the key to understanding its function. There are three major components of the input. First, many regions of the cortex project fibers to these nuclei. These fibers, and those of a second component which arises from the thalamic nuclei (which organize the I/O activity of the cortex), tend to make contact with a few specific neural elements in the basal ganglia. These two input groups may be thought of as specifying discrete patterns of activity to be encoded by basal ganglia neurons; this selection process is like a series of cascaded AND elements, which gate a pattern of activity (or potential activity) to the output lines of the basal ganglia. If they occurred continuously, these outputs could be decoded by lower motor structures into specific movements. It appears, however, that by themselves these cortical and thalamic inputs are insufficient to sustain much activity in light of the strong local inhibition generated by their own action.

The third input component to the basal ganglia arises from a group of nuclei in the mesencephalon which are related to other brain structures (such as the limbic system) which detect the rewarding or punishing quality of stimuli patterns currently being decoded by the sensory systems. This mesencephalic input component has a very different distribution from that of the cortical and thalamic inputs. Within the basal ganglia, each of these axons makes synapses with tens of thousands of neurons; as a result, they cannot specify patterns of activation. This spatially diffuse input is primarily coded temporally rather than spatially. On the other hand, it can exert a widespread gating action on all on-going basal ganglia activity. Thus, this mesencephalic input, which presumably contains information about the intensity of the organism's current emotional state, is capable of sustaining or inhibiting the next phase of the behavior being generated in the basal ganglia. Given the self-quenching nature of activity in the basal ganglia, it is easy to envision a process by which a behavior "suggested" by the cortical and thalamic inputs is continued only if the initial input results produce a sustaining input from the limbic-mesencephalic structures. Strengthening the initial activation pattern might be accomplished by "summing" the third input component with it in order to overcome the self-inhibition. This mesencephalic contribution derived from a "results evaluator" is, of course, ideally situated for this function.

In its most primitive form, this scheme creates a sort of "homing device" which will cause an organism to follow an increasingly intense stimulus, such as odor from food, to its source. That is, as the searching and locomotor patterns generated by the animal result in an increasingly or decreasingly pleasurable stimulus, they are facilitated or eliminated accordingly. Out of this simple feedback guidance mechanism, a host of more elaborate behaviors is developed by evolution and learning. The immense processing power of the cortex aids this process by providing a detailed analysis of the environment and by generating more complex patterns of behavior for possible implementation.

At the present time, we cannot precisely specify the pattern of detailed connections in the basal ganglia which results in these actions. The nature of its operation is inferred from indirect evidence derived by stimulating its inputs or disabling its outputs. These data seem to establish the fact that normal operation of the basal ganglia is essential both to orientation and approach to stimuli, and to the initiation of voluntary behavior and complex learned behaviors such as those which involve anticipatory actions. The convergence on the basal ganglia of processed sensory information from many areas of the cortex provides a source of feedback information which can influence and modify basic plans of action generated by other cortical areas. Damage to the basal ganglia causes a loss of the ability to modify complex actions and judgements on the basis of sensory feedback. (This sort of feedback modification is distinct from the *nonspecific* sustaining action of "emotional feedback" from the mesencephalon.) Finally, as predicted by the model outlined above, damage to the diffusely-connected input component arising from the reward system results in failure to initiate behavior or to orient towards and approach stimuli. Stimulation of this component results in continuation of the immediately preceding behavior or initiation of behaviors appropiate to the sensory stimuli at hand.

There is also a growing body of evidence which indicates that the type of learning called "operant conditioning" may depend on, or even in part occur in, the basal ganglia. This type of learning is established by rewarding or punishing particular behaviors so as to increase the probability of their occurring in the future. To achieve this, all that would be required, in addition to the basal ganglia model which we have described here, is a mechanism by which activity in the diffuse input from the reward system could lower the firing threshold of neurons which were active at the time of this input. No such mechanism is presently known, although its existence is suspected; in our robots it would be easy to include this technique.

X. MODELING THE BASAL GANGLIA

An electronic analog of the basal ganglia model presented here is shown in figure 2.5. (This model does not include the learning function just described.) The essential features are a set of gates to encode the simultaneous inputs from the many cortical regions which contribute to the design of the behavior, a circuit which shuts off the encoded output after a brief delay, and an enabling bus representing the input from the reward system which inhibits the inhibition of active gates. This model is only illustrative, and better ones could be designed to mimic basal ganglia functions. For example, the intensity of activity in the enabling bus should be employed to modulate the intensity of the output.

Considering the very large number of gates required, and that the operation of the system is slow because it requires direction from physical results of actions, in practice it will probably be best to simulate much of the gating and modulation in software on a fast processor. A few relevant principles are worth noting here. The ratio of input to output lines in the basal ganglia is very high. It receives input fibers from many areas of the cerebral cortex (which is by far the largest structure in the brain of man). On the other hand, output neurons comprise less than five percent of the neural complement of the basal ganglia. Clearly a great deal of encoding takes place here: output line permutations are selected by gating an enormous number of inputs.

Consistent with this, the outputs undergo an equally enormous decoding, fan out into the entire downstream motor system, and ultimately

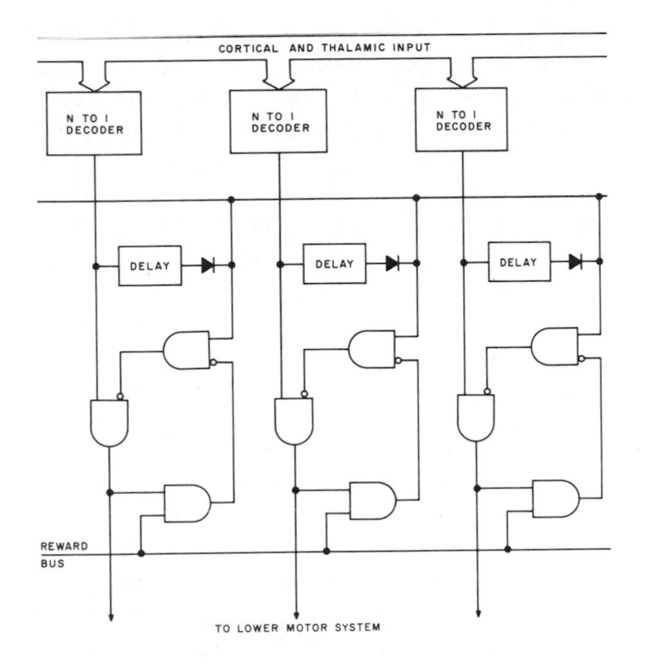

Figure 2.5: *An equivalent circuit approximating some of the functions of the basal ganglia (striatum). Circuits active just prior to the establishment of rewarding inputs will tend to be perpetuated while competing circuits will be blocked out.*

specify the actions of millions of LMN units. The basal ganglia outputs represent a "narrow spot" in the system, through which most of the organism's complex goal-directed behavior must pass. Similarly, the reward system, which gates or modulates input to this information flow, represents the ultimate distillation of the entire sensory world of the organism in relation to reward. The amount of processing going on at higher levels to generate behavior patterns, and the amount required to evaluate their effectiveness is awe-inspiring. Yet, the closing of this complex feedback loop is carried out relatively easily due to interaction at the "narrow points" of the two systems; a simple decision is made to keep going or quit. The need for specific feedback to the behavior generating elements is thereby eliminated. These elements simply try something else which they derive from established hierarchies or generate from similarities with past situations. (We will review this process when we discuss rational thought.)

If we are to provide robots with the ability to modify large-scale behavioral strategies on the basis of evaluation of their effects, or if we wish to provide an operant conditioning capability, it will be necessary to modify massive amounts of information. The most hardware-conservative approach may well be to emulate the basal ganglia system by allowing a simple statement of the evaluative system's reaction to perform a "more/less" modulation of the encoded behavior at a "narrow spot." The behavior generators will try again according to trial-and-error algorithms, rather than by attempting to correct themselves directly. Specific feedback information of a nonevaluative sort, such as position corrections based upon visual observation of a limb, becomes part of the command pattern prior to modulation by the evaluative system. This is effected by simply including these data in the input stream which generates the next output patterns. Given enough processor speed, these inputs could be handled by a software gating system; the effect of the evaluative function could be digitally coded and applied by software rather than by mimicking the brain's analog system.

XI. SYSTEM INTEGRATION

Having looked at the detailed operation of some of the important components of the brain's motor output system, let us conclude by looking at a schematic summary which emphasizes the interactions of the different parts of the system. Figure 2.6 shows the main routes of information flow in the system together with the major controlling inputs. Some of the "black boxes," such as "reward system," will be discussed in later chapters.

One of the outstanding features of the system is that it functions in an organized and integrated way, despite the fact that its parts are in many ways autonomous, and certainly not rigidly synchronized in their operation. A key to this capability is the provision of status information to each unit of the system by each of its neighbors and the ability of each to employ this information in an intelligent way in formulating its own output. A further refinement in this approach is supported by the cerebellum where diverse status information can interact to generate correction information which returns into the main line of the overall process. Wide scale availability of information from special movement-relevant sensory input systems is another unifying feature.

If we leave out the "behavior generating system," which is properly a decision-making system, rather than an executive system, we can discern four major functional portions of the motor system (some structures support more than one function). The first is a system which, when given a high-level command, handles most of the routine traffic and provides automatic elaboration according to established rules. The second is a system which converts parallel statements of action patterns into serially executed instructions to the first system. The third system provides a highly intelligent output terminal which can access the final output elements directly in the service of any of the higher systems on request. The essential feature here is that it is a parallel control for refined special purpose-control, and it is not necessary for most routine actions. Finally, a fourth system provides interaction of the high-level decision-making systems with elaborately processed feedback information; this process generates com-

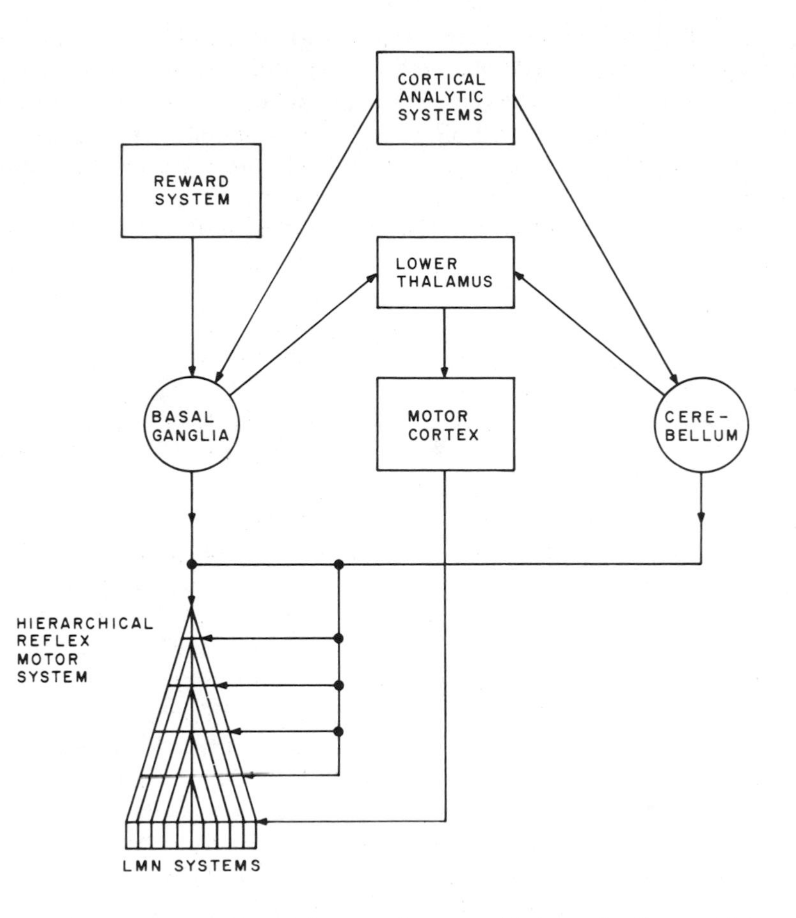

Figure 2.6: *A representation of the overall pattern of information flow and interaction in the several subsystems comprising the brain's motor output system. Inputs at multiple levels modify a "top down" hierarchy of control.*

plex instuctions to the other systems, while continuously screening them for effectiveness.

In this constellation of functions, we find the capability to deal with rapid emergency movements, automatic compensation for imposed deviations, finely graded control under the direction of sensory input, and the execution of arbitrary novel patterns. The organizing principle which seems to best define the system is emphasis upon successively more abstract command functions at higher levels in the system, and a corresponding increase in "situation-free" statements. That is, a high-level element can issue a "walk" command without being overly concerned about the nature of the terrain. This scheme has distinct analogies to high-level programming languages. We shall see the reverse organization in the sensory systems, where detailed information at the receptor level is gradually reduced to powerful statements of object recognition, independent of such details as location, as the information ascends in the system.

Even with all of this elaborate apparatus to direct and coordinate body motion, the problem of movement in the generalized environment remains a challenging one. Despite the massive investment in processing power that the brain has devoted to the problem, we still occasionally fall down. Producing a robot system that even approaches the brain's abilities will be a great accomplishment.

3 The First Analysis of Input

I. THE PROBELM OF CODING

Perhaps the most remarkable feat performed by the organic brain is the creation of useful organization from the flood of data provided by sensory receptors. It is also one of the most difficult tasks faced by the designer of robot systems. Consider the nature of the information which the brain receives about the visual world: patterns of light, varying in wavelength and intensity, are imaged on the retina of the eye by the lens. This illumination produces a complex barrage of neural impulses which flow through several hundred thousand fibers in the optic nerve and activate neurons in numerous brain centers including a portion of the cerebrum called the "primary visual cortex." Obviously, our visual *experience* is nothing like this array of impulses in axons. Instead, we "see" objects, colors, and groupings of objects, which are interpreted in meaningful terms. Our conscious perception of the visual world and of the other sensory modalities is quite a different matter from the stimuli which generate this experience. Although it is not necessary that a robot have "experience" as we do, it will be necessary for them to assemble sensory information into behaviorally relevant elements.

In order to better understand the nature of the sensory processes being performed, it is useful to differentiate the terms "stimulus" and "sensation." "Stimulus" refers to the actual physical event that activates a sensory receptor. In the case of vision, this would be a group of photons being absorbed by retinal tissue. Because the intensity and wavelength of this light are determined by properties of the physical object which reflects it, we often refer to the reflecting object as the stimulus, with the understanding that its action as a stimulus depends on the properties of the light it reflects. On the other hand, "sensation" refers to the mental experience which results from certain kinds of activation of our receptors. There is a close relation between the sensations we have and the stimuli which produce them. Our senses would be useless if this were not so. However, this close relation often leads us to confuse the two, and this is a great error, because they belong to entirely different worlds. A stimulus is an objective occurrence; a sensation is a subjective, mental phenomenon.

To clarify the distinction, consider the sensation of the color red. We all know what we mean when we say an object "is red " or that we "see a red object." A moment's reflection will demonstrate that, strictly speaking, there can be no such thing as a red object. The object's "redness" is solely determined by whether or not it possesses those properties necessary to reflect light of a particular wavelength. If an object reflected light of a wavelength which gave rise to the sensation "red," and we were to somehow change the wavelength between the object and the eye, the object would appear to be another color. Can we say that "redness" is a property of the light? No, because the only relevant physical property of the light is its wavelength, and wavelength is not color. Color is a property of your sensation. Sounds have wavelengths too, and in this case the sensation is interpreted as pitch. Wavelength is only a piece of information which the brain may interpret as it will. In your computer, you could make an analogy with the American Standard Code For Information Interchange (ASCII) code. We use particular bit patterns to represent letters and numbers, but the same bit patterns could just as easily represent something else. There is nothing that inherently requires "01000001" to be interpreted as "A." Similarly, "redness" is simply a convention that the brain uses to label the experience associated with a particular wavelength of light.

The situation becomes clearer as one studies the brain a little further. Sensation is not a result of activating the retina or the optic nerve. If the optic nerve is cut, light falling on the retina produces no sensation, even though the retinal neurons and their axons in the optic nerve are activated. However, artificially activating the visual cortex, to which the nerve used to project, will still produce visual sensations. It seems to follow that sensation is dependent upon the firing of neurons in the visual cortex. Information about specific wavelenghts of light striking the retina is carried in the axons of the optic nerve which arise from classes of retinal cells tuned to particular wavelenghts. This scheme of dedicating groups of axons to carry information about certain wavelenghts may be called a place code. That is, the wavelength information is carried to the visual cortex in terms of *which* lines are active. There is certainly nothing that seems intuitively "red" about which axon carries an impulse. Yet, the sensation of "redness" apparently occurs far beyond the initial retinal encoding process that determines which axons shall be active.

If we accept the relatively modern notion that mental sensations are produced, or at least determined, by the brain as it decodes stimulus-produced activations of receptor lines, according to specified conventions, then the sensation becomes, basically, an example of information processing. The nature of the conscious "experience" of the processed data is a topic we will take up later. For now, our objective is to examine the kinds of transformations the brain imposes on its input data and to ask why a particular transformation, and not another, is useful to the organism in dealing with the environment. The utility of this pursuit lies in the fact that the most likely system for detailed examination of distant objects in an artificial system will use an image-forming system acting on a grid of sensitive transducers. The problem of information processing in such a system is one that the brain has solved.

II. THE IMPORTANCE OF BOUNDARIES

There are about ten million light receptive elements in the retina, and the brain produces a complete analysis of their patterns of illumination about ten times each second. If this were done in a straightforward manner by examining all the possible permutations of ten million bits of information and decoding it against a table of known codes, a tenth of a second of vision experience would be too big a job even for the brain to handle in a reasonable time. In fact, the brain goes to some extremes to reduce the data and some of its tricks are of a quite general utility. The first step in the process is to make a number of decisions about what not to look at.

If an area of uniform illumination is bounded by an area of some other degree of illumination, the constant information within the central area is superfluous. That is, if one had a system that could detect only boundaries between different illumination levels, the center of a uniformly bounded area would not produce a signal. Yet, information about its illumination could be accurately reconstructed by simply extrapolating the illumination level on the inside edge of the boundary across to the next boundary. Any *change* in illumination constitutes a boundary between a lighter and a darker region. Thus, if only boundaries can be detected, extrapolation of levels on either edge of a boundary to the next boundary accurately reconstructs the whole field.

The reduction in the number of points to be considered which is achieved by considering only boundaries is quite large. Think of a 100 by 100 array of retinal receptor cells, illuminated at level A on the right half, and level B on the left. If we were to scan every element in order to reconstruct a picture of this pattern, we would have to examine 10,000 elements. If we were able to scan only those receptors near the boundary between areas A and B, and extrapolate the rest, we would have to look at about 100 elements. In general, the savings are directly related to the area examined. The problem is to locate boundaries without having to search the whole array.

I will discuss the details of how the brain locates spatial boundaries, but first, mention should be made of the next shortcut because their underlying mechanisms are related. Basically, this second trick is to look only at things that change. Aside from the fact that changing patterns of illumination usually imply moving objects, which are important items in the sensory world, special attention to change also has advantages in terms of processing time. The situation is very similar to the preceding one, except that here we must think of change as representing a *temporal* boundary between illumination levels. If we only attend to an element when its illumination changes, and if we always know when it does, we can safely ignore it in the meantime. This is because, by definition, the illumination during the intervening period of "no change" must be steady at whatever level the preceding change brought it to. Thus, it is only necessary to extrapolate the value immediately following a change until the next change is detected.

The eye is sensitive to two stimulus dimensions of light, intensity and wavelength, which we perceive as the sensations of brightness and color. So far we have discussed the two boundary situations, spatial and temporal, in terms of brightness, but the same arguments could apply to boundaries of color. Two areas of equal brightness but different color also must be discriminated. The same mechanisms may be applied to both situations because the brain handles color by providing some receptor elements with differential sensitivity to light wavelengths. For these elements, a change in wavelength effectively *is* a change in illumination. It will either be a change from a wavelength to which the input element is sensitive to one to which it is not sensitive, or *vice versa*. The brain handles the color information simply by recognizing the output of these elements as encoding the wavelength information and interprets it as color. The color boundary problem therefore reduces to the intensity boundary problem. Of course, as species evolve from simple to complex forms, increasing ranges of physical phenomena are captured in the visual system. Only a few animals have color vision; man and other primates have much more sophisticated color discrimination than do other mammals. These facts suggest that a certain degree of processing effort is required to obtain higher color vision and that whatever degree of organization is needed is apparently not worthwhile for most species.

Let us now return to the issue of efficient boundary detection. Detecting temporal boundaries, or changes in illumination with time, is implemented by a process roughly similar to AC-coupling the receptor elements. In fact, AC-coupling of analog-to-digital (A/D) converters, with an appropriate time constant, would be an effective way to model this process in a robot. In the brain, it is simply a property of the receptor neurons themselves, and the details need not concern us. The interesting thing is that the brain uses this AC-coupled characteristic of the neural elements to detect both spatial and temporal boundaries. A selective sensitivity to change, or temporal boundaries, is inherent in the AC-coupling, but a sensitivity to spatial boundaries requires some additional mechanism.

We are all aware of course that the eye moves. We observe it all the time when our gaze turns from one point of fixation to another, or when it follows a moving target. In addition to these obvious motions, there is another which is not detectable by ordinary means. When the eye seems to be at rest, even when you are holding your gaze as intently as possible on a fixed point, there is still a very fine motion with a frequency of about 10 Hz.

The amplitude of this motion, which is rather erratic in its direction, is sufficient to move the image from a fixed object back and forth on the retina by a distance equal to a few times the average separation between the sensitive elements. Those elements near a boundary are swept back and forth continuously from the lighter to darker sides of the boundary at about 10 Hz. This produces a changing signal of the sort to which the AC-coupling property of the system can respond. At the same time, their neighbors further from the boundary, in the lighter or darker regions on either side, are not moved into a region of different illumination. Hence, they "see" an unchanging input, to which they do not respond. Therefore, the receptor elements of the eye act as intelligent terminals which transmit only information about boundaries and changes to the higher levels; this feature allows an enormous savings in the amount of input attention required from higher processes.

III. IMPLICATIONS FOR THE NATURE OF MENTAL EXPERIENCE

Now I hear you say, "Yes, but I can *see* the insides of uniform areas." True, but remember I said your sensations were an arbitrary decoding of the stimulus information, and that the information from areas distant from boundaries was redundant and could be reconstructed by inference or extrapolation. The experience of "seeing" the inside of the area is simply the experience of receiving the appropriate code from the right set of boundary-activated elements.

In the first place, it is relatively easy to demonstrate that you cannot see anything if there is no change. By virtue of some clever optics it is possible to stabilize an image on the retina so that it does not move with respect to the receptor elements, despite the fine motions of the eye. When this is done, the image seems to disappear about a tenth of a second after it is presented. Of course it is still really there on the retina, but your AC-coupled system cannot respond. Now, consider a green disk with a smaller red disk in the center. It is possible to stabilize just one portion of this image in the way we stabilized the whole image a moment ago. If we choose to stabilize just the boundary between the green outer ring and the red inner disk, it should not be possible for the brain to detect that boundary. If this is done, you see neither the boundary nor the inner red disk. What do you see? You see an unbroken green disk all the way across. In other words, if no boundary is detected in the middle, the brain not only does not see the red disk, but it also extrapolates the green all the way across from one outer boundary to the other. Think about it the next time you rely on the evidence of your eyes.

The AC-coupling is not perfect; there is a "DC leak" around it, but the changing-signal-only property of the neurons is enhanced at each step in the transmission process, so that the cells of the visual cortex are found to have almost no response at all to unchanging uniform illumination of the retina. This means that the sensory experience of the interiors of uniform regions is simply what is coded for at the cortex by the byte of information on the boundary conditions. It is not a result of direct translation of retinal illumination conditions on a point-for-point basis into activation of some set of "experience neurons." It is important to grasp this idea, because it points up the fundamental similarity between the brain and the computer. There is no "inner eye" looking out through neural windows. If the encoding process ultimately produced a single neural line which was activated by the sight of a face, then activity in that single line would allow the brain to prepare a response; such an encoded signal would also be a sufficient basis for our subjective experience as well.

The nature of the encoding process for our experience becomes more apparent at levels of the encoding process higher than those involved in the green and red disk experiment. At some level of the process, referred to as "feature extraction," we arrive at a byte of active lines which encodes for some complex pattern. For example, take the repeated patterns of wallpaper: it seems that even for such complex but repetitive patterns, the brain continues its policy of dropping redundant information and carrying forward only information on boundaries. If we look at such a wall, we see a repetitive pattern extending to the ceiling, floor, and the edges of intervening furniture. Suppose we present this same scene to a person who has no vision within a limited region of the visual field. If his injury is at the right level of the feature extracting process, he will report seeing the unbroken wallpaper pattern, including the region within which he is "blind," just as uninjured persons do. It can be demonstrated that his experience of the pattern in the blind region is due to the fact that both he and we are extrapolating the detected pattern across the intervening space between pattern boundaries.

His deficit becomes apparent when we create a boundary in the pattern within his blind region. For example, if we inverted a small patch of the pattern, it would constitute a boundary in the pattern, and we would not extrapolate across it. But if it occurred in his blind region, he could not react to the pattern boundary and would receive the same encoded byte of visual information as before, and claim that he saw an unbroken wallpaper pattern. In an important sense he is blind, yet he has visual experience. You do the same thing; there is a blind spot in the visual field where the optic nerve leaves the retina. You can make small objects disappear by centering them there, but because you can't see boundaries there either, your brain normally extrapolates across it.

If the brain is reducing complex features of a visual stimulus to a signal encoded in a few lines, does this mean that there are things for which we have no feature extractors (encoders), and if so, would we be unable to see them? Apparently this is the case. Experiments suggest that the visual world of simple creatures, like frogs, is quite impoverished. They have elementary as well as complex feature extractors for stimuli (eg: bugs) which are important to their behavior, but as far as anyone has tested, frogs have nowhere near as extensive a set of feature extractors as do mammals.

In theory, it would be possible to have a unique line or coded set of lines activated by every possible combination of activities on the retina, but it would be beyond even the capability of the nervous system to generate that many processing elements. Instead, certain decisions are made as to what things are important to see, and decoding for these is provided. This does not imply that you would not see *anything* when looking at a novel stimulus for which you have no appropriate high-level extractors. At the first level, simple features such as edges, arcs, lines, and spots are extracted. More complex features are extracted from combinations of these. You might be aware only of the activity of the low-level extractors for lines, edges, etc and fail to recognize it as an object, or you might fail to discriminate it from objects which would appear to be identical, but which actually differed in ways which could not be detected by the feature extractors in your visual system.

As an example, it is possible to fool high-level extractors by giving them marginal data. Look at figure 3.1. About 95 percent of the people who see

Figure 3.1. *A white cow with black ears and nose. The head is turned facing you, with the side of the head in shadow. The large white area in the lower right is part of the cow's left flank. Before you "see" the cow, all that appears are white and dark areas. After you "see" the cow, your sensory experience is still the same in terms of the stimuli you are receiving, but your analysis is very different.*

this picture for the first time are able to activate only low-level extractors and report patches of bounded light and dark. It is in fact a photograph of the head and upper forequarters of a black and white cow (facing left, with its face turned towards you) against some trees and a fence. Once you know what to look for, you can nudge the "cow extractors" and get an entirely different experience. Indeed, once you've seen it, it's difficult not to see it. (Don't panic if you can't; about 5 percent never see it.)

IV. THE FEATURE EXTRACTION PROCESS

Actually, there is probably no "cow extractor" *per se,* but rather there exists an assemblage of feature extractors which constitute a code for "cow." Let's look at some of the properties which such high-level extractors should have. The most important one is that they should function independently of position, orientation, context, etc. That is, if we had to have a separate extractor for every position a complex stimulus might assume in the visual field, we would need so many elements that the advantages of the feature-extractor approach would be lost. Next, they should be capable of implementation by learning, so that the available processing elements could be best used to fit the organism's normal visual environment. Third, they should not be limited to spatial forms, but should include detectors for properties such as motion, distance, and other aspects of our visual experience. Understanding how these properties are implemented is a difficult problem, and we haven't a notion of the real number or nature of the highest order extractors in the human visual system. We can examine some of their properties by fatiguing the extractors through prolonged exposure to different types of stimuli and looking at the effects on our visual abilities. In animals, we can follow the process by recording activities of neurons in the visual system during presentations of stimuli to the eye.

From these latter experiments, we have a fairly clear notion of the operation of the lower order extractors, and the process seems easily extensible to higher order features. A general example of the process is a feature extractor which can detect a line segment anywhere in the visual field if it is at one particular angle of inclination to a reference frame. That is, it does not matter where the line is located in the visual field, only that it be a line and that it possesses a certain angle of inclination. This sort of unit appears to be typical of low-level feature extractors of mammalian visual systems.

The basic gating action used is very similar to that of an AND circuit. As I have mentioned, one possible mode of neuronal operation can be implemented by a neuron having a set of inputs which must be "true" in order to achieve firing threshold. In this case, however, we have an AND gate with a safety factor. By this I mean that firing level can be achieved if some percentage of the relevant inputs is active: 100 percent is not required. Think of it as an "ALMOST" gate. (Unlike our conventional computers, the brain continually uses this technique to make "best guesses" rather than precise solutions. Perhaps this is why we make mistakes, but it also provides for powerful inductive leaps which are correct most of the time.)

Connecting a grid of two-input AND gates together, as in figure 3.2, illustrates the basic logic of the scheme. At the bottom level we have a line of receptors. Above these are several additional levels of two-input AND gates, culminating in the top level with only two elements. It is clear that

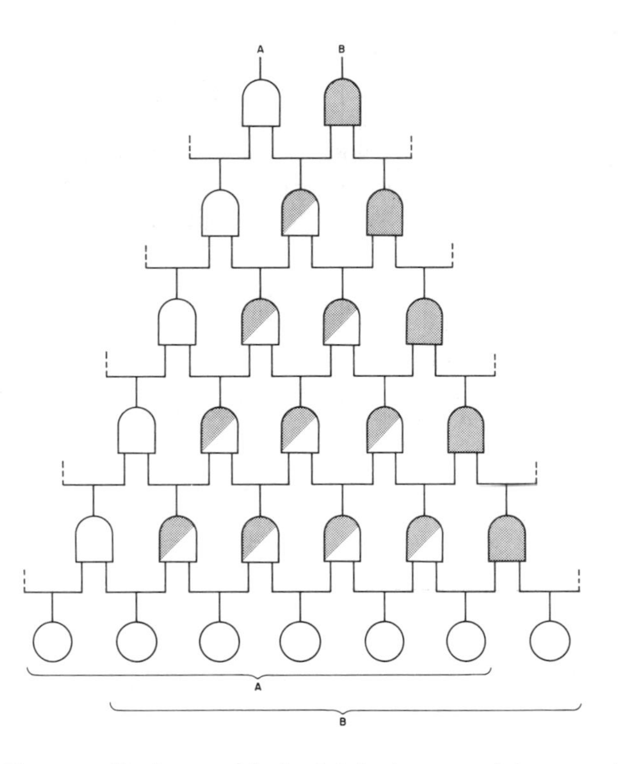

Figure 3.2: *The generalized case of the brain's basic approach to processing input. The two output lines at the top discriminate patterns of input at the bottom that are very similar. In reality, the brain chooses only some of these possible connections to facilitate processing.*

the two top level elements will discriminate between two patterns of activation, in the bottom row, which differ only by one element. Thus, activation of element A encodes the activity of a set of bottom level elements indicated by bracket A, and element B and bracket B represent a different set. If the bottom row represents the output of retinal receptor elements, A and B could discriminate between illumination conditions (A) and (B), which are quite similar. It should be apparent that with enough gates and elements, this sort of general convergence scheme *could* be employed to extract any feature. Because this approach is impractical, the brain adds two principles which enormously increase processing efficiency at the expense of reduced generality. Once the set of retinal activation patterns to be recognized has been selected, specific feature extractors for that pattern are built from the type of logic illustrated in figure 3.2, but the actual gating is modified by the addition of processes called "selective convergence" and "lateral inhibition." The meaning of these terms will become clear shortly. At the lowest levels only a few simple types of features extractors are implemented, whereas at higher levels more complex detectors are built from the simpler subunits. Let us examine the first step in this process.

Within the retina itself, there are several levels of processing which culminate in an output neuron, RGC (retinal ganglion cell), which sends its axon to the brain via the optic nerve. If we record the electrical activity of these RGC neurons we find that their response can be classified into a few basic types depending on the kinds of stimulus to which they respond maximally. Figure 3.3 shows the portions of the visual field which affect the activity of a typical RGC type, and figure 3.4 shows the connections which result in this type of response. We see that this RGC receives excitatory

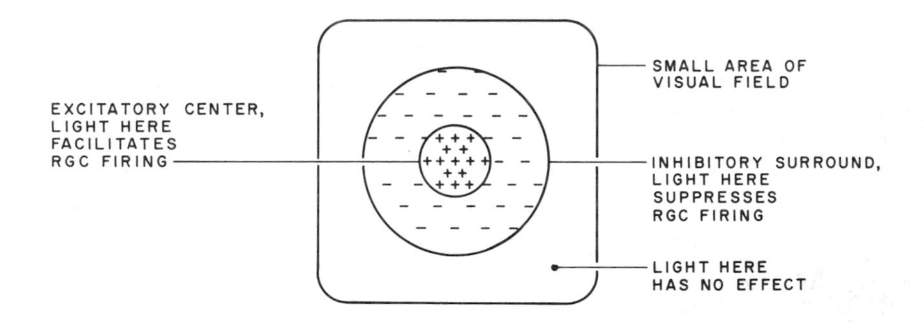

Figure 3.3: *A small area of the visual field, showing the portions which can influence the output of a single, retinal, ganglion cell (RGC) when illuminated. Some visual areas have an excitatory action (+)* and others have an inhibitory action (−) on the RGC's firing rate.

synapses from a small group of receptor elements located in a central spot (+ region) and inhibitory synapses from receptor elements in a ring surrounding this spot (− region). Remember that the type of synaptic effect is the choice of the postsynaptic, receiving neuron, and that the receptor elements in the "inhibitory-surround" area are free to make facilitative connections with other RGCs. When the central spot receives light, it increases the firing rate of the RGC. When the inhibitory-surround is illuminated, it decreases the firing rate of the RGC. (Because the response to a constant stimulus tends to be rapidly diminished, our examples assume that transient stimuli are being employed.) If the entire retinal area which affects our RGC is illuminated, the excitatory and inhibitory effects tend to cancel. Here, as elsewhere in the visual system, there is little response to diffuse light.

Notice that because of the shape of the inhibitory and excitatory regions, a bar of light the width of the excitatory center spot, which extended across the entire active area, would fire retinal elements in both the inhibitory and excitatory regions. This stimulus fires the entire excitatory central region, but activates only a small percentage of the inhibitory elements because it only illuminates the inhibitory-surround in two spots. (See figure 3.5.) The RGC response to a line stimulus crossing the central spot would be strongly positive, but less so than the response to a stimulus which did not touch the inhibitory region.

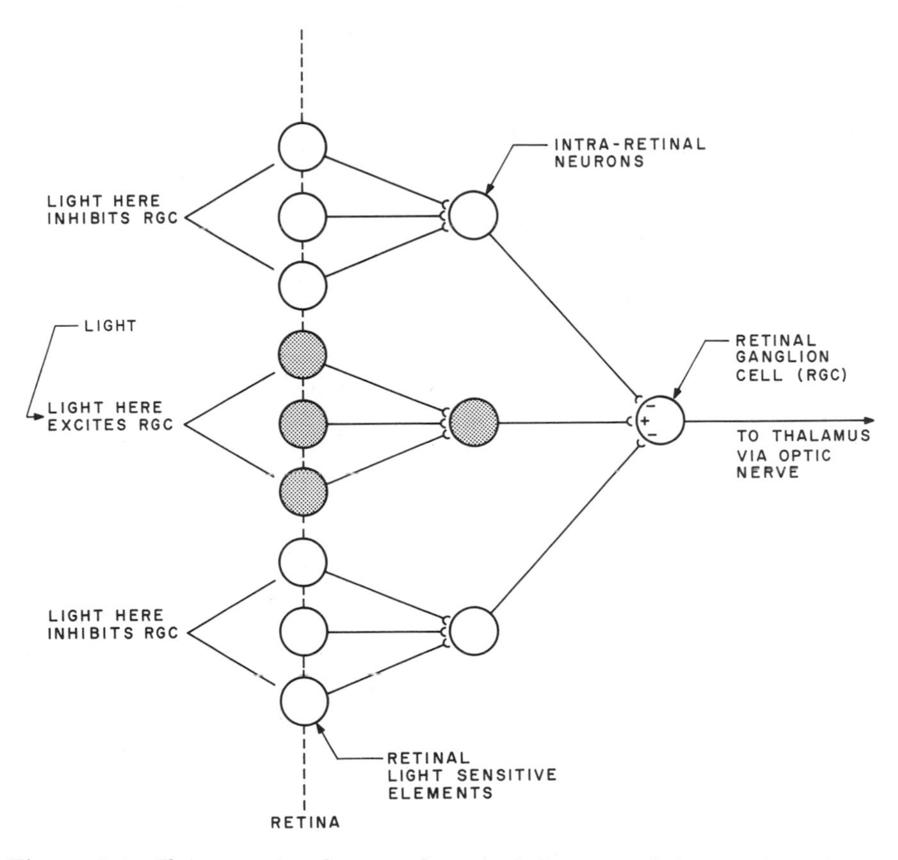

Figure 3.4: *The connections between the retinal elements and the retinal ganglion cell (RGC) which account for the pattern of sensitivity shown in figure 3.3 In this view, the pattern of figure 3.3 would be edge-on as projected on the retina.*

The optimal stimuli for RGCs may be grouped into a few basic types; for example, the receptive field of some RGCs is the inverse of the one above and is composed of an inhibitory center area surrounded by an excitatory ring. The optimal stimulus for this cell is a light annulus with a dark center. I will not discuss these in detail, but will proceed further to pursue development of our location-independent inclined line detector. The next way station, which is one of the targets of the optic nerve, is a thalamic area called the LGN (*lateral geniculate nucleus*). The axons of the

RGCs make synaptic contact with the cells of this nucleus just as the retinal elements made contact with the RGCs. If we record the electrical activity of these cells, while testing for retinal areas that excite or inhibit them with visual stimuli, we find that the cells of the geniculate nuclei have response

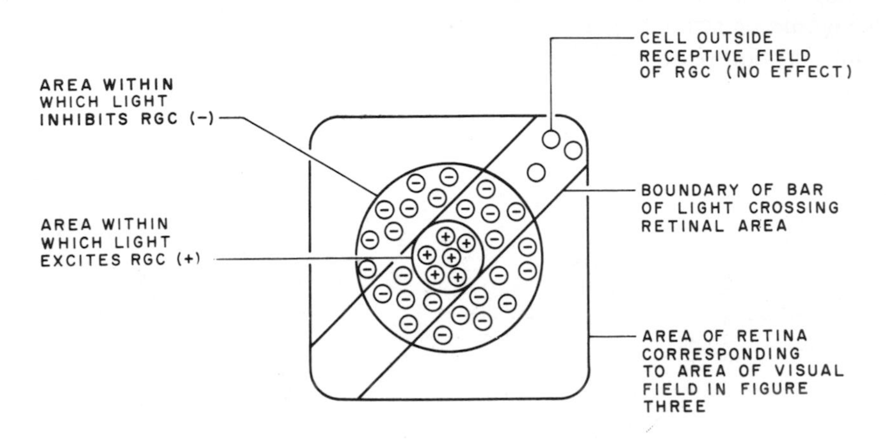

Figure 3.5: *An area of the retina corresponding to the area of the visual field in figure 3.3, showing the retinal elements of figure 3.4 which would be activated by a stimulus consisting of a long line of light.*

patterns rather similar to those of the RGCs, that is, central spots and antagonistic surrounding rings, etc. The active fields tend to be somewhat larger, but it appears that the thalamic cells receive positive inputs from RGCs whose positive centers are close together, so that essentially the same pattern is maintained. This is illustrated for a typical "on-center, off-surround" thalamic cell in figure 3.6. Actually, at this level there is a great

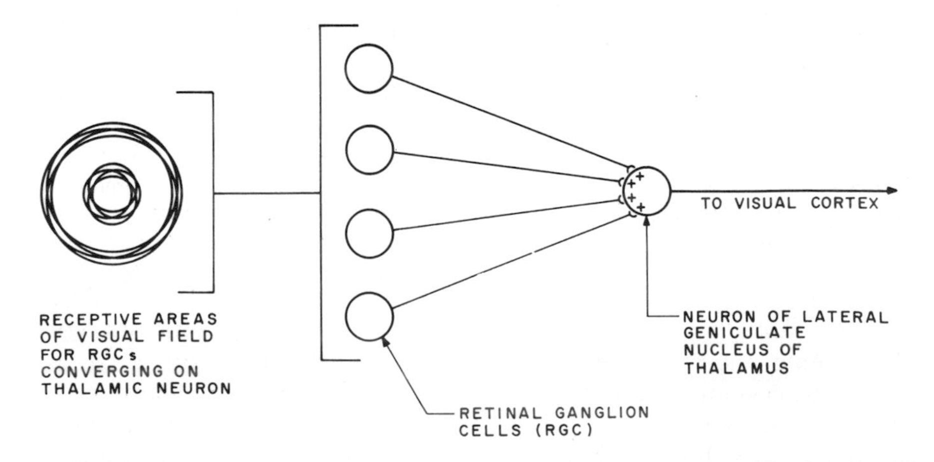

Figure 3.6: *The convergence pattern of retinal ganglion cell (RGC) units on a thalamic cell, showing the spatial relations of the receptive fields of the RGCs which project to the same thalamic cell.*

deal of additional processing which includes not only signals returning from the cortical areas to which the thalamic cells project, but also data from other brain regions that contribute to the analysis of visual information. I will consider these other inputs later, but for now I will follow the line detector system as it projects to the cortex.

The axons of the thalamic cells make synapse on a class of cortical cells known as "simple field cells." These cells typically respond preferentially to stimuli, such as lines of light at particular inclinations and particular locations.

Thus, the optimal stimulus for a "simple" cortical unit is actually rather specific compared to that of a RGC or LGN cell. Usually, it is found that a vertical column of simple cells in the cortex deals with the analysis of some small area of the retinal image and contains a large number of such line analyzers. Each analyzer responds best to a line at a slightly different angle, but all are concerned with the same small area of the retina. Figure 3.7 shows how such a line detector might be constructed from the output of the thalamic cells. Those thalamic cells which have a positive influence on the firing of the cortical simple field cell have excitatory centers in the

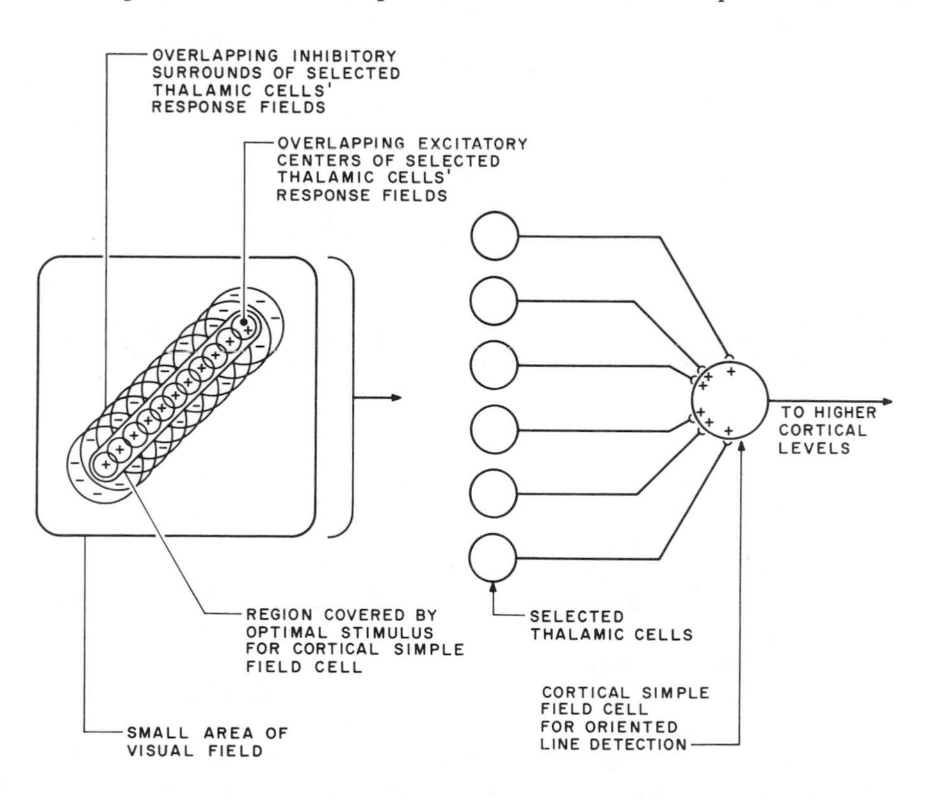

Figure 3.7: *The pattern of convergence of thalamic cells onto a cortical oriented line detector (simple field cell). The spatial relations of the receptive fields of the thalamic cells which project to this cell in an excitatory manner is so chosen that their own excitatory regions lie in a straight line.*

visual field which lies in a line oriented at some angle to the vertical. The optimal stimulus for firing it will be a line which passes through the excitatory central spots of all the lower echelon thalamic cells' receptive fields. Such a stimulus will produce a strong (but sub-maximal) firing in each of the thalamic cells because it will intersect some of the inhibitory territory of each, but such a sub-maximal output from *each* of them is the maximal input for the cortical-level cell. This process of convergence is highly selective; now look what happens if the line is turned at a different angle, or moved to a different position (figure 3.8). If the angle is not aligned

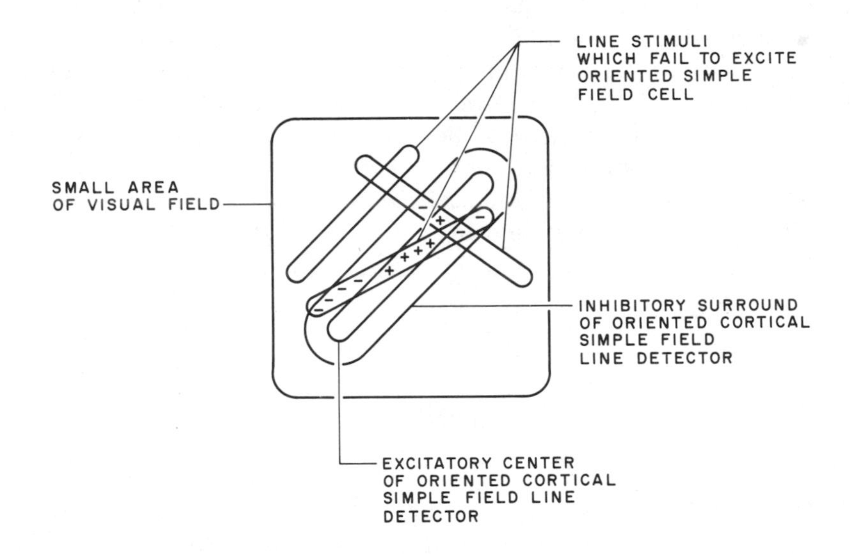

Figure 3.8: *Lines of light which are not optimally placed do not activate the simple field cell since they either do not strike sensitive elements, or strike excitatory and inhibitory areas equally.*

with the line of the "on-centers" one thalamic cell will show a positive response, but the others will not and may have a greater portion of their inhibitory areas activated. In this case there is little input to the cortical line detector. If the line is kept at the correct angle, but moved to the side, it falls off the excitatory region entirely and stimulates completely inhibitory territory. Thus, while our simple field cell can discriminate angle, its response is location-dependent.

The next level of abstraction is apparently provided by the so-called "complex field cells" of the cortex. Complex cells are also stimulated by lines at a particular angle, but in this case, the stimulus may fall within a larger area of the visual field. These cells' response to an optimal stimulus is relatively independent of its spatial location. This response is easily explained if the cell can combine the outputs of a large number of simple field cells with a logical OR function; all of this group of simple cells have the same optimal stimulus orientation, and their receptive fields are spread over a wider area of the visual field (figure 3.9). Again, certain convergence patterns are implemented selectively; in this case the convergence principle is parallelism.

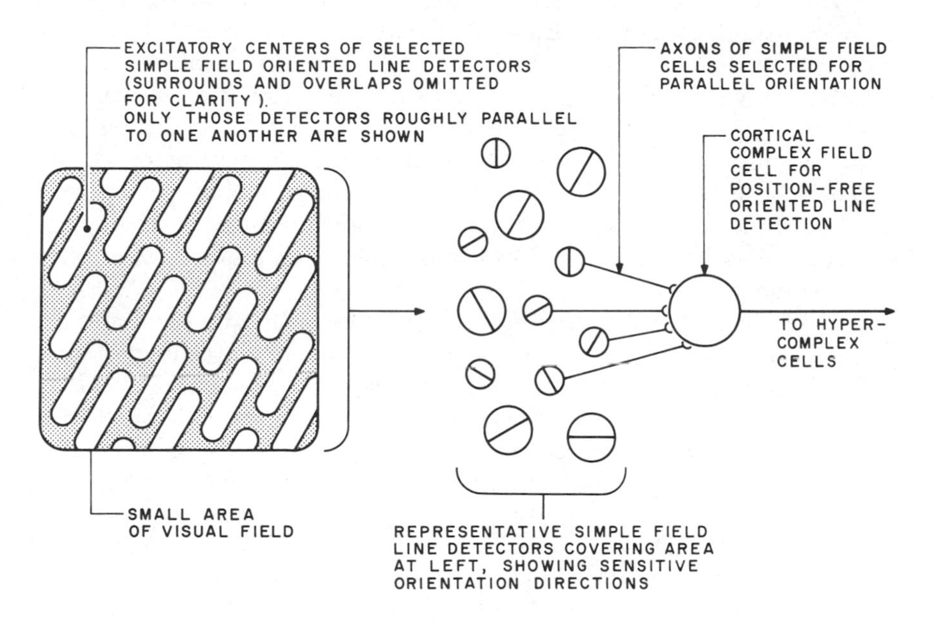

Figure 3.9: *The development of a complex field cell of the cortex which responds to lines at certain angles, but without reference to spatial location over a large retinal region.*

V. SOME FEATURES OF FEATURE EXTRACTORS

In reality, the connections are not so straightforward as I have suggested. As mentioned before, there is considerable interaction between the thalamus and the cortex, and there is profuse interconnection between different layers of cells in the cortex. However, the general principle is relatively simple, although the biological implementation is complex. If one continues this type of convergence operation and combines the outputs of different types of cells, it is clear that feature extractors of any desired degree of complexity could be built. Arc detectors, edge detectors, and numerous other types are already available at the simple field cell level. At the level of more complex feature extractors, which may be in regions of cortex outside the primary visual area, it is very difficult to determine the effective stimulus for a cell simply because of the enormous number of stimuli to which it might respond. For example, in a monkey's brain recordings were made from a cell which responded only to the outline of a monkey hand. For our purpose it is not necessary to specify all types of feature extractors which exist. The particular set that was most useful to a man's brain would probably differ from the most useful set for a robot brain. It is sufficient to see the principles by which the feature extractors can be constructed. Let us review these.

First, it is clear that only a small subset of the possible combinations of receptor elements are actually encoded by higher level cells. Features which are useful, elementary building blocks are encoded by convergence of low-level cells selected on the basis of their spatial relations of their retinal receptive fields. Higher level cells then employ selected combinations of these simple features to form extractors for more complex features. Instead of employing the generalized encoding scheme of figure 3.2, the brain has chosen the more limited but more economical approach of omitting some of the connections and creating specialized gates by selecting particular branches from the preceding levels. This is what is meant by "selective convergence."

Second, the response of the higher echelon cell is frequently fine-tuned by a process of lateral inhibition. That is, lower echelon cells frequently have inhibitory projections not only to adjacent cells but also to the neighbors of the higher echelon cell to which they send excitatory projections. The arrangement of these inhibitory projections may help a higher order cell discriminate between stimuli which are similar enough to the target feature to generate some degree of excitation. Thus, in the simple cortical field cell line detector in figure 3.8, if a line is presented whose orientation differs only slightly from the optimum angle, it would still excite several "on-centers" of thalamic cells and thereby cause considerable response in the corresponding cortical cell. This difficulty is surmounted by the fact that in crossing the first cell's line of on-centers at a small angle, the stimulus also crosses a large amount of territory which inhibits the thalamic cells' output. Thus, a stimulus which is slightly skewed still generates some excitation, but simultaneously produces a lot of inhibition. If the angular discrepancy is large there is little problem. Of course, a stimulus which is suboptimal for one detector may be ideal for another; those higher level cells which receive optimal stimuli will send inhibitory signals to suppress the activity of those higher level cells receiving less than optimal stimulation.

Thirdly, how much excitatory input will be required to fire a higher echelon cell? Recall that the neurons do not function strictly as AND or OR gates, in that only a certain percentage of inputs being excitatory is required for firing to occur. This "ALMOST" gate principle is one of enormous power, and I shall have more to say about its application to intelligence in a later chapter; but for the moment, look at what happens in the sensory system if we adjust the amount of excitation required for firing. If all lines had to be active as in a conventional AND gate, we would have a perfectly accurate system, like any good conventional computer. We would also have a slow and inflexible system. To have all the inputs properly set up, we would need to wait for perfect alignment of the image; the object would have to be close to the eye for high resolution and would need appropriate illumination to prevent any marginal situations. Such a system would be accurate, but your ancestors would never have reached reproductive age if they had to use such a scheme before decoding the stimulus as a wolf. On the other hand, if we let the system be sloppy and fire feature extractors when only a small number of relevant input lines are active, we will achieve fast results with a lot of errors. In particular, we would be unable to make fine discriminations between similar stimuli which would have activated many lines in common.

This kind of error is easily demonstrated. Briefly flash a picture of a circle with a small piece missing on a screen, and your subjects will report that they saw a complete circle. Only if they examine the image longer will they be able to discriminate the broken circle from a complete one. Clearly, both modes of processing have their uses, and it would be nice to adjust the excitation requirements of the "ALMOST" gates to suite the task at hand. In the brain this is done by axons from control regions outside the sensory system which make diffuse and widespread contact with large numbers of sensory processing elements. These inputs do not carry specific visual information, but by excitatory or inhibitory action they can bias the processing elements towards or away from firing threshold. This influence increases or decreases the amount of excitatory input which is required for firing to occur. When this process is driven beyond normal limits, as with various drugs, the feature extractors can be biased so close to firing that little or even no input is required. The result is a variety of visual distortions and hallucinations. [*As an interesting visual experiment, sit in a perfectly dark room for about 30 minutes. You will notice that your experience is one of tiny black, gray, and white flecks of visual sensation which apparently are derived from spontaneous, non-stimulus-driven activity of the retina and higher visual systems. What function this serves is completely unknown....* **ed**]

A fourth point worth noting is that the organization of this sensory system resembles that of a pipelined processor. As soon as the cells of any echelon have fired in response to the current state of their inputs, higher echelons begin dealing with that fact, while the earlier echelon begins to respond to the next state of its inputs. This process is not clocked; rather, results trickle through as fast as they can. Depending on the levels of convergence required, some things take longer to recognize than others. There is no need for the system to proceed in lock-stepped stages like a real pipelining system, because information can be siphoned off the line at any stage as well as be passed to the next. If you need to catch a fast moving object, you can respond to information about its position, which is encoded fairly early in the process, without having to wait for a detailed analysis of its surface markings, which will be derived by extensive processing.

I have emphasized the development of a particular feature extractor to clarify the process involved. The emphasis on selective convergence should not obscure the fact that each lower echelon cell's outputs usually go to many higher echelon cells. Further, these lower level outputs may be involved in the extraction of entirely different features at each of the higher echelon cells to which they project. It is not the case that we have a grand convergence which starts with a ten million bit byte of retinal elements and gates itself down to a few high-level cells. Rather, we come out the other end of the process with a "byte" containing even more lines than the input byte. The difference is that the bits in the lowest input byte represent the spatial patterns of illumination on the retina in a simple point-for-point code. On the other hand, each bit in the output byte of the system represents the occurrence or nonoccurrence of a complex pattern of features in the visual world, and can be used as direct inputs to higher processes. Thus, the input byte and the output byte of the visual system each contain the same basic information: the content of the visual world. However, in the output byte, the information is recoded so that each bit represents highly useful pieces of information about the patterns which occur among the input bits. Referring again to figure 3.2, the real situation would be one in which there were as many cells at the top of the figure as at the bottom, with each convergent tree leading to a top level cell containing many elements in common with other convergent trees, just as the two shown do.

VI. THE TEMPORAL BYTE

So far I have discussed the processing of spatial patterns of retinal illumination. There are many other qualities which must be processed. One that deserves special mention here is the encoding of intensity information. This is done in the brain by use of the analog information in the cell's "temporal byte." That is, each line carries one bit in the "spatial byte" which encodes the existence of some set of conditions at the retina related to *which* cells are activated. The rate of firing of the line encodes, in pulse-frequency analog form, information about the strength of that activation. For low echelon cells, this rate is essentially information about the intensity of the light falling on the receptors. At higher echelons in the sensory system, it is information about the "degree of certainty" with regard to a cell's identification of a feature. This information is derived from both intensity and spatial information, because either higher pulse rates or more lines active will increase the firing of the cell. This is an example of how brain combines digital and analog information in a single decision process. The nature of the information which is being encoded by intensity of firing at the higher levels of the process may be better understood by applying the "degree of certainty" concept to the lowest levels, where the temporal byte represents light intensity. Obviously, the low-level element has the greatest degree of certainty that it is being illuminated when it receives strong illumination. At higher levels, the number of inputs and the activity of inputs can trade off with respect to being able to drive the receiving cell; this approach is generally appropriate, because the degree of certainty about the existence of the feature to be decoded is increased if the inputs are "very certain." In general for the brain, "He who yells the loudest has the most to say." Because cells don't have egos, it works.

In any realistic approach, using present day hardware, this aspect of the system would probably have to be modeled using a byte of several bits in place of each single line in the brain. "Which byte" would be equivalent to "which axon," and the bit pattern would carry the information carried by the temporal byte on the axon.

VII. THE FEATURE EXTRACTOR AS A PROM

Given that the number of conceptual features into which the visual world could be subdivided is virtually unlimited, whereas the number of available bits in the system's output byte is merely enormous, how does the brain decide which features to encode? The simplest features are undoubtedly the result of evolutionary selection and are hardwired by the time birth occurs. The remaining features are probably developed in response to the type of visual environment in which the animal grows up. For example, there is evidence that if a kitten is exposed to a visual world containing only vertical lines for a certain period of its development, its visual cortex will be rich in line detectors with a near-vertical orientation and poor in detectors for other orientations. Apparently this pattern persists throughout later life. It seems similar to a programmable read-only memory (PROM).

VIII. FEEDFORWARD

Finally, I should mention some types of non-visual input to the process that carry very specific correction information. Try this experiment. Look across the room while moving your head from side to side. Notice that the world seems to stay still, even though you are moving its image on your retina. Now move the image around on the retina in a different way. Place your fingertip against your lower eyelid and lightly jiggle the eyeball while looking across the room (keep the other eye closed). Notice that this time the world seems to jump around as the image is moved about on the retina. Why the difference? In both cases the image is moving around on the retina. The answer is that compensation for the movement of an image caused by moving the head is an everyday problem for the brain in its interpretation of the visual world. It solves the problem by using feedforward information from the motor nuclei which control the movement of the head and body to correct the interpretation of the relative motion of image and retina. Because you don't usually jiggle your eyeball with your finger, your brain has never developed a mechanism to pre-correct for doing so, and you see the motion. There are more subtle non-visual inputs to the processing too, such as your motivations, but these are a subject for later chapters.

IX. MODELING THE FEATURE EXTRACTORS

Now the difficult part. How might we model such a visual system with current digital technology? As a start, let's examine what would be required of a "brute force" approach if we did not care what it cost. It would seem that the most straightforward method would be to have a set of microprocessors at each echelon to model the activity of each of the elements at that level. Because, with straight digital techniques, we would have to code intensity on a byte of several bits length, each lower echelon input element talking to an upper echelon unit would have to present a byte rather than a line. This means (say) eight lines for each converging step and each plane of lateral inhibition, instead of one. Each processor would then accept a number of bytes from elements at a lower level, which it would process according to a small read-only memory (ROM) program; a number of bytes received from its lateral neighbors, would also figure in the result. The ROM program would determine the type of response of the "cell," and its output would be a byte on a bus that ran to a number of yet higher echelon processors and laterally to its own neighbors. If we really wanted to model the brain's operation, this would all be conducted with handshaking logic and the processors would all have their own private clocks. Each processor would continually compute the result of whatever inputs it had available at any instant and output the result. When the input from any of its information sources changed, the output would change.

With a processor and a read only memory (ROM) to represent each cell, such things as the weighting of percentage input from an "ALMOST gate" action and the continuous alteration of output on the basis of the output of lateral neighbors, are simple. This system would be fast, powerful, and incredibly expensive. Let's say we opted for a minimal system which ran from a 64 by 64 grid of photosensitive elements. Further, let's say we want to keep the ratio of input to output at each echelon approximately uniform, so we wind up with about 4,000 highest echelon feature extractors. That means the system can recognize 4,000 different complex stimuli. Then let's say we want to carry the analysis to a depth of 5 echelons. (That determines the complexity of the stimuli which can be extracted by the highest level. Remember that the brain only took 4 echelons to get to the complex field cell. Hypercomplex cells can handle some very advanced extraction problems.) At this point however, we are talking about 20,000 processors. Even at 8008 prices, that's not exactly cheap.

Now suppose we try to trade speed for cost. The system just described obviously runs much faster than the real brain. A first step might be to have single processors at each echelon doing the work of many, even all, of the 4,000 elements at that echelon. Suppose that it is possible to update the output of a single simulated neural element in 100 microseconds. That means we could do all 4,000 simulations in each echelon in about half a second. That's not bad; it's still pipelining the processing from echelon to echelon, so the system would see a picture updated every half second with a (1/2 second number of echelons) delay between the stimulus event and the final analysis. Even if we pulled some information off the pipeline at an early stage in order to allow rapid reaction, however, it's still too slow for real-time work. (If you ever have the equipment at hand, try playing catch in a room illuminated with 2 per second strobe light flashes. At any flash rate below 10 per second the task becomes difficult.)

Two complexities also appear when we try to update serially the simulated elements of an echelon. One is that the program for each element is different, which makes our ROM considerably more complex. The other is that the output of each element in the array depends in part on the current output of its neighbors, including the ones you have not reached yet in the current pass. With only one pass across the array per update, a "lateral lag time" error would be introduced. Correcting this with the simple expedient of iterative passes takes too long. Furthermore, you have to carry some information in scratch pad. How far do you want elements to be able to interact laterally? For most purposes, a few elements away might do, but for some tasks, the brain converges outputs from widely separated elements. Lateral interaction among these is probably best ignored in our hypothetical simple system. Presumably, some optimization could be found in which several processors simulated each echelon; each one could handle a number of elements serially.

A different approach to trading speed for cost would be to have all your available processors simulate the elements of a single echelon, store the result, switch programs and simulate the next echelon, etc. This way you get through each echelon faster because a complete update of an echelon is divided among more processors, and fewer serially simulated elements per processor means you finish more quickly. However, with this scheme you lose the pipelining feature of the system, because a new input byte has to wait until the last byte gets all the way through before the system can start to deal with it by simulating the first echelon again.

Of course these notions do not exhaust the approaches to the problem, and I didn't promise to solve it for you, but they do illustrate some of the difficulties we will have to overcome. Perhaps ultimately, the most efficient approach will be specially designed chips with arrays of gates under the control of a processor which could alter their characteristics. This would require the development of chips which are just beginning to be designed, but in the long run this technique would be the cheapest approach to a feature extractor network.

The best approach may not involve replicating the detailed features of the brain's processing steps in recoding the sensory input. However, what does seem worthy of study is the general logic of the brain's approach, which includes such items as: ways of eliminating redundant information, ways of using selected feature extractors as building blocks at each stage of the perceptual process, elimination of each level of restrictions such as position on the generality of the feature encoding line, and the use of the "ALMOST gate" concept to provide continuously variable levels of stringency in the encoding process.

4 Some Further Types of Initial Input Analysis

I. SPATIAL FREQUENCY ANALYSIS IN THE VISUAL SYSTEM

In the preceding discussion, we focused on only one aspect of the visual system's analysis of the patterns of illumination on the retina. We might refer to this procedure as the "geometrical" feature extraction process. Knowing the complexity of processing available to the brain, we should not be surprised to find that more than one approach may be employed; in fact, this seems to be the case. Up to this point, what we have seen is a fairly straightforward procedure for recoding certain features as patterns of activity in some sets of neurons. I say "straightforward," because the process, although complex, is one which can be easily grasped in terms of the obvious spatial properties of the stimulus on the retina. I will now consider a process which is thought to be used by the visual system, that will require that the reader has at least an intuitive understanding of the concepts of spatial frequency and of Fourier analysis.

To begin, it is necessary to understand the term spatial frequency. This quality is similar to the kind of frequency with which everyone is familiar, except that the denominator of the "cycles per second" is not time, as in acoustic or electrical frequency, but rather the denominator is some spatial dimension (eg: cycles per centimeter). As an example, think of the alternating pattern of light and dark bars that constitute a grating. If the bars are of equal width and equally spaced, such a pattern is the spatial analogue of a square wave in time. Like a square wave, it has a characteristic fundamental frequency. If the bars did not have sharp edges, but rather shaded gradually into white and then back to dark, we would have something similar to a spatial analogue of a pure sine wave. In fact, it would be possible for the brightness of the bars to change with position exactly according to a sine function, and we would then have a spatial sine wave of some single frequency. The spatial square wave has a fundamental frequency and higher harmonics, just like its temporal counterpart. The two kinds of frequency are mathematically identical. In the mathematics of frequency, whether the "t" stands for time or for distance along some spatial dimension is not important. Now if this is true, it ought to be possible, at least in principle, to deal with spatial frequencies in all the same ways that we are accustomed to dealing with temporal frequencies; filtering would be one example. (How do you build a spatial filter? Consider a comb.)

In terms of visual processing, it is interesting to examine what can be done with Fourier analysis and temporal frequencies, and then try to apply the same procedures to spatial frequencies. Most people who work with electronics are familiar with the fact that various kinds of periodic waveforms can be broken down into a collection of component sine waves which is the so-called "Fourier series."

Given a slight extension of the method, it is also possible to describe any waveform, periodic or not, as a similar collection of component sine waves. It is not even necessary to know the mathematical formula of the waveform to be analyzed, because the method may be applied empirically to any waveform. This means that it is possible to represent *any* function of time as a set of component sine waves of specified frequency, amplitude, and phase relation. Thus, an arbitrary function of one variable, such as time or distance, may be represented by an equivalent function of frequency, that is, a function which gives the amplitude of the component sine waves along the frequency dimension. (See figure 4.1.)

It is very useful to be able to do this. For example, in brain research, it is often of interest to know how much power there is in each of the frequency components of the composite electrical activity which can be measured from the surface of the skull (the electroencephalogram, or EEG). This "power spectrum" is easily found by performing a Fourier transformation on the composite waveform, and then for reasons of mathematical convenience, squaring each of the components of the frequency axis. Or one can perform mathematical filtering of a signal by taking the Fourier transform, which gives a representation of the function in terms of frequency, subtracting the undesired frequencies, and retransforming the new frequency function back to the original function of time or space. The result is the original waveform with the undesired frequencies filtered out. This points up two aspects of the Fourier transformation of a function that make it very interesting from the standpoint of image processing. The first is that the transformation is unique and directional. That is, for every function in time or space, there is a single specific counterpart function on the frequency axis, and methods exist for interconverting one into the other. The second, as the example of mathematical filtering illustrates, is that some operations are easier to perform on the function when it is represented in one form than in the other. Thus, this technique allows us to do things which are not intuitively obvious, such as removing a particular spatial frequency component from an ordinary image.

These properties account for the frequent use of Fourier transforms in computer programming. Quite apart from any relation to analysis of

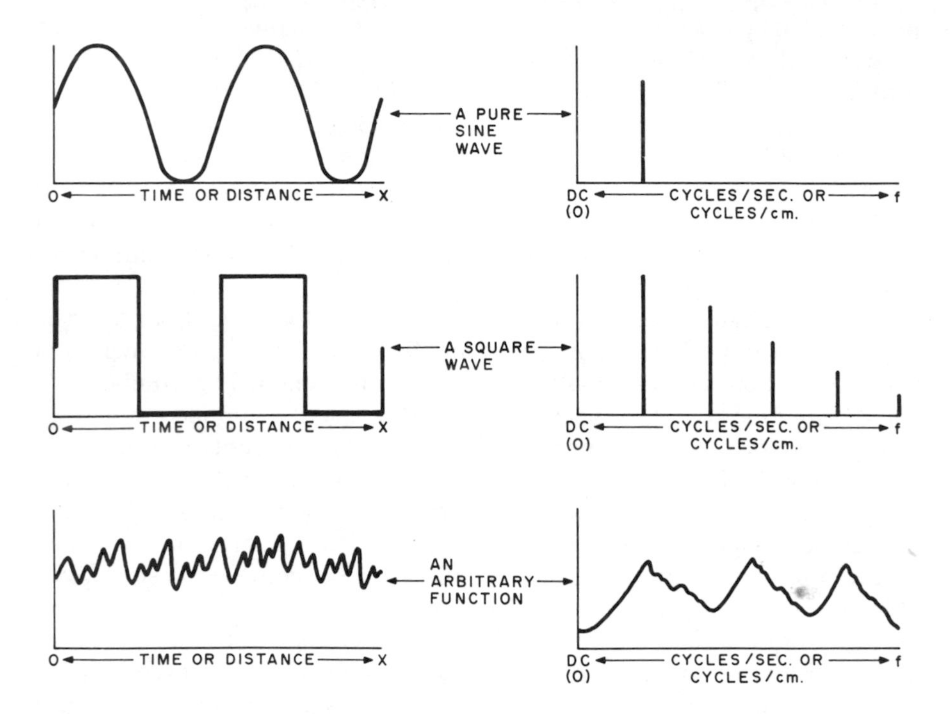

Figure 4.1: *Representation of three different functions in both the "normal" manner (time domain or space domain) and in the Fourier transformed frequency domain. A sine wave, a square wave, and an arbitrary function are shown as they would appear in these two kinds of representation. The same information is present in both cases, but is encoded differently. The differences in method of encoding may be useful in some kinds of operations on these functions. (The frequency domain is represented here only by the graph of its amplitude component; a complete representation would include a second graph indicating the phase relations of the different frequencies.)*

waveforms, Fourier analysis makes it possible to do things such as taking the convolution integral of two functions by simply multiplying their Fourier transforms and reconverting the result. As a result, a great deal of effort has gone into developing fast Fourier transform algorithms for computers. (It is also possible to produce Fourier transforms of time functions by analog means, and when applicable it is generally a very fast method.)

These developments have suggested to some workers in the field of visual perception that the brain might employ methods similar to Fourier methods, applied to the spatial frequencies present in the visual image, as a means of pattern identification. The essence of this idea is that a line of illumination intensities on the retina will have as its Fourier-transformed counterpart a unique set of spatial frequencies. The technique is easily extended to two dimensions, but becomes harder to visualize. Such a representation of retinal illumination might offer several advantages in terms of analyzing the patterns which combined to produce an image; here we have the spatial analogy of the mathematical filtering process, described above, which offers a simple way of identifying individual frequency components which are otherwise distributed within a waveform. Might it be, for example, that a "spatial frequency feature detector" could be formed by arranging for a cell to respond to some particular set of amplitudes in a Fourier-transformed representation of the visual scene?

In an experimental approach to this question, the response of the visual system to different spatial frequencies has been studied by attempting to fatigue the postulated spatial frequency extractors. This can be done by illuminating the retina with a particular spatial frequency derived from a grating. If there are cells which respond to this spatial frequency, they should be strongly stimulated by this procedure. Because brain cells become less sensitive, or habituated, to their optimal stimulus after strong stimulation, it might be expected that after being exposed to such a fatiguing process, we would see the world minus those spatial frequency components that normally would be encoded by the habituated cells. To see a simple example of the kinds of effects that this operation produces, try the experiment outlined in the caption of figure 4.2.

Astonishing, isn't it? This kind of experiment shows that habituating those cells which apparently detect one spatial frequency can certainly affect our perception of stimuli that are composed of different spatial frequency components. But, can we show that the brain is in fact using this information in the fashion we suggest?

A very interesting recent test result indicates that this may be the case. It is possible to predict how the brain should combine certain kinds of images if a Fourier process is being used. To do this, one can perform a Fourier transformation of a visual stimulus into spatial frequency components, separate the amplitude and phase information from the components, and present only one type of information to each eye. Because the more complex visual processes rely upon information derived from both eyes, we may stimulate each retina independently and expect that the two patterns of activation will be merged at a higher level. Note that this stimulus paradigm is very different from presenting portions of an image divided on the basis of geometry. The two sets of spatial frequency components don't look at all like two halves of the original image. Nonetheless, if the brain is capable of analyzing images in terms of their Fourier components, we might predict that a higher level feature detector would still achieve the correct result. Experimental evidence indicates that it does. This constitutes fairly remarkable support for the spatial frequency hypothesis of feature extraction, because it is difficult to see how a geometrically-based feature extractor could function given the highly abstracted stimuli employed here.

Additional properties of the visual system's spatial frequency-based processes may be quantitatively evaluated by similar experiments: questions of interest include system band width, the number of discrete fre-

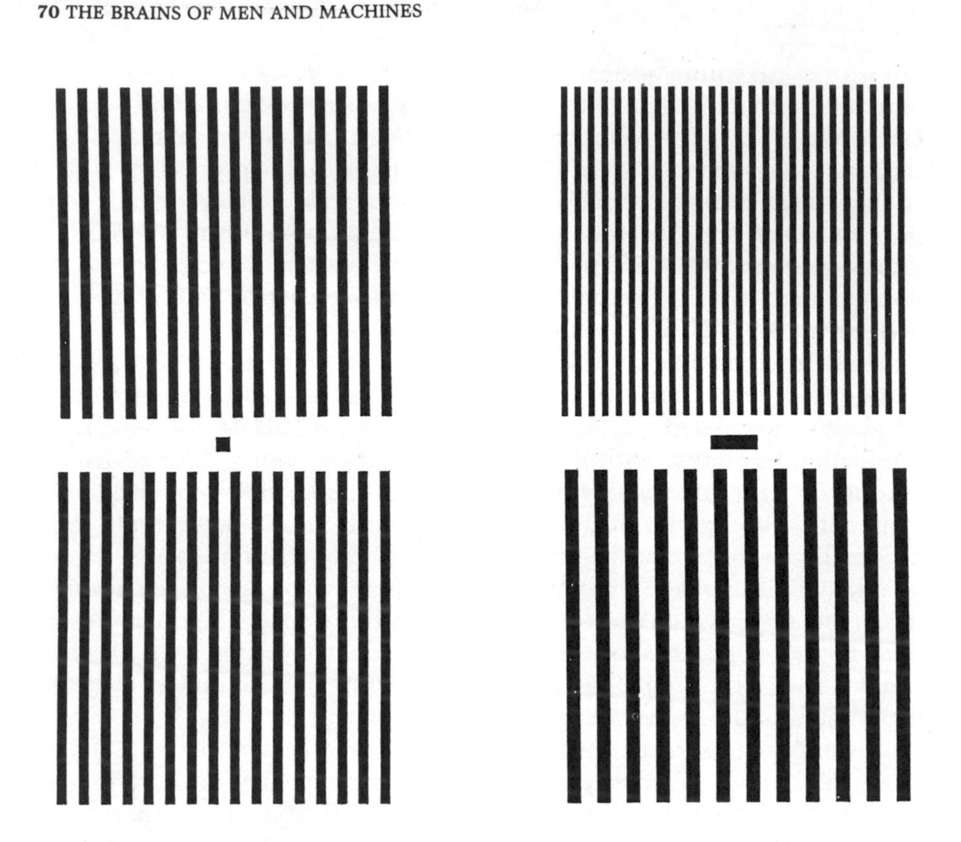

Figure 4.2: *View this figure from about 2m and fixate the small square. Note that the upper and lower grating appear the same. Now view the rectangle allowing your gaze to wander back and forth along the rectangle for about one minute. Again fixate the square and note the change in the appearance of the gratings which are previously identical. (Adapted from "Size Adaption: the New Aftereffect" by C. Lakemore and P. Sutton, Volume 166. Copyright © by The American Association for the Advancement of Science.)*

quency "channels" to which the system is sensitive, and the selectivity or sharpness of tuning of each channel. These may be approached by threshold detection measurements and by observing the various interactions of different spatial frequencies. The results of these psychophysical experiments not only illustrate a number of interesting properties of the human visual system but also are consistent with the large volume of data derived from physiological studies in animals.

Some of the most interesting results of this experimental approach are as follows: first, in the process described above it appears that high spatial frequencies are more efficiently analyzed than lower frequencies. Below a frequency of about three cycles per degree of visual angle, the brain is an inefficient detector. This may suggest that the spatial frequency mechanism is particularly important in the analysis of fine details that involve high spatial frequency, whereas the geometrically based feature extraction process is more important in the analysis of broad shapes and outlines. This dichotomy of function may explain why the fine, repetitive details of surface texture are more easily detected than are individual features of the same size. (Remember, however, that although the experiments are usually done with gratings which cover a wide angle, *any* image, even a single edge, has a characteristic frequency spectrum in the Fourier domain.)

Another finding of interest is that the system is quite sharply tuned; that is to say, the process is highly selective and can detect fairly small differences in spatial frequency. Experiments indicate that spatial frequencies more than an octave apart do not interact strongly; this finding suggests the existence of independent detectors for the two frequencies. The exact bandwidth of these detectors, or channels, is still in doubt; some experiments indicate that their tuning may be considerably sharper than an octave.

Finally, it appears that the different channels are subject to surprisingly linear additive processes. Although the Fourier process is linear, this is striking because there is no particular reason why the brain could not use some nonlinear process of its own that in simple cases might approximate the results predicted from a linear, Fourier-analysis-based model. So far, the agreement between theory and experiment is rather good. Using Fourier theory and the experimentally measured sensitivity of the visual system at several discrete spatial frequencies, it has been possible to predict a detection threshold for more complex objects, such as bars or lines, whose spatial spectrum is the sum of many component frequencies. These predictions are in good agreement with experimental results. Related experiments seem to demonstrate that other properties, such as an object's width, cannot be used to predict a detection threshold. That is, there do not appear to be width selective channels whose sensitivity may be used to predict the detection of a complex object.

Complex feature detectors might respond to particular activity patterns among sets of neurons which are sensitive to particular spatial frequencies. Such higher order detectors would look for a particular pattern of activation or "signature." That is, a feature would be identified according to the values of particular components in an object's spatial frequency spectrum.

Having considered some of the concepts underlying the spatial frequency hypothesis of feature extraction, we may look to the architecture of the brain to see how the neuronal structure of the visual system could support the experimental results associated with this paradigm. To begin with, we may compare the response of cell types with which we are already familiar to different spatial frequencies. For example, if we stimulate the retina with a single spatial frequency (derived from a grating which has periodic alternation of light and dark regions), we find that the response of the retinal ganglion cells is very specific; that is, there are large numbers of cells which respond only to narrow bands of spatial frequency. This kind of response is quite appropriate to the first stage in a system which transforms an image into its constituent spatial frequency components. These cells do not show any preference for stimulus orientation. Even though moving gratings produce optimal response, it should be noted that spatial frequency is independent of grating velocity; therefore, the RGC response is not derived from some simple consideration such as flicker.

At the level of the thalamus, cells in the lateral geniculate nucleus also show a selective response to gratings which generate a limited range of spatial frequencies; these cells show no preference for any particular grating orientation will elicit maximal response. As was the case in our initial discussion, the best response from a simple cell is obtained by moving assume that whatever additional processing is done at this stage is related either to interactions with other systems, or to some aspect of the stimulus which is more complex than the characteristics suggested by geometry or spatial frequency.

When we record from cortical units, using the stimulus paradigm described above, we find that simple field cells again show a preference for lines at a particular orientation to the vertical. For each cell a specific grating orientation will elicit maximal response. As was the case in our initial discussion, the best response from a simple cell is obtained by moving a grating perpendicularly to its lines. What is added by the spatial frequency paradigm is the finding that simple field cells also have a preferred range of spatial frequencies, which is relatively independent of the grating's velocity across the receptive field. Like the cells lower in the system, these cortical units seem to be "tuned" to a particular passband of frequencies. This may be understood in light of the previous discussion of geometric feature extraction without having to add any novel "circuitry" to the simple field cell. All that is required is to make a further restriction in the selective convergence of thalamic cells. That is, a specific simple field cell will receive input from thalamic cells whose maximal response not only falls within the same spatial frequency passband, but also is derived from receptive fields that have a unique spatial relationship.

If the selectivity seen in these simple field cells underlies the apparent spatial frequency specificity observed in human psychophysical studies, we would expect these experimental results to vary with stimulus orientation: in fact, if a grating of a particular orientation is used to fatigue a spatial frequency "channel," the detection threshold for other similarly oriented gratings in that passband is raised for a time; if the grating orientation differs by only a few degrees, the habituated "channel" does not seem to effect the threshold. This result indicates a high degree of orientation, as well as frequency specificity in the simple cells' response to these stimuli.

It appears that simple cells encode a description of the location of the relevant portion of the stimulus, its orientation, and its contribution to the spatial frequency components of the visual stimulus. If we make the assumption that simple field cells in a cortical column have a distribution of frequency passbands, just as they have a distribution of preferred stimulus orientations, then each column not only encodes a geometric analysis of that portion of the field, but also provides a multidimensional Fourier analysis of the field broken down by spatial frequency components.

In selecting their inputs, higher order cells might apply a selective convergence principle which could specify spatial frequency components at particular stimulus orientations, and thereby assemble a set of Fourier components which corresponded to any given spatial distribution of intensity.

The fact that the Fourier domain system seems to extract only the higher-frequency components of a spatial pattern suggests that this means of recording the stimuli would be particularly useful in the analysis of fine details in which luminance changes occur over small spatial displacements. Given the (relatively) large size of the receptive fields' "on" centers, it could be difficult to encode those stimuli having high spatial frequencies by using a geometric analysis paradigm.

All this is well and good, until we recall that RGC cells, at the lowest level of analysis, have a selective response to spatial frequencies. How can this be? No one knows for sure, but there are some interesting points to consider. One might account for the RGC's differential sensitivities to spatial frequency if the ratio of excitatory (or inhibitory) center diameter to inhibitory- (or excitatory) surround diameter varied from cell to cell. For example, a large central region would not be able to discriminate between spatial frequencies which could place more than one cycle into the receptive field. Conversely, a small center region with a (relatively) large surround could not differentiate low spatial frequencies, because the same illumination level would be falling on both the center and its antagonistic surround. Whenever areas of even illumination exceed the average receptive field size, as occurs in the case of low spatial frequencies, the RGCs will not respond.

This type of reasoning is consistent with RGC response characteristics which I discussed previously, and it could explain the RGC's spatial frequency selectivity to stimuli derived from a repetitive grating. It is not clear how such a system might perform generalized Fourier analysis of a nonrepetitive intensity function.

This suggests that while the lower level elements of the visual system demonstrate spatial frequency selectivity in response to stimulation by repetitive gratings and similar stimuli and higher level cells may exhibit spatial orientation frequency sensitivity, we will have to look to even higher levels of the system for a Fourier-domain-based explanation of certain experimental results. I shall return to this topic after the operation of higher level perceptual analysis, which occurs outside the primary visual cortex, is considered.

Some of the software-based attempts at artificial perception and pattern recognition have employed a Fourier-domain approach, and there are numerous reasons why this method is attractive. For example, information about an object's spatial displacement in the visual field could be carried in the phase information of its Fourier transform. An operational advantage of this fact is that one could choose to ignore such information and instead look at only the "power spectra" of such objects (produced by squaring the transform). This technique is useful for seeking patterns independently of their spatial location.

The problem with such an approach is the large amount of processing time required by traditional serial processing machines to perform Fourier transforms. It is not difficult to build analog integrated circuit Fourier analyzers, and there is considerable activity in this area. A combination of these techniques with advanced parallel processing capabilities would seem quite promising. This is an area where we might profit from an advance in our understanding of the brain's perceptual machinery.

II. THE ANALYSIS OF MOVEMENT

It is clear that one of the major goals of a pattern recognition system is the ability to detect a pattern independently of its spatial location. Otherwise, an object would be defined according to its position, and a given object, appearing in different places, would seem to be a unique entity at each location. The brain goes to some lengths to develop location-independent feature extractors. On the other hand, knowledge of an object's location is also necessary, and as we have seen, this information is available from the lower echelon, place-specific units which contribute to the operation of the higher level location-independent units.

However, a problem occurs when we examine the requirements for the visual system's analysis of movement. This process is a fundamentally important aspect of the visual system's ability to provide information relevant to dealing with the environment. Specific rates and directions of motion of objects are required to compute trajectories, and even the existence of any kind of motion is an important property that may distinguish unimportant and important objects. To take an obvious example, predators move.

It appears that the detection of motion has been a matter of great importance in the development of the visual system, and even the most primitive of such systems has special features that serve this function. Indeed, in many cases motion is one of the essential features, along with shape, size, etc that is required for stimulus "recognition" by a complex feature extractor such as the "bug detector" of the frog's brain.

The reason why detection of motion may require a separate system is that the processes described so far which are aware of location and shape would be very inefficient as motion analyzers: movement would have to be calculated from changes in the temporal distribution of activity among the location-sensitive, lower-echelon units (which activate the location-independent feature extractors). This is not a simple problem, because its solution requires a knowledge of which lower level units are being affected by which objects, and at this stage, the perceptual process does not associate specific input lines with a particular object. A fundamental design philosophy of the nervous system seems to be to avoid such a commitment wherever possible in favor of dedicated, parallel-processing of relatively abstract features.

In the case of a fundamental visual feature, such as motion, we might expect to find that specialized detectors appear at the retinal level. With respect to the temporal properties of their response to a stimulus, there are (at least) two populations of RGCs. These two types are referred to as X and Y cells, which have "sustained" and "transient" responses respectively.

The sustained, or X cell type, corresponds mostly to the kind of RGC cell which we have considered up to this point. It is characterized by a small, circular receptive field, with a well-defined center response and an opposing surround. These cells are most common near the fovea, or central area of the retina which is used during visual fixation. The transient, or Y cell, receptive field is similar in organization, but its dynamic characteristics are quite different; Y cells have larger receptive fields and become more common in the peripheral retina which observes the edges of the visual field.

The terms sustained and transient are derived from the observation that the sustained cells, while giving the type of "AC-coupled" response discussed earlier, do show a maintained "DC leakage" following the initial strong response to a change in illumination. On the other hand, transient cells are characterized by a much more pronounced degree of AC-coupling. Thus, X cells respond well to static illumination gradients over small spatial areas; Y cells respond to changing illumination over larger areas. With respect to the neuronal organization underlying the receptive field properties, it appears that the manner in which primary receptors are summated to produce the excitatory or inhibitory regions of the receptive field is different in the two types of cells. In Y cells, increases in illumination intensity tend to increase the size of the receptive field's center relative to its surround. This finding suggests that there are primary receptors of relatively high threshold, located in the retinal region related to the surround, which may be dynamically included in the organization of the receptive field center. The antagonistic "surround" effect depends on a relatively low threshold of the units in this region and may be diminished as higher threshold units come "on line" and contribute to the receptive field's "center" at higher illumination intensities.

The consequences of this architecture become apparent when one examines the symmetry of the receptive field's response. Because the receptive fields are circular, one might suppose that if a light/dark boundary bisected the field, the additive properties of the excitatory center and the inhibitory-surround should ensure that the cell's response will be constant if the light and dark halves of the bisecting field are transposed. In the case of the sustained cell type, this prediction is correct; reversing the light and dark halves has no effect. However, in the case of the transient cells, there is a distinct response to this type of change. This is due neither to any asymmetry in the geometry of the cell's field, nor to the sensitivity distribution within the field, because there is no preferred orientation of the bisection.

Thus, the transient cell responds well to changes of the first derivative of illumination patterns in its receptive field; this response is well suited to a detector designed to recognize motion. Apparently, the relative contribution of a portion of the receptive field to AC-coupled excitatory or inhibitory effects is dependent on illumination level. Even though these properties are symmetrically placed in a given receptive field, manipulation of the excitatory/inhibitory balance may be affected by asymmetries in the pattern of illumination. As we have seen above, there is some evidence to suggest that this is the case with the transient cell's receptive field organization, in which the output of a particular receptor will be coupled into an inhibitory pool at one level of illumination and into an excitatory field at another level.

There are a number of reasons for believing that transient and sustained cells contribute to different aspects of vision. Sustained cells appear to contribute to the analysis of geometrical features, whereas transient cells contribute to the detection of motion. We have seen that the sustained cells are sensitive to fine detail, as might be predicted from the more precisely defined geometry of their center/surround antagonism. Their sharply tuned sensitivity gradients are well designed to differentiate fine changes of illumination with position. In fact, the sustained cells can respond to position phase in the finest gratings resolvable by the retina, whereas the transient cells respond only to the onset or offset of motion of such stimuli. Thus, the sustained cells show a fine discrimination of the geometric aspects of spatial intensity distribution, but only a coarse discrimination of its temporal aspects; the transient cells show a coarse sensitivity to geometric aspects of spatial intensity distribution, but a fine sensitivity to its temporal aspects.

Further, there is evidence that the sustained and transient RGC fibers contribute to pathways within the nervous system which are functionally and anatomically distinct. The sustained cells have relatively smaller (more slowly conducting) axons, which project to the LGN of the thalamus in the fashion described earlier, and thence by relay to the visual cortex. The transient cells give rise to larger (more rapidly conducting) axons which also follow this route, but in addition travel to other subcortical structures involved in oculomotor coordination. The more rapid conduction of transient cell axons is what would be expected of a system concerned with temporal resolution. Further, it appears that at the thalamic level, the separation of transient and sustained pathways is maintained, so that a given thalamic LGN cell receives fibers either from sustained or from transient type RGC cells. As a result, the thalamic cells also possess the characteristic sustained or transient type responses to events in their receptive fields and preserve this distinction in their interaction with higher centers. At the level of the LGN, there does exist a strong inhibitory compiling between the X and Y cell population. That is, a rise in activity in the Y channel will inhibit the X channel. This fact may contribute to the experience that when we make quick eye movements the visual world does not smear, and our ability to detect fine detail decreases during rapid eye movements. Again, it is important to remember that both sustained and transient cells show minimal responding to a truly fixed stimulus that has no relative motion to the retina despite the constant small movements of the eye. What gives the transient cell its unique character as a motion detector is its strong response to stimuli that change in time even if the ratio of inhibitory and excitatory regions created by the new pattern of illumination remains the same as before.

As mentioned earlier, at the retinal level, we find a difference in the spatial distribution of sustained and transient cell types. While both types are found in all locations, the transient types are much more prevalent in the peripheral retina. This distribution corresponds to the fact that sensitivity of the visual system to motion is better at the edges of the visual field than it is at the center. However, acuity for form falls off markedly as we move outwards from the center of the visual field. Consider the advantages of such a scheme: first, it is an imperative of survival that sensitivity to important stimuli, such as those which move, be maintained at the visual periphery. If these attention-getting stimuli initiate an immediate orienting response, which brings higher resolution form analyzers to bear, it is not necessary to provide extreme spatial acuity at the periphery. Indeed, it would be difficult to do so because the geometric analyzers require a degree of exact focusing of the eyes which occurs only after an object is fixated and its distance is determined. On the other hand, the motion-sensitive transient cells do not have sufficient geometric acuity to profit from sharp focus.

Second, in terms of cybernetic considerations it seems to be efficient to use a scheme which orients a spatially restricted battery of high-acuity analyzers instead of evenly distributing a much larger number of these high-resolution elements across a non-orienting detector field. This fits nicely with the fact that relative to the large number of geometric, fixated objects which require detailed analysis, there are probably only a limited number of moving stimuli which require the same degree of acuity.

The distribution to subcortical structures of transient cell axons to subcortical structures other than the LGN nucleus of the thalamus is also consistent with this view. A principal target of the transient cells is a structure in the mesencephalon known as the superior colliculus. This is an ancient visual center which carries out visual processing in primitive vertebrates that do not possess a cortex. One of its functions, then and now, is to provide input to the nuclei of the lower brainstem and upper spinal cord that drive the muscles of the eyes and neck. These are important in the visual-orienting reflex, which brings an attention-getting stimulus to the center of the visual field. One of the most effective stimuli for eliciting this orienting reflex is a motion at the periphery of the field. It is almost impossible *not* to orient to an unexpected motion seen out of the corner of the eye.

Motion information is also important to higher perception and may be a distinguishing feature of objects. Indeed, one of the principles of perceptual organization, called "common fate," is that objects which move together belong to a single perceptual entity. Whether the objects move against a fixed background, or the background moves and the objects are fixed doesn't matter; only the relative motion is relevant.

Therefore it is not surprising to find that the output of LGN Y transient cells influences the response of simple and complex field cells in the cortex. Some simple field units not only have a preferred orientation and type of stimulus, but also require that the stimulus moves in a specific direction. Of course, other simple field units are quite happy with stationary targets. There are also cells which may represent higher order feature extractors for motion itself; for example, in another area of the thalamus, which receives fibers from both the visual cortex and the superior colliculus, there are cells which respond to discontinuous motions of various sorts.

It is not difficult to imagine that the motion-related properties of cortical feature extractors involve selective convergence of transient-type populations of thalamic cells, but the generation of directional selectivity is something of a puzzle. The basically symmetrical nature of the RGC field, even in the case of transient cells, does not permit a simple convergence scheme, which could not only explain a preferential response to motion in one direction but could also provide inhibition for motion in the opposite direction. Yet, direction sensitive elements are reported in the thalamus, and according to some reports, even at the RGC level. A number of possible models have been developed, most of which require the addition of a time-delay element which projects to cells lying in the preferred direction and combines with the direct, undelayed input to that area. One such model is shown in figure 4.3. While there is no positive evidence for this

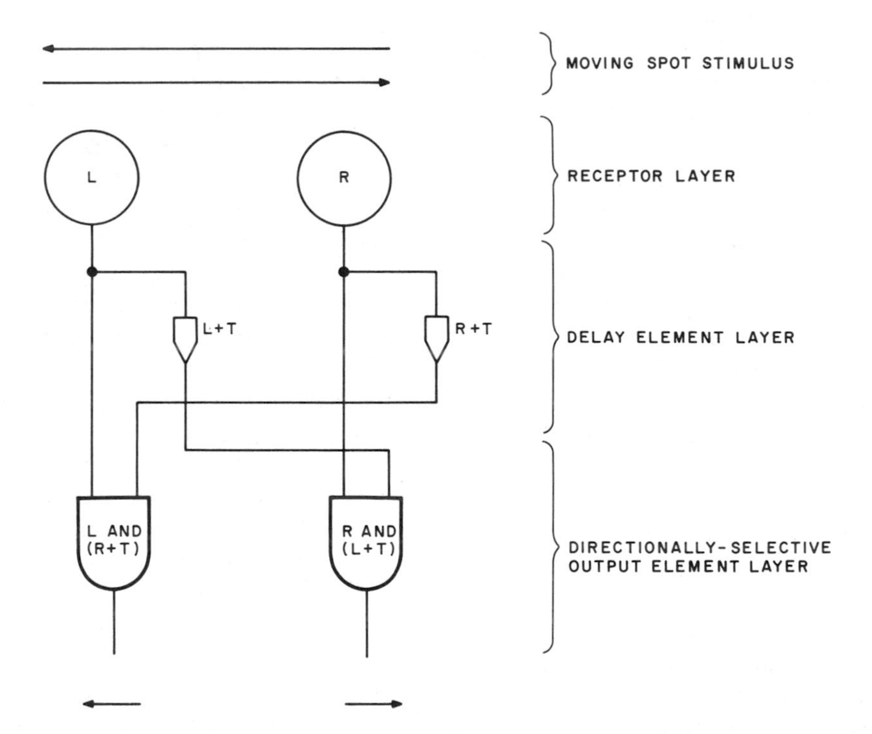

Figure 4.3: *A schematic diagram illustrating a possible mechanism by which retinal elements might display a sensitivity to direction of movement of a stimulus. The appropriate elements exist in the retina, but it is not known that they function in just this fashion.*

mechanism, existing connections within the retina could support this interaction. In any event, it is a workable model for artifical systems and might lend itself to being implemented by surface acoustic wave (SAW) or charge-coupled technology.

There are a number of convincing perceptual experiments which demonstrate that the visual system treats motion as a separate qualitative entity. For example, it is quite possible to manipulate stimuli in ways which affect the perception of motion and geometric forms independently. Because the transient cells respond poorly to either high spatial frequency or low contrast, it is possible to present stimuli which are moving quickly in terms of their geometric contours, but which are not capable of stimulating the motion detection channels; they therefore appear to be stationary. The opposite situation is also possible. Many people are familiar with the so-called "waterfall" effect; in this case, a stationary background surrounds a constantly moving stimulus, such as a waterfall. This constant motion can fatigue the relevant motion feature extractors. After one views such a scene for a few minutes, looking at another surface will give the impression that the area of the visual field previously occupied by the waterfall is "flowing" at the same rate in the opposite direction.

A more convincing demonstration of the independence of geometry and motion can be constructed with the aid of a phonograph turntable, tape recorder reel, or similar rotating surface. Watch it for a while with the eye fixed at the central hub. A minute should do. Then, while keeping the eye fixed at the same spot, have someone stop the rotation. The object will now appear to rotate in the opposite direction. If the gaze is now shifted quickly to some textured surface, a curious "twisting" movement will be seen, which demonstrates that the effect is quite independent of the geometric nature of the visual stimulus which fatigues the motion extractors. It is interesting to note that the effect can also be produced, although less strongly, by observing the motion with one eye and then using the other eye for the visual aftereffect. This indicates that at least some of the fatigue is occurring in motion extractors above the retinal level.

Finally, you can also demonstrate the perception of "pure" motion without being aware of the object's form. This demonstration relies on the poor coverage of the periphery of the visual field by geometric line detection channels and its good coverage by motion sensitive channels. Look straight ahead, and fix your gaze on some convenient point. Then, holding up the first two fingers of each hand, and holding your hands at arms length, move your hands to the extreme periphery of the visual field. With a little experimentation, you can find the point at which your fingers just disappear at the extreme edge of your field of view. Now, when you have arranged it so that you cannot see your fingers while looking straight ahead, wiggle them slightly. The motion is instantly detectable, and at first you have the impression that you can in fact see the fingers again. However, most people will find that there is an identifiable zone on the periphery of the visual field where they can actually see the motion of the fingers without being able to see the object which is moving. This "ghostly" motion is rather disconcerting and is an interesting perceptual experience. Note also the strong tendency to orient the eye towards the motion.

III. VISION IN THREE DIMENSIONS

Almost everyone is familiar with the notion that depth perception is primarily derived from the existence of binocular vision. In reality, binocular vision is only one of many factors which the brain interprets to estimate an object's distance. These factors include the degree of surface texture, illumination, knowledge of an object's real size, interposition of objects at different distances, apparent motion of the object, relationship to the background, observer's movements, etc. Clearly, when we cover one eye our depth perception is impaired, but does not vanish. Many species do quite well at depth perception even though their eyes look in opposite directions and there is very little overlap of the left and right visual fields.

You can see the effect of some of these cues by covering one eye, and holding still. Notice that there is still depth to the visual world. Now, move your head from side to side, while still keeping the eye covered. You will notice an increase in the apparent depth of the scene as apparent motion relative to the background is added to the list of available cues. Finally, when both eyes are opened, the contribution of binocular vision can be appreciated.

Obviously, many of the other cues which are employed, such as knowledge about the real size of an object, require levels of processing that are not available to the primary visual system and probably represent processes in perception that occur at higher levels of the system. However, in the case of binocular vision, there seems to be a clear basis in the lower levels of neural hardware for the interpretation of visual depth. The principle behind binocular depth perception is simple trigonometry. The brain receives an image from both eyes, and in order for the two images to exist on corresponding parts of the retinas, the eyes must be crossed to a degree determined by the fixed distance between the eyes and the distance to the object in question, which varies. If the degree of angular deviation of the eyes required to make the two images of an object coincide is known, the distance will be encoded uniquely. This mechanism is most important for near distance, where the deviations are large relative to the object's distance. In the far distance, small changes in angular deviation correspond to large changes in object distance and other cues are most important. There are two possible approaches here: depth information could be inferred from the degree of ocular convergence ("eye crossing") required for the image of the object to register on equivalent parts of the two retinas, or it could be determined by the degree of disparity in the retinal positions of an image for a given degree of ocular convergence. The brain probably makes use of both mechanisms.

The brain has a simple measure of the degree of ocular angular deviation in the output of the pontine and medullary "oculomotor" nuclei which drive the eye muscles. The trick is to know when the two images are being received from equivalent parts of the two retinas. In an attempt to explain how this could be done, we need to know in detail how the data from a particular retinal location is passed to the brain. On the way from the retina to the thalamus, optic nerve fibers cross over in such a way that those from the left half of each retina are routed to the brain's right hemisphere, and those from the right half go to the left hemisphere. Thus, each hemisphere receives two independent sets of data, one from each eye from the part of the visual field with which it is concerned. Recall that the visual cortex was organized functionally into vertical columns, each of which dealt with the analysis of a small area of the visual field; within each of these columns, the input from one eye is dominant (giving rise to the name "ocular dominance columns"). Moreover, there appears to be a regular alternation among the columns with regard to which eye is dominant. Thus, every column has a neighbor which analyzes the same part of the visual field, but the analysis is dominated by input from the opposite eye. When the two columns have matching analyses, it indicates that the eyes' convergence is correctly marking the distant of an object in that part of the visual field. It is not difficult to imagine cells which would "AND" this information and fire only when two adjacent columns have identical output. Such a mechanism might signal that an association was to be made between an object and the distance encoded by the output of the eye muscle drivers. You can readily observe this "double image" process. Hold up your thumb and look at a distant scene beyond it. If you maintain this fixation while paying attention to the thumb, you will see two thumbs.

However, there are aspects of binocular vision that require a different type of analysis. In particular, there are certain experiments which demonstrate the ability of the eye to generate an illusion of depth on the basis of small binocular discrepancies; given a set of random dots whose retinal position has a small binocular disparity hidden within a field of other random dots that have no binocular disparity, one will see the first set "float" or "sink" above the background. In this case, there is a very pronounced perception of differing depths, yet the eyes are in fact converging to the same extent for both portions of the stimulus. This indicates that differences in retinal position between the two eyes must also be used to generate depth information during constant degrees of convergence.

An aspect of receptive field organization within the ocular dominance column seems to provide an explanation of this ability. Some cortical cells receive input from binocular receptive fields which are displaced horizontally for a given degree of ocular convergence. When the image presented to the two eyes has a retinal disparity which matches this displacement, the two receptive fields will sum their activities; when the respective images do not have this critical displacement, the two receptive fields will interact in an inhibitory manner. Thus, there are a series of cells within each cortical column which are sensitive to specific degrees of spatial disparity between the corresponding portions of the two images. This allows the elaboration of a spatial depth channel which extracts depth information on a point-by-point basis in the visual image, even when ocular convergence is held constant. In each column, there are cells tuned to specific values of depth.

Cells which are inhibited by a mismatch of their preferred degree of retinal displacement and the two retinal images of a given visual stimulus provide a basis for the suppression of out-of-register images. This fact could be important in the creation of a single image out of the two slightly different views provided by each eye.

In a recently developed robot system that can drive a car at 30 miles per hour and avoid obstacles, a depth perception system has been employed which seems similar in principle to the scheme above although its practical operation is rather different. In this system, two television cameras view a scene from different positions and produce slightly displaced images as a result. In this arrangement, the difference in "position" is coded as time, because the image is scanned sequentially at a constant rate. In this case a depth channel consists of a device which searches for coincidence in the two images at a specified time lag. Whether the image shift is encoded in terms of a time lag in a scanned system, or as a distance on a receptive surface in a non-scanned system, is really a matter of engineering convenience.

IV. A COMPARISON OF VISION AND HEARING

Having examined in some detail the processes underlying the initial processing of information in the visual system, I would now like to discuss (in more condensed form) the operation of the initial stages of processing in another modality — hearing — before I attempt some generalization on the brain's approach to the initial stages of sensory processing.

The fundamental data in the auditory system are temporal patterns of frequency and amplitude. From this information, the auditory system is required to differentiate sounds of significance to the organism and to determine their spatial locations. In many respects, the basic task is fundamentally different from that performed by the visual system. However, I hope to convince you that despite the differences in processing which, at the lower levels of the system, are related to the physical nature of the stimuli, the higher levels of the auditory system organize the data in a way which is reminiscent of the higher levels of processing in the visual system. The significance of this observation will be discussed in a subsequent section.

To begin with, the transduction mechanism for sound consists of a pickup, the eardrum, which converts pressure waves in the air to mechanical vibrations in a fluid-filled chamber (the cochlea) and performs mechanical impedance matching between them. Within this chamber, a membrane of tapering width is set into motion by the vibrations from the eardrum and takes the form of a traveling wave. Due to the mechanical properties of the membrane, the maximum displacement produced by this traveling wave occurs at a different point for each frequency. Where the membrane is displaced, it produces firing in the fibers of the auditory nerve that originates in cells which are stimulated by it, and the frequency of firing is proportional to the degree of membrane deflection. I need not discuss the exact mechanism by which this transduction occurs; it is probably of little interest to the designer of artificial systems. The important point to note here is that one qualitative component of sound, pitch, or frequency has immediately been reduced to a "place code." That is, pitch is represented by which group of neurons is stimulated, not by their firing frequency. Intensity is represented by how frequently a particular pitch-encoding group of neurons fires. (There may be some exceptions to this statement in the initial stages of processing for very low pitches, but these tones are still assigned a place code before ascending very far in the system.)

The initial stages of auditory processing, prior to the cortex, are not as simply laid out as the visual system is. There are a variable number of neuronal relays which may introduce delays into the system that are important in the analysis of temporal patterns. Also, there are several way stations in the auditory pathways which provide significant interaction between the ascending fiber tracts arising in each cochlea. The interaction may encode phase differences which are important to auditory localization. As in the visual system, the auditory nuclei project to subcortical centers which have auxiliary functions.

At the earliest levels of auditory processing, which occur in the brainstem, one finds mechanisms which are reminiscent of the RGC receptive field. For example, there are neurons which are rather sharply tuned to narrow bands of frequency; that is, these neurons are excited by sound frequencies within a restricted range, but are inhibited by broad bands of other frequencies.

Such a compound excitatory-inhibitory tuning mechanism is probably required to provide a sharp frequency response, given the limited selectivity that could be produced by the cochlear traveling wave mechanism alone. However, once achieved, this finely resolved spatial coding of frequency is not retained as a characteristic of neurons at high levels of the system. Rather it seems to represent an initial processing stage which is required if higher order complexities are to be resolved on the basis of power spectra analysis. Pitch perception appears to be completed early in the analytic process. As might be expected from this conclusion, removal of the auditory cortex has no effect on the ability of animals to distinguish pure tones, although the ability to distinguish more complex features of acoustic stimuli is lost.

In another subcortical auditory center, the inferior colliculus, there are neurons which seem to be selectively sensitive to those stimulus properties which allow determination of the spatial location of the sound source relative to the organism. Here we find cells which are optimally activated by particular delays between the time of arrival of sound at each ear, but which are quite unconcerned about the particular frequency or intensity of that sound. Other cells in this same structure show a sensitivity to differences in preferred degrees of intensity between the two ears, but show no preference for absolute level of intensity. Still others are sensitive only to sounds with a continuously changing relation between the signals from the two ears (ie: a motion detector in auditory space!).

At the level of the auditory cortex, early studies indicated that there was a "tonotopic" arrangement of the neurons; that is, the frequency response spectrum was mapped across the surface of the auditory cortex in much the same way that the spatial map of the retina is preserved across the visual cortex. However, it now appears that there is little if any significance to this mapping, except perhaps for some convenience in establishing interactive connections. For one thing, there are now known to be at least three such maps in the auditory regions of the cortex. Further, most cortical cells do not even respond to well-defined frequencies any more than visual cortical cells respond well to simple light stimuli. It seems more likely that the neurons of the auditory cortex extract specific features from the spatially coded frequency and phase relations encoded by the neurons at lower levels of the system by means of selective convergence and lateral inhibition.

A cat which has had its auditory cortex removed will easily be able to discriminate three different tones, but will be unable to discriminate two different "melodies" (ie: orders of presentation), composed of those same three tones. When one looks at the kinds of stimuli preferred by individual neurons of the auditory cortex, one finds cells which are sensitive to rather broad bands of frequency (like location-independent visual cells, these auditory cells are frequency-independent). Three quarters of them do not respond to a simple pure tone stimulus. On the other hand, these units do tend to show very specific preferences for more complex properties of sound such as frequency or amplitude modulation. Some of these cells may show a preference for sounds with a particular range or direction of modulation. Others may require sounds which have particular attack or decay times, or some ratio of attack to decay. Yet more complex types may respond to the degree of envelope symmetry, or to order and timing of certain components of a sound. There is some evidence that there may be a mapping of units in two dimensions by frequency and intensity, so that particular combinations of these two properties may be selectively extracted and enter into yet higher order analysis.

In species ranging from frogs and birds to monkeys, a number of studies show that there may be units in the auditory cortex which are selectively responsive to species-specific vocalizations. Most animals employ a variety of specific vocal signals that constitute a rudimentary language; these calls serve specific functions of social interaction such as warning, mate-calling, territorial claim, etc. They are not really equivalent to human language, because the meanings are highly specific, and they appear to be inherited rather than learned. The stereotyped nature of some vocalizations makes them good examples of complex features which could be detected by wired-in units of the auditory cortex. These units probably represent the highest level of genetically-determined feature extraction. Similarly specific visual cues subserve vital social functions in animals, and one might expect to find analogues in the visual cortex, although I am not aware of any studies in this area.

The transformation of complex temporal waveforms into a spatially-mapped frequency spectrum by the chochlea and its associated nuclei is another example of analog Fourier analysis. It would seem that this process alone would be sufficient for the perception of pitch; however, there are pitch phonomena, resulting from combinations of complex sounds, which seem to require a more elaborate explanation.

For example, "resultant" pitch phenomena can be obtained by presenting two different complex sounds independently; to each ear their combination into a subjective pitch is occurring at a higher level of the system. Several models which might account for these phenomena suggest that in the auditory system, pattern recognition is based on a search of the Fourier spectra from particular stimuli in much the same way that the visual system seems to use spatial frequencies.

There are several other senses which feed the central nervous system; some of them, such as smell and taste, are less understood in comparison to the visual and auditory system. In the case of touch, it appears that principles similar to those used in vision and hearing may be employed. For example, cells of the somesthetic cortex, which receive sensory input from touch receptors in the skin, have a spatial receptive field on the surface of the skin which possesses the familiar center-surround structure seen in the visual field of the RGC. Senses such as taste and smell do not seem to have cortical representation, although their projections onto subcortical centers have been investigated; for these two senses, it appears that there is a spatial representation of sensory quality associated with specific receptor fibers, whose rate of firing encodes quantity. As one ascends in the system, various combinations of these features are converged to create higher order units which have greater specificity. At the present time the existence of selective convergence and lateral inhibition in such systems is uncertain.

V. SOME GENERAL PRINCIPLES IN SENSORY PREPROCESSING

So far, we have been concerned with understanding the operation of the brain's techniques for the initial analysis of its input data. By "initial analysis," I mean that processing of primitive information that takes place in a genetically or developmentally determined manner; this level of analysis is based on current receptor activation and is confined to a single sensory modality. This analysis, which I have rather arbitrarily taken as extending from a receptor to the primary sensory cortex, can be thought of as having the same relation to higher order perceptual processes as a preprocessor has to a central computer. As shown, the architecture is rather different but similar principles may apply; the central machine relies on the preprocessor to provide it with an easily digestible summary of relevant aspects of the information and to delete that which is inessential. Intelligent terminals prevent the central processor from being burdened by lower level activity. Except for rather general specifications, the preprocessor is allowed a free hand in its area of concern because the central machine knows the general algorithms which are being applied. It would not do to have the preprocessor experiment on its own or to take orders from other parts of the system.

Further, as I will presently attempt to show, there appear to be principles of coding with which the primary receptors' diverse codes for widely differing physical properties are progressively reduced. In all probability this is the internal code of the central mechanism—the code with which it develops the perceptual world model constructed out of the totality of past and present sensory input.

I have dealt primarily with the visual system because it is relatively well understood, and because its operating principles are of some interest in their own right to the cybernetic architect concerned with artificial perception. Certain general notions developed in that view have been shown to apply in the case of the auditory system as well. For initial considerations of the operation of the higher perceptual apparatus, it may be useful to examine the idea that these fundamental principles of sensory analysis are independent of modality and to explore the implications which such a view has for the nature of the central process.

First, let us review the notion that stimulus "quality" is represented by a place code ("which" neuron), while stimulus intensity is conveyed by a frequency code ("amount" of neural activity). At the lower levels of the sensory systems we have examined, this concept has served us well. Visual location and color are encoded in a neuron's identity, while light intensity is encoded in its firing rate. Similarly, sound frequency is encoded by a particular neuron in the auditory nerve, and loudness is indicated by how rapidly it fires. As selective convergence was employed to generate new sensory qualities, such as inclination, the degree of firing of the "extractor" for that quality was representative of the intensity of the stimulation.

In the case of a higher order feature detector, the interpretation of the "intensity" of stimulation deserves more careful analysis. How does one conceptualize "intensity of inclination?" Previously we have spoken of "degree of certainty" in this regard, and more explictly, it would seem that in a case such as this, the firing rate of the extractor is really encoding the degree of match between the cell's input pattern and its "ideal" input pattern. That is, the neuron is essentially "tuned" to a particular pattern of inputs, high on some lines, low on others, and its firing rate declines whenever the input pattern starts to depart for any reason from the optimal one. Given a detector for a line, at a particular location with a particular inclination, we see that the firing rate would be less than the optimal if the line deviated from the desired inclination, or if the position of the line were displaced from the optimal position, or both. The evidence suggests that most such cells also have inputs which define other desired properties of the line, such as color, movement, etc. The properties are assembled by an appropriate convergence of detectors. Thus, the cell is tuned in a multidimensional space, and firing rate encodes the "degree of fit" between the stimulus, as characterized in that space, and the cell's tuning curve.

Conceptualizing such a feature detector as a multidimensionally tuned system (essentially, a vector in a multidimensional perceptual space), allows one to make use of the relatively broad tuning of the cells we have been considering; and this has some very interesting implications for the nature of the central coding scheme of the brain. However, before considering this, let us examine a few other examples of sensory input to see how this model of a tuned detector applies to the general case of sensory analysis.

In the case of color, we find that in the retina there are three classes of receptors which are broadly tuned to the three primary colors of human vision by virtue of particular pigments which absorb various wavelengths of light. The filter curves are broad enough that a stimulus of almost any wavelength will produce some activity in at least two, and probably three, color types. Thus, we should think of the system's response as being defined by the *relative* activities of these three types insofar as we are talking about quality (wavelength) and by the absolute levels of activity in the three types when we are talking about quantity (intensity).

Therefore, it is no surprise to find that the next step in processing of color information is a set of neurons which respond to particular *balances* of activity among the primary color receptors. These are the so-called "opponent process" cells of the color vision system, which have excitatory inputs from one primary color receptor and inhibitory inputs from another. The result is a cell that is "tuned" to a particular combination of activities among primary color receptors.

Next, consider the case of kinesthesis—the sense which defines joint angle for the brain. One might expect that there would be neurons which fired at progressively higher rates as the joint was opened further and further. However, such a scheme would present the system with the difficult task of making fine discriminations of the firing rate of the cell in order to decode the joint angle. It appears that the brain chooses instead to let angle be a quality, and encodes it in terms of which neurons are firing. We find that the kinesthetic receptors which service a particular joint each fire most strongly when some particular joint angle is reached, and their firing rates fall off on either side of that angle. Thus, they constitute a set of tuned receptors for particular angles of the joint, and their firing rates encode the degree of match between the actual angle and their own preferred angle; at any angle, there will be some group of receptors that is active.

The breadth of the kinesthetic receptors' tuning curves implies that fine discriminations of angle would have to be made by establishing "angle detectors" through the process of selective convergence, both inhibitory and excitatory, to produce a pattern of activity in a population of inputs which corresponded to the desired quality. That is, any angle would activate several broadly tuned receptors to varying degrees and not activate others at all. The relative activities could define the angle as finely as desired simply by polling more of the broadly tuned detectors. Similarly, temperature is a quality encoded by populations of cells tuned to respond maximally in specific temperature ranges, rather than a quantity of heat-stimulation encoded in the firing rate of a single cell type.

What is seen in all of these cases is the establishment of feature detectors which are based on a common formula. That is, place code defines the degree of match between the activation of the relevant ultimate receptors and the unidimensional or multidimensional tuning curve of the detector. It does not matter whether the quality in question is inherent in the physical nature of the stimulus (as in color and the wavelength of light) or is a "derived" quality extracted at a high level of the system (as in inclination). Neither does it matter whether the quality is a geometric property such as location, in which changes in the quality are encoded at the primary receptors in terms of which ones of some population of receptors are activated, or a non-geometric property such as color, in which changes in the quality are encoded in relative activities of an unchanging population of receptors.

For some modalities, such as color, temperature or joint angle, there are not many features to detect, and this stage of representation is arrived at early in the analysis in subcortical structures. For example, color processing appears to be complete at the thalamic level, even though the output of this color detector machinery may then rise to the cortical level as one dimension of input to multidimensionally-tuned higher order detectors. Other qualities, like the various features of inclination, angularity and curvature, which are important in the ultimate perceptual analysis of form, require more stages of processing for their derivation and are represented at higher levels of the system, such as the visual cortex.

To summarize this position, it is apparently the case that the brain defines its perceptual qualities in such a way that the nature of the quality is encoded in which cells are active, and the rate of their firing encodes the "goodness of fit" between the stimulus to which their receptors are exposed and the feature to which they are tuned. This process is an evolutionary development of the encoding process at the receptors, where firing rate is correlated with stimulus intensity.

This view of the feature detector would not be of much use if we were to conceptualize the system as continuing the process to the point where there was a specific neuron for encoding the presence of each and every possible feature. That is, if we thought of the system as consisting entirely of convergent trees of selected cells which culminated in a single cell and expressed, by its activation, the existence of some precise pattern of activity in the receptors from which its particular unique convergence selections are derived, then the breadth of tuning in the lower echelons would simply be indicative of an incomplete process.

This sort of reductionist representation of the sensory world, in which there is a specific cell for every possible perception, is a very unlikely model of the brain's internal code. In the first place, the number of "ultimate" cells required would be unthinkable, even by the brain's standards. There would have to be a separate cell to encode every possible state of the sensory world which you could discriminate. In the second place, the brain is constantly losing neurons. Like other cells in your body, they die all the time, but unlike other cells in your body, they are never replaced. Your brain is shrinking at the rate of about a thousand cells per hour. Clearly, this is a system that needs a lot of redundancy. Or, the system is able to run a relatively constant program in a varying supply of parts. If every discernible sensory event were encoded in a single, highly complex multi-modal feature extractor, and sufficient redundancy were provided to account for the brain's apparent insensitivity to damage and cell loss, the problem would become completely ridiculous.

The implication of these considerations is that the brain does not, indeed could not, encode perceptions of the world in the form of single active lines for each perception. The alternative is to encode them as some pattern of activity in a population of neurons. If that is done so that the set of discernible sensory events is encoded in the set of possible permutations of activity of a large number of neurons, both the quantity and redundancy problems become manageable. The distinction is exactly that between coding a number by a bit position in core or coding it by the arrangement, or permutations, of bits in a small byte. Numbers from 0 to 65,535 can be encoded in the permutations of the bits of a 16 bit byte. The alternative would be to encode the numbers as the presence or absence of a "1" in one of 65,536 bits. The evidence suggests that the brain employs a permutation code for the same reasons that computer engineers do.

The advantages obtained by the brain's simultaneous parallel processing and nonsynchronous operation are especially well suited to this kind of encoding scheme. Any functional unit of the brain can have access to any "bit" from any part of the perceptual code at any time independently of any other functional unit which may be looking at the same information. (I am not implying a random access capability; a functional unit would have to be wired for the bits it was going to access, either genetically, or perhaps developmentally during the perceptual learning process.) It is as though you had a large number of parallel processors working on some part of the perceptual problem, and each is able to access the collection of bits it needs from core at the same time.

If this is true, why have feature detectors and selective convergence in the first place? Clearly, all the information on the retina is encoded in the pattern of activity of the population of retinal cells. Why not have the decision-making apparatus pull whatever bits it wants that encode something? Of course, the answer is that it is a trade-off between hardware and processing costs, just as in the input preprocessor case in a conventional system. It is too expensive in terms of hardware to reduce every possible sensory state to a single active bit, and it would require much processing to have the central machinery attend to every raw bit on the retina. The compromise is to have the central processor (or in this case processors) look at activity patterns in a smaller group of neural lines that are already reduced to a set of commonly useful features, such as angles, edges, lines, arcs, etc, and deal with these encoded patterns in the equivalent of a smaller byte

The system is free to carry the feature extractor process to whatever lengths are useful. In his excellent discussion of these points, Erickson (see bibliography) makes the observation that we would expect to see large numbers of rather general-purpose feature extractors corresponding to the most common sorts of stimulus elements encountered and a few very specific extractors where they proved useful. We would not expect to see many cells devoted to highly specific feature recognition, because they would be seldom used and it would be wasteful. Curiously, the prediction of this argument, which is borne out in observation, is that you would expect to find the most specific feature extractors in the most primitive systems.

If one considers creature such as a frog, it is clear that it has a very limited repertoire of behavior, and that these are controlled by a very limited number of stimuli. The system has only a few basic decisions to make. It does not need to discriminate a variety of different sensory inputs, but only has to recognize a few. It has little use for large numbers of general-purpose feature extractors and concentrates its limited hardware resources on the most efficient possible extraction of a few highly specific stimulus features. These are all it requires for the handling of its limited behavioral repertoire. Experimentally, in frogs' eyes we find very specific detectors for bugs, and in frogs' ears, very specific detectors for mating calls. Everything else can be pretty well lumped under "undifferentiated obstacle"—go around, or "undifferentiated danger"—flee, according to whether it is large and non-moving or large and moving. At the level of the mammal with a large behavioral repertoire, we need something like a general-purpose perceptual analyzer, and we find a visual cortex with many more feature detecting cells. However, the majority of these cells are less specific.

This range of levels of specificity can be conceptualized in our previous terms by saying that the central machine's processors may access a selection of data bytes from the collection of bits representing the outputs of the preprocessor's feature extractors. Some are long bytes of rather nonspecific feature detectors that encode specific information in a large number of permutations, while others are short bytes (in the extreme case, one bit) that encode specific information in fewer permutations. The short bytes save central processing, but are expensive in terms of preprocessor hardware in their production. The fact that evolution has found the tradeoff to favor complex detectors for simple brains suggests that we might want to consider this guideline in the development of our robot's systems. The earliest, most behaviorally-limited robots should have a small number of carefully selected specific feature detectors in their perceptual systems.

Let us return to the question of broad versus narrow tuning in the multidimensional detector's input space. This concept is directly related to the number of "lines" which are required to specify the full perceptual range of the quality in question. In the case of a simple, one-dimensional quality such as color, it is clear that the permutations among three broadly tuned detectors are sufficient to specify the range of colors seen at each image point. At the other extreme, a detector for a specific object at a specific location (if there were such a detector) would have extremely sharp tuning in many dimensions. If it did not, it would lose its specificity.

When we examine an actual feature extractor of the primary visual cortex (eg: the line inclination detectors), we find that the tuning is rather broad. The specificity is not so broad as the spectral tuning of the color receptors, but still is much broader than that which would be required to support our ability to discriminate lines at different angles. This suggests that within each small visual area subserved by a set of orientation detectors, the quality "inclination" is encoded as a pattern of activity, or permutation, among a relatively large number of these rather non-specific units. The more narrowly tuned the detector elements are, the greater specificity they have, but then more of them are required to cover the full dynamic range of the quality. A retinal color detector will be satisfied by absolutely anything that has the proper location and color. A more narrowly tuned cortical line detector requires a more extensive and specific set of stimulus properties for satisfaction, but many more of these elements are required to cover the same small retinal area with regard to that quality.

An extension of this method of encoding perceptual information is the idea that the breadth of the tuning curve for a multidimensional detector need not be the same in all dimensions. For example, at high levels of processing, we tend to find units which may be narrowly tuned for form (object specificity), but are broadly tuned for location. I have already mentioned auditory cortex units which were narrowly tuned for qualities such as directional frequency modulation, but are broadly tuned for absolute pitch. Such units could retain the generality of broadly tuned units with its attendant savings in hardware, while giving the advantages of narrowly tuned units with their savings in further processing costs; this is the real import of the location-free feature detector.

Presumably, the central machinery could mix together any combination of broadly and narrowly tuned units to drive further processes, but in doing so it would be committing that part of its apparatus to increasingly narrow functions. In effect, it would be creating a further step of extraction. The considerations we have just examined suggest that the central machinery will prefer to keep its units generalized and to allow the rise of specificity through particular momentary patterns, generated in broad "bytes," by the interactions of several broad types of relatively nonspecific elements. Narrow bytes of specific elements probably have limited use in special-purpose fast response systems.

In fact, this feature of differential breadth permits one of the most important properties of the brain's perceptual coding scheme. This is the parallel addition of different perceptual analyses to the total code. Simple qualities, such as location and pitch, may be extracted at low levels of the input system and placed onto an "input bus" where higher levels can make use of them. More complex features may be extracted by higher level input units which have access to these "bits," but their outputs do not replace these earlier bits on the input bus; they are added to them in parallel. This means that the central machine gets a final input "byte" which has bits that come from lower levels of input processing and which encode qualities such as location without regard to content, and other bits that come from higher levels of input processing and which encode content without regard to location. The higher level units are broadly tuned with regard to qualities such as location and narrowly tuned with regard to content. The reverse is true for the lower level bits.

There are many advantages to this type of system. First, there is savings in hardware; you don't need to have a different higher level unit for each location of its particular specific feature. The low-level location code, which provides the higher level extractor's inputs, is passed along to yet higher centers together with the output (if any) of the higher level extractor, and the whole combination is the code for the particular object at a particular place. The central machine does not have to waste time analyzing the set of location codes for different patterns representing content; that has already been done, if possible, by the higher level extractor, and the higher level extractor's output does not need to convey location information. This presages the existence of two perceptual "channels," one for recognition of content without knowledge of location, and one for awareness of the spatial framework of the world without dissection into categorized objects. Normally these two aspects of our perceptual experience function so closely together that we are not even aware that they are separate processes.

The second advantage is more fundamental. Suppose that we didn't care about hardware costs; we could have a higher level extractor for every location of every stimulus, no matter how complex. Nonetheless, we would still want to design our system with parallel addition of features as they were extracted. This is because any stimulus pattern which failed to fire some complex feature extractor would then be lost. The information would simply disappear. You might devise some kind of error routine, like the ones you use when the input doesn't match anything in the lookup table, but that's not satisfactory when there's no one to read the error messages. Think about your experience with the cow picture. If that worked for you as it was supposed to do, what happened was that you got a lot of information from low-level extractors put onto the bus, and the central machine had access to them, but some set of feature extractors for encoding "cow" didn't fire because of inadequate input. When they finally did, you got the perception of a cow *added* to the perception of lines and shapes that was already there. Nothing changed in, or was added to, your perception of the basic simple content of the visual field, but something else was added to it in parallel. If we were using serial decoding, nothing would have gotten through until the "cow recognizer" finally fired!

Thus, the advantage of adding features in parallel to the perceptual code is that the entire "byte" is available at any given instant to the higher centers. This is immensely important in a machine designed for a generalized environment. When this design principle is employed, fine tuning of high-level extractors with regard to properties encoded at lower levels is superfluous because their outputs would then encode that information, and it would be redundant. It is also wasteful because many more high-level extractors would be required. Broad tuning of high-level extractors for such lower level properties means that their outputs do not encode those properties, which is unnecessary, and as a result, fewer of them can be used for the same property. Hence, system resources can be devoted to increasing the number of properties dealt with by high-level extractors on a fine-tuned basis.

For some reason, this picture of the nature of the central code is disquieting. It is somehow much easier to associate a subjective perceptual experience with activity in a single specified cell than with something so abstract as a pattern of activity in a group of cells. There is, however, no reason to prefer one over the other as a logical basis for subjective experience. In fact, there is no suggestion of a way to link the mental experience to the nature of the correlated physical brain event; in either case, one is as good (or bad) as the other as far as any *a priori* basis in logic is concerned.

If the perceptual code given to the central machine consists of permutations of a large number of lines which, on the whole, are rather broadly tuned, and if further processing is represented by patterns generated from interactions between equally broadly based systems of cells and these input lines, it follows that the interacting systems, and the output system to which the result is delivered, must be of similar organization in terms of data representation. That is, we expect that the "top end" of the final output processor has a large number of broadly tuned input lines (for representing actions), and that movement specificity results from a specificity in the pattern of activation of these lines. This approach is different from the notion that a particular movement is the result of the command from a particular high-level motor system cell fanning out to the muscles.

This position is consistent with experimental evidence and with the considerations of cybernetic efficiency that we applied to the input system. (If it were not so, the cell at the top of a movement heirarchy would essentially be the final and most specific detector of the input system!) In computer terms, the output processor is handed a control byte, which it decodes to produce the desired motions.

The final representation of the data from the input processing system occurs in the internal code of the central machinery. Erickson has called the process "parallel population coding," and the implications of this approach for the design of artificial intelligence systems are many.

5 The Higher Perceptual Processes

I. THE SECONDARY SENSORY CORTEX

The preceding chapter has developed the basic ideas which seem to underlie input preprocessing in the central nervous system. Although receiving different kinds of physical stimuli and requiring very different types of receptor processes, each of the sensory systems seems to transform its information into a common perceptual code; perceptual "qualities" are derived from stimulation and are represented as activity in particular neural lines. In a modality, the full picture of activity is represented by permutations of activity among these lines. However, there is still an enormous gulf between the kinds of information produced by this process, at the level of the primary sensory cortex, and our usual subjective perceptual experience of the sensory world.

The first level of visual sensory organization is sufficient for encoding simple features such as lines and colors, but it does not explain how one may discriminate an object or other unified perceptual entity from its visual background. This sort of operation requires additional processing which includes not only further analysis of patterns in some modality such as vision, but also the information with data from other senses and from memory. These steps result in "recognition" of a set of patterns within the stimulus with a relationship to one another, defined in terms of a conceptual object, and may include the identification of the object as a member of a class.

In this chapter and the next, I will examine the outlines of this object recognition process. Although we are not yet in a position to specify the actual neural connections in very great detail, there is enough information available to enable us to see what sort of processing is going on and to speculate on the ways in which it is being accomplished.

Figure 5.1 shows the principal sensory areas of the human cerebral cortex. Notice the locations of the primary visual, auditory, and somesthetic cortices. These areas, with which we are already acquainted, are bordered by a series of "secondary" sensory cortices that are involved in further processing of sensory inputs after their analysis by the primary sensory cortices. In turn these secondary areas all border on the "middle parietal cortex," which appears to be where information from many modalities is combined with memory and verbal processes. The arrangement of the areas in itself suggests some sort of progressive merging of sensory into perceptual and then multimodal perceptual structures; data moves inward from the surrounding primary projection areas to the middle parietal cortex. While this is undoubtedly a gross oversimplification, it will do as an initial conceptual aid in visualizing what is occurring.

The study of the receptive field properties of sensory cells has been extended to secondary and tertiary visual sensory cortices. The results are consistent with the type of organization already discussed in the primary visual cortex; ie: the cells in these regions are found to continue the feature extraction process begun in the simple field cells and complex field cells of the primary cortex. In these secondary cortical areas, the majority of cells are termed "hypercomplex " and can be divided into high-and low-order types.

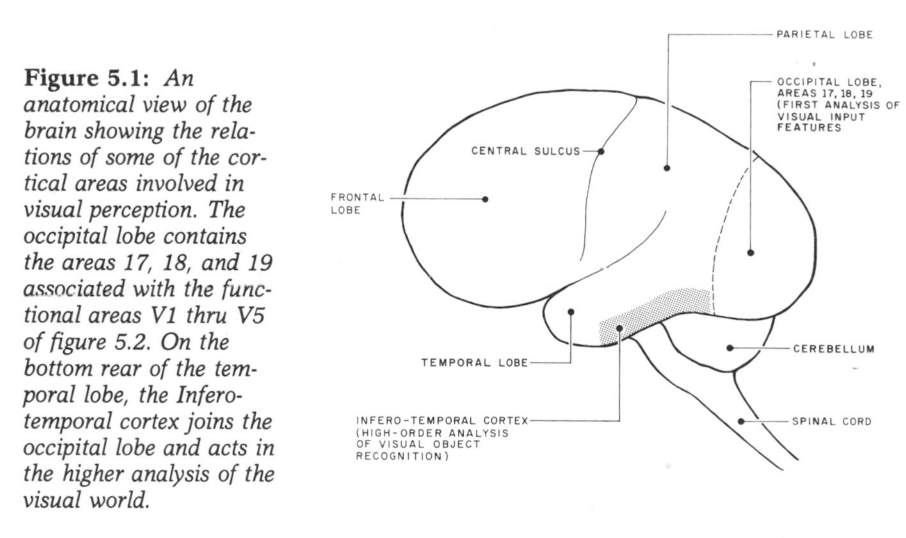

Figure 5.1: *An anatomical view of the brain showing the relations of some of the cortical areas involved in visual perception. The occipital lobe contains the areas 17, 18, and 19 associated with the functional areas V1 thru V5 of figure 5.2. On the bottom rear of the temporal lobe, the Inferotemporal cortex joins the occipital lobe and acts in the higher analysis of the visual world.*

It is typical of these cells to have a very narrow tuning (in the sense of the preceding discussion) for various aspects of spatial form and very broad tuning for qualities such as location. In many cases, the orientation quality becomes broadly tuned as well, so that some cells are responsive to a particular complex spatial feature without much restriction on its preferred orientation within that cell's receptive field. This is analogous to the way in which complex field cells show an independence of location. A similar "detuning" of hypercomplex cell responses in other dimensions is exemplified by cells that are sharply tuned to particular spatial forms, but show no preference for light on dark or dark on light relations between the form and its background.

The forms preferred by these cells include singly terminated lines (ie: a line segment which is required to end within the cell's receptive field at one extremity) and doubly terminated lines (a break in a line segment). Edges (boundaries between areas of opposite illumination) are also selected for singly and doubly terminated properties, and with these features, "corner detectors" and similar highly complex form extractors can be created. At this level, the thrust of the tuning process seems to favor detection of singularities in the features already extracted by the units in the primary sensory cortex. These kinds of singularities are known to be of great importance in perception, and in all likelihood represent another step in the search for boundaries which seems to dominate the perceptual process.

The existence of these sorts of cells in the secondary and tertiary sensory cortices is not surprising, and it is not necessary to belabor the point that their responses can be understood in terms of the same principles of selective convergence and inhibition which we found useful in understanding the lower level cells of the system. However, by itself the existence of hypercomplex receptive fields does not tell us much about the functions which are performed in these higher regions of the sensory cortex. The evidence from experimental or accidental damage to visual sensory cortex in animals and men suggests that although vision is lost when the primary cortex is damaged, more subtle changes result from damage to the higher levels of the cortex.

The blindness which results from destruction of the primary visual cortex suggests that it is part of a common pathway for fine pattern vision through which the data for all related perceptual functions passes. I say fine "pattern vision," because there is some evidence that after destruction of the primary visual cortex certain kinds of visual ability remain, such as the ability to fixate the eyes on an "unseen" target. This probably reflects remaining activity in some of the mechanisms supporting the visual orienting reflex. As far as the ability to "see" in the usual sense, the subject is blind.

However, destruction of *secondary* visual cortical areas does not result in total blindness, but causes the loss of certain kinds of visual perception. This buttresses the suggestion that perception is not a simple serial process passing from lower to ever higher levels of abstraction through a serial, linear system. If this were true, breaking the chain at any point would produce absolute blindness, as occurs when the fiber tracts between the eye and the primary visual cortex are severed. Rather, it appears that from the primary visual cortex or even from the subcortical level upward, perception is a branching, parallel process.

There are a number of functions performed on the data from the primary sensory projection areas which are associated with the secondary areas of that modality. One of these is the formation of a spatial "map" of the "features" which the primary cortex has provided. In other words, each small area of the primary cortex "knows" what sensory features are present in its own small region of authority, (eg: a line at this or that angle) but in order to form a unique representation of an extended object or scene, these elements must somehow be associated in terms of their spatial relationship. This must be done by a system which has as its province either the entire sensory field, or at least that subsection which contains the object in question. Basically, this is the process that allows the recognition of an object by its outline or contour. Of course, objects have contours in tactile as well as visual space. Our perception of a unified "space" containing an object is the result of the unconscious synthesis of visual, tactile, and other spaces which is a later part of our perceptual process. However, even within a modality, the outline and location of an object in that modality's space must first be determined.

In light of the earlier discussion of parallel population coding, we would not expect to find many "ultra-hypercomplex" feature detectors responding to particular objects. Rather, we would expect to find that complex objects create patterns of activity among populations of cells which could be interpreted elsewhere in the system. For example, following corneal transplants to correct lifelong absence of pattern vision due to cataracts, the patient can initially distinguish squares and triangles only by counting the corners. It is only later that he learns to "see" them as being different perceptual objects as we do. This acquisition of a perceptual skill probably reflects the existence of an initial perceptual construct such as the "corner detectors" just discussed and the subsequent development of an "object-level" entity through the acquired recognition of patterned, organized activity in these elements.

II. ACTIVE PROCESSES IN OBJECT PERCEPTION

Studies of the secondary visual cortex reveal at least two types of processes, "active" and "passive," involved in the recognition of spatial objects. The passive process is essentially what we have been considering all along that is, the analysis of the input presented to the receptors. In the active process the system utilizes additional information which can be gained by correlating the changes in receptor orientation, which it can command, with the resultant changes in input pattern. The passive process encompasses everything the system could do to a scene which was presented in a brief flash that did not allow time for active examination processes; it is clear that such a presentation can be effectively analyzed, because people have no trouble recognizing shapes presented in this fashion. However, it is also true that the system can do much more with an active process, which may be of special importance in the perceptual analysis of novel patterns; the passive process is at its best with previously analyzed patterns.

The existence of a processing mode other than the passive one is shown by the fact that shapes can be recognized when they are presented, one part at a time, through a small hole in a screen. This indicates that an object recognition system exists which can handle the sequential presentation of information at the same retinal location. Similar findings apply in the case of touch: an object can be recognized by simply placing it statically against the skin, but recognition is greatly enhanced if an active process is permitted in which the receptor surface is allowed to "explore" the object. In both tactile and visual examination of objects, the fingers or the eyes tend to follow the contours of objects (with special attention to corners, edges, etc).

The advantage of such a process would appear to be as follows: if the analyzing device has access to the program which moves the receptors or at least to information which encodes the spatial motion of the receptors, then a concatenation of that information with the successively encoded states of the feature extracting units (which cover a small retinal area responsible for a region in space) could uniquely encode the spatial nature of the object in its entirety. In other words, there is no need to produce a spatial synthesis of the results of feature analysis from many different retinal locations. This scheme would seem to be preferable for some kinds of perception. In particular, it would be well suited for the separation of objects from their surroundings, because objects are usually characterized by boundaries in space; these boundaries take the form of continuous contours which could be tracked by the receptor under the guidance of edge detectors and corner detectors in a specified small region of the visual field. This process could provide feedback to the motor systems moving the receptor and allow accurate scanning of a boundary.

What evidence is there that such a process occurs in the brain? In both the visual and tactile secondary sensory cortices, the majority of hypercomplex units are selectively sensitive to moving stimuli and have preferred directions of motion. In both systems, these cells are also anatomically associated with the systems that control receptor motion. In the secondary visual areas, there are "motor eye fields" where electrical stimulation produces eye movement. It is curious to find motor areas so far from the classical motor pathways unless one can understand their presence in relation to some surrounding region. For example, one finds regions for producing eye movement in the frontal cortex which seem to be related to voluntary changes in direction of gaze. The eye movement centers controlling the visual orienting reflex in response to novel, peripheral auditory and visual stimuli have already been mentioned. In the secondary visual cortex, there are motor eye fields which do not serve either of these functions, but which appear to be related to the visual reflexes which follow moving targets in the visual field and also function in the reflexive tendency of the eye to "scan" along continuous lines in viewing an image. Similarly, in the tactile areas, there are neural mechanisms that are intimately related to the reflexive grasping and palpating motions which form the basis for active exploration of an object. The active and passive mechanisms of tactile perception even seem to be represented in different regions of the secondary tactile cortex. A lesion in one portion of this area will impair active tactile perception, without interfering with passive methods or reducing tactile sensitivity.

There is evidence that the active exploration of a bounded figure may involve eye movement only when the size of the object exceeds the size of the area of high acuity at the center of gaze. However, this does not restrict the active process model to such objects. Experiments indicate that the point of visual "attention" moves around on the examined object, even when eye movement is not involved. There is considerable evidence that processes exist in the higher levels of the perceptual apparatus that modify the operating area of perception within the visual field without recourse to eye movement. This, in fact, is probably one meaning of the poorly defined word "attention." It may be that eye movement is an auxiliary mechanism brought into play whenever an attention shift exceeds a certain degree of spatial displacement. In principle, the inputs which direct such attention shifts could be treated identically with eye movement information in generating a location code. The attention shifting inputs might well be the primary basis for generating the location code. Eye movement might be a secondary process, transparent to the user, which is called in whenever the demands of the system call for attention shifts which exceed the spatial bounds of good activity with the primary process.

One important benefit of an active perceptual coding mechanism would be that the code from different modalities could include a common element of spatial position; this would enable the association of data from different sensory modalities derived from the same spatial stimulus point. Such an ability is an important task in the development of a unified perceptual world model, and one which characterizes an advanced brain. It is well established that the final perceptual representation of the world by man is a unified multimodal model which permits description in terms of any modality, and in which information gathered from one sensory system may be transferred to the desciptive language of another. Information acquired through tactile input may be used to identify objects by vision and vice versa. Interestingly, this kind of cross-modal comparison appears to be effective only if active processes have been used during the exploration of an unknown object.

The essence of this processing method, which gives a unified representation of an object, is the ability to represent it as a serial set of codes, each containing both the "what the scanner saw" code and the "where the scanner was" code. The chief advantage is that if the scanner is driven by feedback from a local edge-follower, the boundary of the object is automatically coded without necessity for complex decision processes to decide what belongs with what. The chief disadvantage is that these sets of codes must be acquired serially in time, because a visual or tactile scanning is a sequential process. On the other hand, the passive process is a spatially parallel process in which the information arising simulaneously at various points on the receptor surface must somehow be further analyzed in terms of "what is part of what." However, it does not follow that the final representation of the two forms of analysis must be different. A series of code elements, acquired in the time-serial form via an active perceptual process, could be represented in parallel spatial form. These elements would not even have to be scanned in their original order for retrieval or further processing. When a visual perception of an object is acquired by a forced, active process (ie: when it is only presented a piece at a time), it is important for the *acquisition* of the perception that the proper order for serial presentation of the stimulus elements be observed, but the final perceptual content can be described in any order. What most likely occurs is that the encoded perceptual material from either active or passive processes is similarly encoded and subsequently recoded by combination of the two codes and combination of information from other modalities into large multimodal perceptual units. This sort of recoding appears to require cortical mechanisms beyond the secondary sensory cortices.

III. THE PASSIVE PROCESSES AND THE HIGHER PERCEPTUAL CENTERS

The passive aspect of object recognition relies on the same basic inputs from columns of feature analyzers in the primary cortex as does the postulated active process, but there are at least two major lines of evidence concerning ways in which it is accomplished. An understanding of both of these proposed mechanisms (which are not exclusive) requires that we consider the higher visual cortical areas as a functional entity.

Therefore let us get an overview of what is being done, and where. If we regard the secondary visual cortex as the ''front end'' of this mechanism, it is possible to follow its operation in a more or less direct fashion through several intervening stages, which culminate in the forward portion of the ''infero-temporal cortex.'' (See figure 5.1.) It is possible to draw a simple scheme for information flow in this system on the basis of the anatomical connections between the areas involved; an outline of the major pathways is presented in figure 5.2. In this scheme, the area labeled ''V1'' corresponds to the primary visual cortex, and ''V2'' thru ''V5'' correspond to the ''secondary visual cortex'' which we have just been considering. ''ITC'' denotes the infero-temporal cortex which is apparently at the end of the processing sequence.

I would like first to call your attention to the fact that there are essentially two distinct pathways from the retina to the ITC. The first of these is the ''main route'' that we have been following, which runs from the retinal ganglion cell (RGC) of the retina to the lateral geniculate nucleus (LGN) of the thalamus, to the simple and complex cells of Vl, and thence to the several secondary visual areas and to the ITC. There is also an essentially independent route through various subcortical centers. You will recall that in our discussion of the detection of movement, I mentioned that there exists a pathway from the RGCs of the retinal level motion detectors to the mesencephalon; these terminate in the superior colliculus (SC) which is connected with the nuclei in lower centers that move the eyes. We observed that this motion-sensitive channel was important in the ''orienting reflex'' to moving peripheral stimuli. It is also the case (see figure 5.2) that

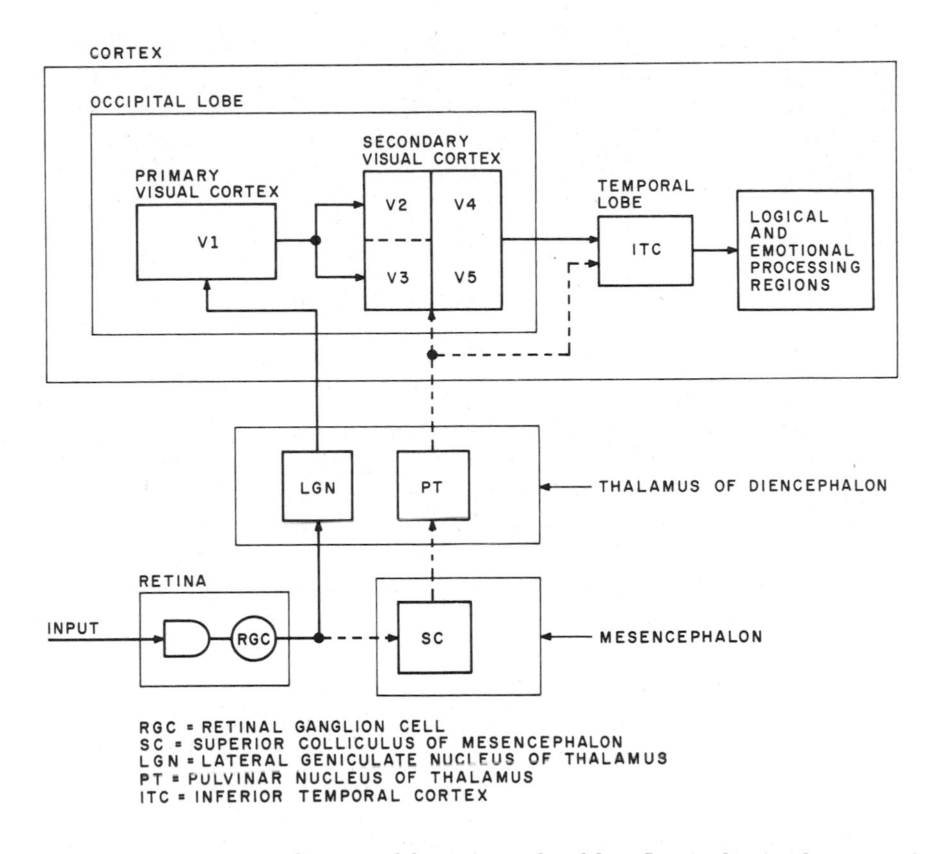

Figure 5.2: *A summary diagram of the major paths of data flow in the visual perceptual system. Two major pathways, which diverge at subcortical levels are apparent. The functions of these two paths are discussed in the text (redrawn from Gross, et al.)*

the SC sends an additional projection directly to the ITC, that bypasses the rest of the "main route" of the visual input. In addition to its direct retinal input, the SC also receives an input from the primary visual cortex, V1. The structure labeled "PT," or pulvinar, is a thalamic nucleus. Although it has no direct retinal connections, it does receive heavy input both from the SC and from V1, which gives it access to a great deal of preprocessed visual information. This structure also projects to the ITC and to the secondary visual cortices as well.

This second pathway, supplementing the "mainline," is a subcortical system which receives information from early stages of the visual system and passes its outputs to the later stages of the primary system. This architecture suggests that the alternate subcortical pathway may be parallel processing information which is different than that processed within the mainline system, but which is needed by it for its operation. The evidence suggests that the mainline system is concerned principally with the analysis and categorization of the *contents* of visual space, while the second system is responsibile for the analysis of global *location* in visual space.

This second system may be responsible for the phenomenon of "blind sight" which can occur in certain kinds of blindness. As I mentioned earlier, destruction of the primary visual area, V1, leads to blindness, and the patient reports no visual sensation. This is consistent with our view that the mainline pathway is the input system required for conscious visual experience as we understand it. At the same time, however, both experimental results in monkeys and results from cases of human disease and accident seem to indicate that some kind of visual ability remains.

Animals or persons with this kind of deficit frequently show the ability to avoid obstacles or to reach for objects accurately. A person with such a visual problem may say that he has absolutely no visual experience remain ing in all or part of his visual field, yet when asked to "guess" at the nature or location of an object in his region of blindness, he may show surprising accuracy. He will still deny that he had any experience of sight and will usually claim that he was "just lucky." The accuracy of object discrimination achieved in this fashion is inferior to that performed with normal vision, but the ability to locate objects in visual space is quite good. Both kinds of residual visual performance can be improved with practice, and the possibility of training this residual "blind sight" is under study.

The evidence suggests that this ability depends on the operation of the secondary visual pathways. If it is true that these structures are the major locus for parallel processing of movement and location data, it is not surprising that the ability to localize objects in space is retained. In cases of cortical damage, at least a minimal ability for pattern recognition would be required by this system in order for it to function as a location detector. Its inputs to those higher centers important in the *recognition* of visual input could then account for the retention of some ability to discriminate as well as locate objects, and lesion of the secondary visual cortex indeed eliminates "blind sight." What is fascinating is the implication that the conscious experience of sight depends on the operation of V1. Apparently the operation of the higher visual identification centers enters consciousness as the abstract knowledge of what is there with or without the raw visual experience. This is an example of parallel addition of successive perceptual analysis to the perceptual code; it is a result in accord with common sense, because even a completely unrecognizable random pattern with no objects in it still gives rise to a visual sensation.

Unfortunately, at the present time little is known about the details of the operation of this second visual system, and we shall have to infer its functions from our knowledge of its influences on the mainline system, to which we now turn our attention.

The secondary visual areas, V2 and V3, retain the retinal map. That is, there is a point-for-point mapping of activity on the retina into activity in these cortical areas in terms of spatial relations. This is true of the primary area V1 as well, and the result in V2 and V3 simply reflects the fact that the outputs of V1 are themselves mapped onto V2 and V3 in a point-for-point manner. However, in this case the map is duplicated twice: once in the upper portion of the secondary areas, and once in the lower. Thus, the outputs of V1's feature extractors are sent to two different areas in the secondary visual cortex.

The results of lesioning these two areas indicate that each region performs different functions. Destruction of the upper cortical region eliminates the ability to perceive visual spatial relations among objects, but does not interfere with pattern discrimination or visual acuity. On the other hand, lesion of the lower cortical region produces serious deficits in pattern and object recognition. This is another indication of the splitting of the analysis of "what" from the analysis of "where" in the visual system. To my knowledge this fact is not established, but we would expect that the "where" system would be concerned with the interaction between the location-retaining second input pathway and the "object-identification" information coming from the mainline pathway. In this fashion, we construct a perceptual "framework of space" which incorporates visual location (including depth) as a quality bound to other qualities of the field, but not necessarily analyzed into objects and categories. The lower portion seems to continue the familiar process of abstracting content from location in the pursuit of object identification freed from the details of location and orientation. The construction of visual spatial relations by the other system is an important part even of this process, for some spatial relations are important in object identification and are not just "spatial noise."

Beyond areas V2 and V3, point-for-point mapping of the retina is lost, so there is no longer any relation between where on the retina the visual stimulus occurs and the location within the cortical area where the resulting neural activity occurs. It seems that at the level of V4 and V5, the object abstraction process has reached the point where the *nature* of the stimulus rather than its position is important in determining the cells which are activated. This remains true at all subsequent levels. It would be a reasonable guess that at the level of V2 and V3, one processing route is continuing the object recognition process, while the other area is preparing a location code which could be appended to the code for that object, thus freeing the system from the constraint of encoding location by cortical area. This would be an oversimplification of the total role of the location encoding system, because it also has inputs to the later non-mapped systems of the cortex. Its functions there involve the dynamic restructuring of receptive fields to cover particular locations, rather than location detection *per se*. This restructuring also occurs as part of an apparent feedback process from later to earlier stages in the analysis, so the later system must know the locations of the objects in order to know which elements of the earlier mapped stages it is appropriate to address.

When lesions are made in the ITC cortex, we find that the disrupted processes are at the highest level of function that one would want to distinguish as being "purely" perceptual (as opposed to memory or reasoning in the visual mode). As in the case of the secondary visual cortex, lesions here do not alter acuity, nor do they interfere with object recognition or spatial relations as does damage to the secondary visual cortex areas. In keeping with the finding that every area of V1 projects connections onto every area of the ITC via multi-element relays through the secondary visual cortices, lesions in the ITC do not produce any deficits in circumscribed regions of the visual field. The kinds of deficits which do occur seem to have to do with the ability to recognize objects as members of categories or to learn discriminations among similar objects. These are the fundamental perceptual processes of generalization and discrimination, which were defined long before the operation of the visual system was studied.

Discrimination is essentially the ability to differentiate between two similar stimuli. That is, it may be necessary to assign objects to two entirely different categories in regard to your actions toward them on the basis of small differences in their otherwise similar appearance. Generalization is the process of treating perceptually different objects as members of a class for purposes of behavioral response. In order to accomplish these functions, it is necessary to classify objects in terms of perceptual categories, and this is the function that seems to be disturbed with ITC lesions.

The ITC is divided into a rear portion and a forward portion. The rear portion receives inputs from the secondary areas V4 and V5. It projects to the forward portion, which in turn sends its outputs to areas of the brain which are concerned with emotional and rational response and with memory. Apparently, the deficits in discrimination and classification are not because of failures in memory or judgement processes, even though learned discriminations are involved, but in that the effect is limited to strictly visual functions. Discrimination in other modalities is not affected. In the posterior region, the deficit appears to be a failure to discriminate between patterns or objects, while in the forward portion, the deficit seems to involve a failure of perceptual classification. I say "appears to" in these cases because most of the information comes from animal studies in which an animal's perceptual abilities must be inferred from his behavior. Verbal reports from injured humans are useful for generating hypotheses, but often an injury is not precisely localized in the region we are studying.

There have been a number of possible interpretations of the substrates for the functions described above. At the moment, the one which seems to have the most general acceptance is that the rear portion of the ITC generates models of visual content (which may be improved through a feedback process), while the anterior portion classifies these objects into categories for rational or emotional response, for putting into memory, or for use as templates in recalling stored information.

In these terms, we might suppose that the reorganization of perceptual data, which took place at the moment you perceived the cow as an integrated object, was the operation of the rear portion, while its identification as a member of the class "cow" was the function of the forward portion. If this analysis is correct, the rear ITC would be functioning to form a model of the object which would include some parts of the scene, while excluding others which belong to different objects. The forward portion would then be involved in categorizing the object into a class on the basis of its prominent features. Obviously, this latter function involves memory processes in all but a few simple cases, and this region probably functions as the interface between visual perception and visual memory. It is appropriately connected anatomically for this function, as we shall see.

In general, discriminations which involve comparisons of objects are inhibited by damage to the rear portion of the ITC, while classification of objects into groups is interfered with by lesions to the forward portion of the ITC. Neither lesion affects the ability of the animal to recognize objects.

Discriminations generally involve the *relative* comparison of objects, while categorization tends to involve *absolute* properties. Visual memories and emotional reactions to visual stimuli are in terms of absolutes and categories rather than comparisons. You remember seeing a tall, dark man, not something which was more manlike than something else, but someone who was a bit taller and a shade darker than "usual." Moreover, as long as we exclude verbal processes and deal strictly in terms of visual memory and judgement, the evidence indicates that people employ relatively few categories in classifying objects; this paucity of classes could be encoded by a "short byte" which excludes irrelevant detail. (I shall return to this point in discussing the uses of spatial frequency coding.)

The suggestion is that the rear portion of the ITC represents the final processing stage of the visual image for detailed examination and current response. The function of the anterior portion may be to classify or categorize these stimuli into a more compact code of significant features for employment by memory storage (which may be why visual memories are not detailed as a rule) or by rational or emotional processes, which, in most cases, need categorical rather than detailed information.

As a general hypothesis, the following order of events occurs: retinal images send information which diverges into a pattern analysis system represented by the flow of information from the RGC to the LGN to V1, and then through several paths in V2, 3, 4, and 5, to the ITC. Simultaneously, another pathway which looks at spatial relations moves through subcortical routes including the SC and the PT. This system has interactions with the pattern analyzing system at most steps along the route. The major interaction in terms of the construction of spatial relations among visual stimuli occurs in a portion of the secondary visual cortex.

Following the pattern analyzing pathway, we find a progressive abstraction of content from situation at each level, until, at the ITC, the analysis of the object as such is completed and free from the constraints of particular conditions of presentation. As a final step, the object is further abstracted from the details of this particular occurrence of its type and presented to the rest of the brain as an instance of a general class.

At each level, the content of conscious experience is enriched by the results of processing at that level. Thus, random patterns give visual experience without meaning or recognition because the highest level of successful processing is V1. More organized patterns may be bounded areas which are experienced as objects with spatial relations to one another. This level of experience seems to be added at V2, 3, 4, and 5. V5 is probably the highest level at which analysis added anything to the cow picture before you saw it as a cow. The dependence of the "blind sight" phenomenon on the secondary visual cortex suggests that this added dimension of perception is really added as an abstract quality, quite independently of the "sight" experience which is the lowest level of conscious processing. Next, the rear portions of the ITC region assemble this information into a model of a visual object with meaningful relations among its parts. The object is "recognized." This step also enters consciousness as an abstract qualitative experience. In terms of what you actually *saw*, nothing changed during your perceptual reorganization of the cow. What you experienced was probably the "feeling" of a successful model building process by this area. Finally, detail is sacrificed for generality when the anterior portion of the ITC sends to your memory or rational processes the information that a cow was encountered. Next week, you will have only the haziest notion of what the cow actually looked like; if asked to draw it you might make all sorts of reasonable, but inaccurate, changes in the cow. This indicates the development of the perceptual code to the "semantic" level, which is discussed in Chapter 9.

IV. SOME DETAILS OF PASSIVE PROCESS OBJECT RECOGNITION

Accepting this as a general picture of *what* is occurring in the higher perceptual processes and more or less *where* it is happening, how does the brain function at the cellular level? The first process to be considered is basically an extension of the selective convergence process which we examined in V1, with a few new additions. I have already described similar receptive field and tuning characteristics of the neurons in the secondary visual cortex. However, when these properties are examined in the case of neurons in the infero-temporal cortex, however, some startling differences appear. The most notable is the size and placement of the visual receptive fields.

In accord with the finding that all portions of the preceding retinally "mapped" cortices project to all portions of this area, it is found that the receptive fields in the ITC tend to be large and centrally placed. An average ITC neuron has a receptive field of about 1000 square degrees of visual angle, although some are as small as 100 square degrees. Moreover, all of the receptive fields include the central area of the visual field (called the "fovea," where visual acuity is highest) which is centered on the object of visual attention by automatic eye movements. This is not surprising, because the postulated "pattern extracting" portion of the visual system, following the mainline route, has its retinal detectors concentrated in the fovea, and thinly distributed elsewhere. The "secondary" visual system, for movement and location, is more uniformly distributed and favors the periphery of the visual field. This is one reason why vision at the center of gaze is very sharp, but peripheral vision is not. The system achieves economy by putting its available pattern processing machinery into a relatively small space to achieve high resolution, and by moving the eyes in order to bring the object under scrutiny into this most acute region. Thus, in that portion of the system most concerned with object identification, it is expected that the detectors would cover this region. What we apparently see here is a region whose cells cover the central area of vision in the same overlapping fashion that the simple field cells within a column of the primary cortex cover the small retinal area allocated to their column.

The cells in this region seem to respond only to visual stimuli; ie: it is not an area where multimodal sensory processing is undertaken. Most units have tuning characteristics which include multiple qualities such as size, shape, contrast, orientation, color, and movement. Some have requirements in many of these dimensions; some are very broadly tuned in most. More than half of the units here are indifferent to the absolute nature of the contrast. That is, they do not care whether the stimulus is light on a dark field or dark on a light field. Most of them are more sensitive to moving stimuli than to stationary ones. Many of these characteristics can be understood in terms of the sensitivities, already examined, of their input elements in the preceding cortical areas and probably do not represent any new level of processing on the part of these units. Most of the units are found to respond well to line stimuli, and, again, this may simply reflect the preference of their inputs.

The problem here is the number of possible stimuli that *might* be optimal for the cell, but simply were not tried. That is, a V1 simple field cell will give some response to a spot falling on the bar which defines the area of its excitatory field, simply because the spot excites some of its LGN input units. It would respond better to a line, if you tried one. This becomes more than just a theoretical point in the infero-temporal cortex, because some cells have been found (usually accidently) to have optimal responses to very specific stimuli. For example, parts of a monkey's anatomy have been found to give very specific and pronounced responses in some units in monkey brains.

This observation presents a problem of interpretation. The evidence seems to support the notion of coding advanced earlier in which the ultimate code will consist of a few highly specific units extracting important features, with many more generally tuned units encoding most stimuli by permutations of their activity as a group. This would provide a superficial fit with the findings. The advance in processing in the ITC would then be primarily that of maintaining a group of generally tuned units that covered the whole of the region of fine vision, which could produce a location-free object code. Nonetheless, given the existence of at least some highly specific units here, it is not possible to exclude the idea that there might really be an encoding scheme based on reduction (by selective convergence) to object-specific cells for every different kind of object. It might be that we simply have not found the optimal stimulus for most of these cells. One possible argument against the view is the fact that the simple optimal stimuli discovered for the seemingly simpler units remain the same in all parts of their large receptive fields.

At this point we must introduce the finding that a cell's receptive field properties are not always static, but may be subject to dynamic reorganization. This is not to say that the physical connections of the cells are changed from moment to moment, but rather that there appears to be a mechanism for changing the weighting assigned to each of the inputs of a feature extracting cell. It is probable that the mechanism employed is that of pre-synaptic inhibition. You will recall from the discussion of the "basic neuron" in Chapter 1 that axons can make synaptic contact with other axons, as well as with dendrites and cell bodies. Such contact often occurs just before the termination of the target axon on a cell body and serves to inhibit the ability of that axon to transmit its message across its synapse. This form of inhibition acts to selectively turn off particular inputs to a neuron, rather than to lower the neuron's general firing tendency (as in the case of normal post-synaptic inhibition). A graded pre-synaptic control can be achieved by partially reducing an axon's contribution to firing the following neuron.

It is apparent that applying this kind of control to the inputs of a feature extractor could be used to alter the shape and location of its receptive field, if there were a large body of inputs from which to select. It is even possible that the "selectivity" of selective convergence is actually established more by selective inhibition of unwanted inputs than by anatomical selection of desired inputs. If this were true, the function of the feature extractors would be under dynamic control rather than statically determined. We know that the size of receptive fields can be altered to some extent by various sorts of motivational or emotional states, but this could be accomplished by just altering the firing threshold of the feature extractor cell and would not require actual dynamic control of receptor field geometry in the fashion postulated here.

There is evidence that this kind of control is exerted as part of the perceptual process in very specific ways. For example, it has been reported that the axis of orientation of area V1 line detectors is not fixed, and that it can shift in a compensatory fashion under conditions of head tilt. Presumably, this would be accomplished by reflexive feedback mechanisms from vestibular and kinesthetic areas to area V1.

From the standpoint of perceptual processes, a potentially more exciting finding is that electrical stimulation of the rear portion of the ITC may cause changes in the receptive fields of area V1. If this is true, it suggests that the ITC may exert complex feedback control over the feature extraction processes which generate its own inputs. It is possible therefore that the perceptual model building process is an active feedback process which involves restructuring of the input feature tuning until a match is found between a possible analysis of the input and some known category of stimulus.

According to this hypothesis, the process of object recognition would begin with a "standard" configuration of the receptive fields of V1. If a good match was not found to a category in visual memory by the ITC, that area could employ feedback control to shift the feature extraction process in the preceding analytic stages so as to optimize extraction for some other object class, in ways suggested by the nearest "category similarities" found in visual memory. This process would continue until a satisfactory classification of the object was achieved, or until the object was accepted as unrecognized.

This type of action might explain the response time exhibited in ITC neurons, which is considerably longer than could be explained on the basis of propagation delays through the preceding stages. This long latency would be understandable if the process of "hypothesis formation" by the ITC consists of a cascading feedback process controlled by units with the most successful match, which brings more and more units into play as the process finds more successful matches of data and memory template. This feedback scheme would explain not only the long time taken to perceive figures in perceptually difficult situations, but also the way in which they become readily apparent once the perceptual solution is found. It would also explain why certain ambiguous figures, which can be perceived in more than one way, tend to be seen either one way or the other in their entirety, rather than as a perceptual "mixture." Thus, a particular perceptual organization for one part of a figure establishes a paradigm that tends to organize the rest of the field. Context and prior experience exert strong effects on perceptual organizations, and the mechanism hypothesized here could explain this effect in terms of the successive classification schemes which were tried in the search for a successful model. In the case of a novel object, the superiority of the active scanning process in perceptual object discrimination could reflect the passive process's lack of adequate classification templates.

The receptive field organizations of the ITC neurons also seem to be variable. By cutting the fibers which run from the preceding visual cortical areas to the ITC, it may be shown that this "mainline" route is the essential pathway for visual form discrimination. Without it, form discrimination is lost. This leaves the question of the contribution of the secondary subcortical pathway which also projects to the ITC. (See figure 5.2.) When this subcortical input is cut and the receptive fields of ITC neurons are plotted, it is found that the type of optimal stimulus object for the neuron is unchanged, but that the boundaries of the receptive fields are expanded enormously, so that all of the neurons appear to cover the entire visual field! Given the findings which implicate the subcortical system in the analysis of visual location, this suggests the availability of an apparatus for directing the object identifying apparatus to any particular location in the visual field. The geometry of the connections suggests that this must be occurring by an inhibitory input (probably a pre-synaptic one) to the fibers reaching the ITC from earlier cortical stages. The ones arising from unwanted locations in the spatially-mapped earlier stages would be inhibited and the ones reaching the ITC neuron from the location of interest would be free to enter feature recognition data. This mechanism is reminiscent of the "focusing" of attention in which concentration on a particular small detail can effectively eliminate perception of other portions of the visual scene, so long as they do not trigger the automatic attention and orienting mechanisms.

One suspects that this input from the subcortical systems usually operates more or less automatically; perhaps this process is directed by inputs from retinally mapped cortical regions, which act as feedforward to the "dynamic tuning" feedback method of perceptual model building just described. The ITC has reciprocal connections with the forward parts of the prefrontal lobe of the cortex which are to be involved in voluntary attention processes, and it may be that conscious direction of visual attention operates through this mechanism, as well as by bringing the object to the center of the visual field. It has been established that the location of attention can be dissociated voluntarily from the locus of visual gaze direction. However, the two usually operate synergistically, so that attended objects will be studied with the optimum acuity.

The fact that in many experiments, the receptive fields at all levels are generally static is an artifact caused by the use of anesthetized animals. When similar experiments are performed in a conscious, actively attending animal, the receptive fields are dynamic. Many of the perceptual deficits seen after ITC lesions could be explained by a failure to attend selectively to appropriate details of the scene.

V. A DEMONSTRATION OF THE INFLUENCE OF PERCEPTUAL MODELS

A process such as this dynamic model building scheme, whether based on the selective convergence mechanism or not, is probably required to explain the phenomenal degree to which perception reorganizes the subjective world to be in harmony with the perceptual model currently in use. The following demonstration is difficult to do on the first try, but it illustrates as vividly as any I have ever seen the way in which our perceptual organization of the world can be influenced by adherence to a prior perceptual model. I was once having lunch with a professor of philosophy who inquired whether or not I thought that our assumptions could determine our perceptions. I whipped this one out, and I think from his reaction that I may have come close to destroying his belief in the physical world.

All that is required is a sheet of paper and a pencil. Fold the paper as illustrated in figure 5.3 so that it can be set on the floor in front of you in the form of an "M," as shown. Now sit or stand where you can comfortably observe the object from a distance of three or four feet. Arrange it so that the "M" cross section at one end faces you, and you are looking down the length of the folds. It will help if you initially close or cover one eye and perhaps defocus very slightly to reduce the influence of depth perception cues.

A moment's reflection will show that the two-dimensional image on the retina is, like a shadow, ambiguous with regard to the orientation of the object in the third dimension. The retinal image of the object as it now sits is identical with that which would be produced if the near and far ends were swapped, and the paper were sitting partially upright at an angle to the floor as illustrated in the diagram. For the first part of the demonstration your task will be to "see" the object as being in this second position rather than it its true orientation. Some people can do this instantly, some never can. Most take a little trying. It has been my observation that creativity helps, but I really don't know what makes it easy for some and hard for others—something to do with perceptual rigidity I suppose.

In any event, stick with it until you get it. You will know unmistakably when you've got it right: the transition is an example of that sudden total perceptual reorganization I was referring to earlier. After you've got it, you will find that a little practice at making it stand up and sit down will help you to hold the perceptual organization in the upright position as long as you please. This in itself is interesting, but the best is yet to come. After you can hold the image comfortably in its new orientation, try rocking your shoulders slightly toward and away from it, or from one side to the other, so that your head moves with respect to the object. You may find that as soon as you try this you will lose the new perceptual organization, but again a little practice will help. What you will experience when you can hold it while moving around is the sight of the object twisting and turning to follow your every move like a tracking radar antenna. Everything else looks normal, but there it sits in the middle of your floor moving as though alive, and keeping itself oriented towards you. If you return to a normal perception of the object, of course, nothing of the sort is seen.

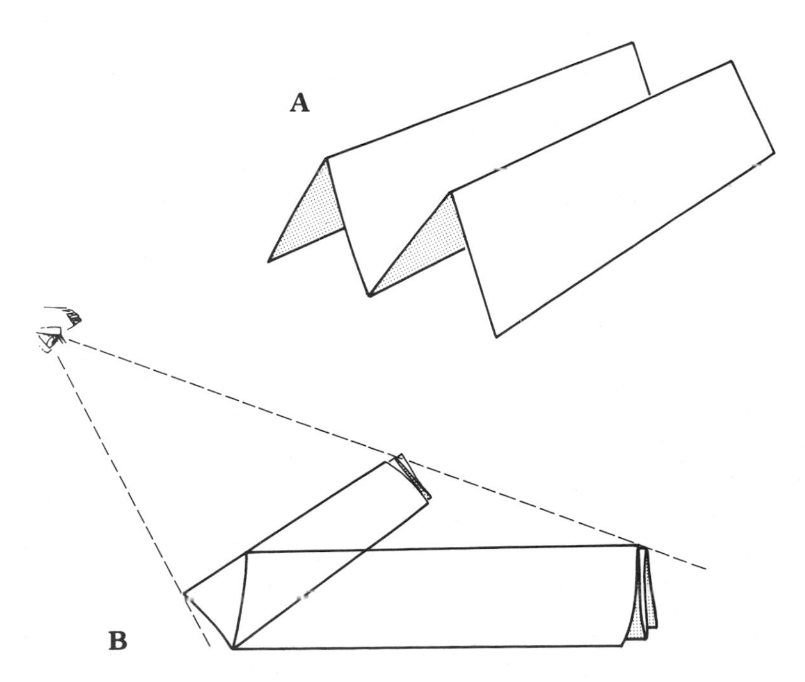

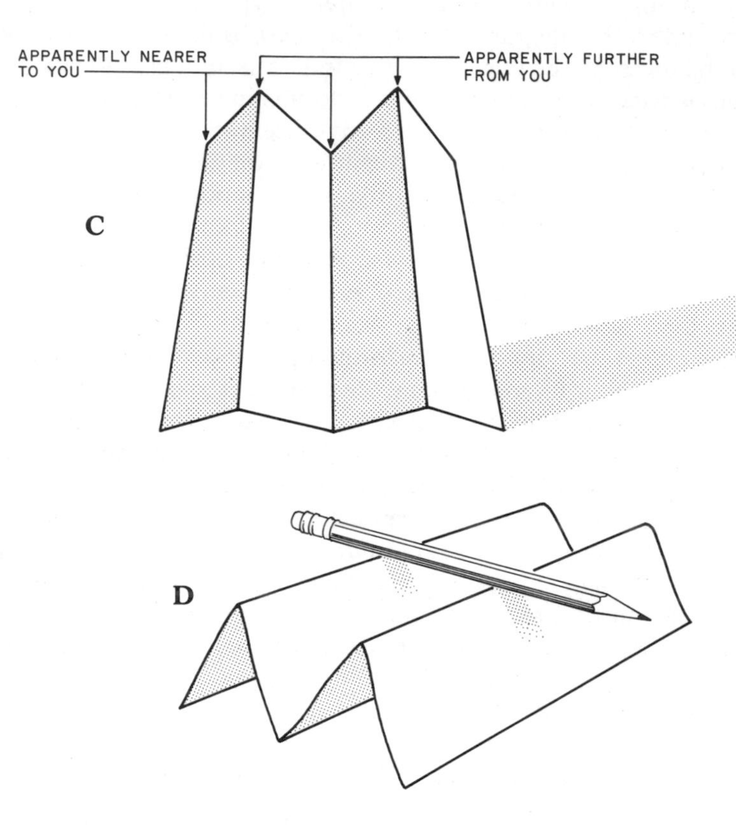

Figure 5.3: *A demonstration experiment illustrating the dependence of perception on the perceiver's assumptions about the nature of the world. The initial task is to view the object in a false position after reducing depth cues and reinterpreting the ambiguous information contained in its outline (see text). When this is accomplished, certain compensatory adjustments in other preceptions are seen. These include apparent motion of the object to "track" the observer and a pencil floating in the air.*

a) Fold the object into the shape shown. An ordinary sheet of typewriter paper works well. Make the first fold parallel to the short side of the paper, across its middle.

b) The principle behind the ambiguity inherent in the figure's geometric outline. Either orientation satisfies the requirements of the outline on the retina if other depth cues are disregarded. A shadow for example is perfectly ambiguous in this fashion.

c) An attempt to depict the object as it appears when in the false "standing up" perceptual interpretation. The effect is sudden, complete, and unmistakable. Keep trying until you know you've seen it.

d) The proper orientation of the pencil for the "floating pencil trick" (see text for a discussion of this effect and the preceding movement effects).

Your perception of the object as standing up, rather than lying down as it really is, requires you to make a certain false assumption about its orientation. Because this assumption is not incompatible with the sensory information if depth cues are minimized, the perceptual apparatus can accept this imposed model and put together a coherent visual picture. However, once it locks in to this perceptual model it will try to reinterpret everything else in terms of the same construct. When you move, there are changes in the relation of the parts of the object and the parts of the background which are incompatible with the imposed model. As geometry would have it, these changes are compatible with the currently operating (though false) perceptual model, given the assumption that the object is moving to compensate exactly for your motion. Rather than abandon the model, which is probably locking itself into the system by meeting the criteria of successful object classification, the perceptual system serves up a perceived motion of the object. The perceptual mechanism isn't bright enough to reject motion by inanimate objects, and your rational processes are going along with it for the sake of the experiment, so the perceptual mechanism computes and produces in perfect detail the necessary motions to keep the relations of object and background compatible with the model of the object.

If you want to see how far this sort of thing can be carried, play with it until you are capable of easily maintaining the perceptual change while moving, and then place a pencil on the object across the width of the "M," as shown in figure 5.3d. When you repeat the experiment, the perceptual system isn't using a false model of the pencil, so it cannot move with respect to the background and must appear in its proper location. The object supporting it is seen as incompatibly oriented and in motion. The poor perceptual system resolves this in the only way possible: while the folded paper goes through its gyrations, the pencil hangs motionless and unsupported in the empty air in front of it!

It would probably require a couple of hours with a hand calculator and a trigonometry text to compute the apparent motions that have to be imposed on the false perception of the paper in order to compensate for your motions within the constraints of the object and background relations which are givens. Astonishingly enough, not one bit of this scene, even with the pencil, is in contradiction with the retinal image. You may have noticed, depending on your lighting, that the light and shadow areas on the paper are incompatible with the true lighting situation when it is seen in the upright orientation. If so, you probably also noticed that the paper appeared to glow mysteriously with its own light. This is another example of the adjustment of the perception to both the retinal facts and the assumed model. All of this, however, was performed by preprocessors to your conscious experience and handed to you as accomplished fact.

It is clear that the perceptual mechanism has at its disposal enormously powerful machinery for locking onto a model. In this case, the model was one suggested by your conscious intervention, but given that you had only to conceive of it in very general terms and not bother yourself with the details of the motion and other effects, it seems quite plausible that the same machinery could do an equally efficient job in locking into an internally consistent interpretation suggested by a partial match between an attempted model and a category from visual memory. The real-time handling of all this is testimony to the power of the brain's parallel organization even in the face of slow components.

VI. SPATIAL FREQUENCY AS A HIGH-ORDER FEATURE

An alternative to the continuation of geometric feature extraction by selective convergence as a means of object identification and model building in the higher visual cortex is another version of the Fourier-like process which was discussed in relation to the lower visual centers. You will recall that there was considerable evidence that certain perceptual processes, which were difficult to explain by other means, could be easily accounted for by assuming that the brain employs Fourier analysis in the domain of spatial frequencies in its analysis of visual input. Thus, visual exposure to a square wave grating fatigues the brain's ability to respond to odd harmonics of the square wave grating, even when these are subsequently presented individually in the form of pure spatial sine wave gratings. Even harmonics are not affected. The only easy explanation of this is that the brain processes the spatial frequency components of the spatial square wave by some Fourier-like process which separates them into independent frequency channels. We saw evidence that in many cases the simple field cells did show a rather narrow tuning for particular bands of spatial frequency when presented with various gratings. At the time it was noted that, although it would be relatively easy to imagine a mechanism for the spatial frequency selectivity of these cells when shown a repetitive grating type of stimulus, a mechanism for a true Fourier analysis of an arbitrary spatial waveform at the lower levels of the system was not apparent.

The model to be presented next provides a means by which just such an analysis could be carried out at the level of V1 and above, provided that the V1 level cells can show spatial frequency sensitivity to repetitive stimuli. Of course the results of purely perceptual experiments do not show where in the system the Fourier analysis takes place.

Consider an array of V1-type oriented line detectors arranged as shown in figure 5.4. This array represents a one-dimensional sample of luminosity values across the horizontal axis of some small area of the visual field. This is equivalent to saying that the firing rate of each of the units represents a value of luminosity as a function of its position along this axis. We could say: $L = f(Rd)$, where L is the luminosity of the image at point d along that axis, and R is the firing rate of the cell covering the region centered at position d. This sort of function is entirely amenable to Fourier transformation, which, if it could be accomplished, would give a description of the luminosity distribution along this axis which contained all the information in the above function, but which was stated in terms of the amplitudes and phases of a set of spatial frequencies. Such a description of luminosity may be retransformed into the original description of luminosity in terms of position. The two are simply different ways of describing the same data, which are interchangeable, but which have different languages that suit them to different sorts of manipulation.

For the moment, assume that it is possible for the brain to accomplish

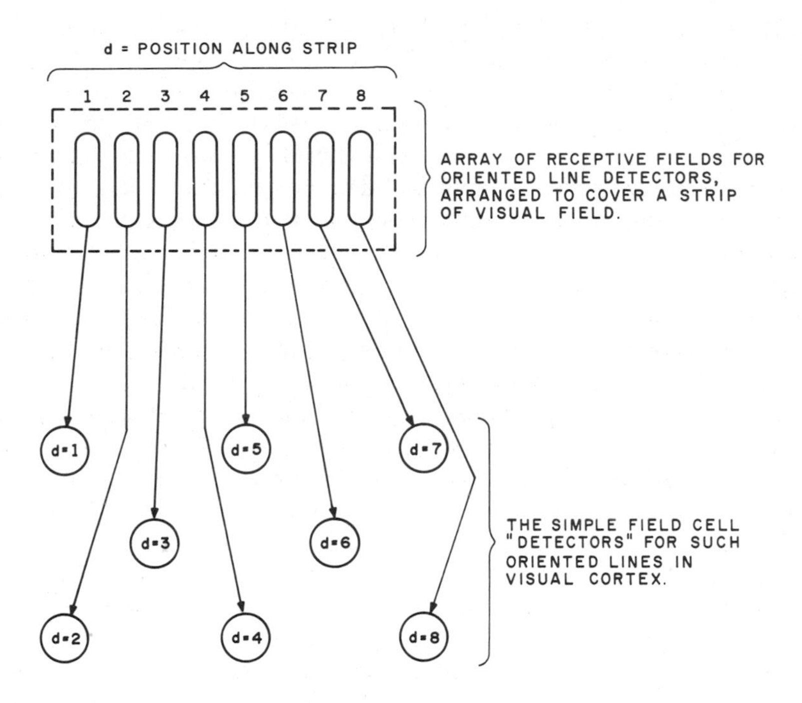

Figure 5.4: *A collection of simple field cells selected so that their receptive fields are ordered as a strip of parallel lines perpendicular to the long axis of the strip. Such a collection of cells would encode a description of luminosity as a function of position along the strip and hence of the one-dimensional, spatial luminosity function along the strip. Such a function can be subjected to Fourier transformation to generate a spatial frequency description of this function.*

such a transform of the luminosity function encoded by a strip of parallel receptive fields. If this is possible, then it is also possible to take a similar transform of the luminosity function along any randomly chosen line crossing the same area. Thus, we might envision several such strips crossing the same area at different angles as shown in figure 5.5. If we took the transforms of enough such strips, it would obviously encode the luminosity distribution in the entire two-dimensional visual field. Once the possibility of the transformation of the luminosity functions of the individual strips is admitted, it is a simple matter to show that the two-dimensional Fourier transform of the spatial luminosity function for the area covered can be constructed in a very straightforward manner by forming a composite array of the individual, one-dimensional transforms.

Such a process for constructing the two-dimensional transform is only approximate, because of the finite number of strip orientations and the finite slit width of the line detectors. However, for any given degree of approximation to the two-dimensional transform, it is possible to calculate how many orientations and what slit widths would be required. As you have probably noticed, the hypothetical strips covered by parallel slit detectors at each orientation look rather like the assemblies of simple field cells which were presented in an earlier section as the probable set of inputs to the V1 complex field cells. Both the number of such preferred line orientations found among complex field cells within a column covering a specific area and the observed slit widths of the simple field elements are what would be required to give an adequate approximation to the two-dimensional Fourier transform of the spatial luminosity function of the area. ''Adequate'' means sufficient to allow a resolution of the spatial luminosity function comparable to the visual acuity actually observed for the different areas of the retina. This means that a usable two-dimensional Fourier transform of the spatial luminosity function within each area does in fact exist, encoded in the firing of these simple field cell units. Whether or not the system is capable of using information encoded in this way is another matter.

The notion that it might requires an examination of the over assumption of a moment ago that a one-dimensional Fourier analysis of the spatial luminosity function along a single strip of oriented-line detectors was possible. The model which we have used earlier for the wiring of the location-free complex field cell is really an integrator along such a strip. (See figure 3.9.) The essence of the Fourier transformation is just such an integration with spatial frequency included as a function parameter.

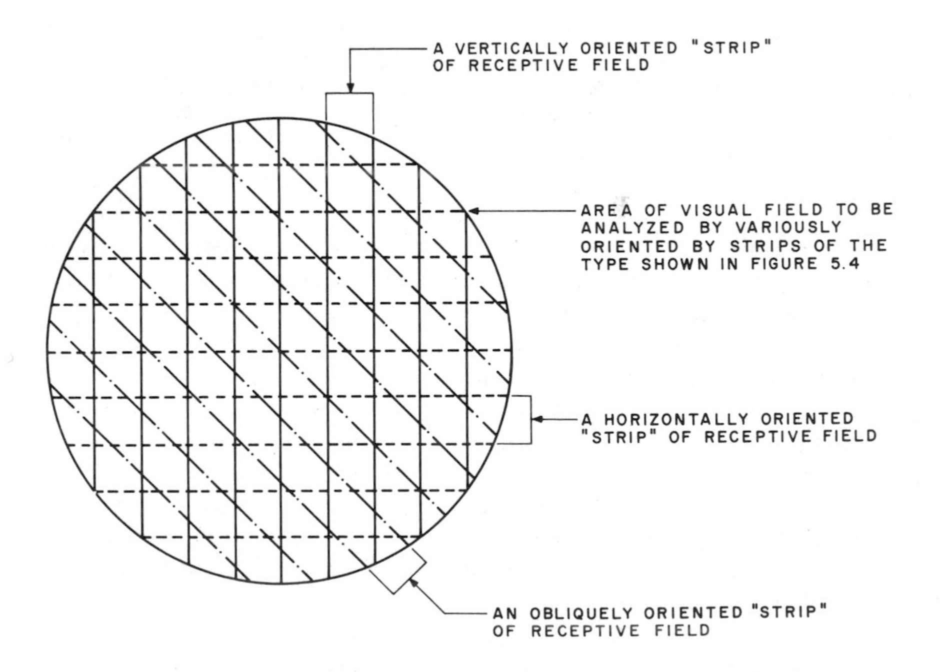

Figure 5.5.: *Parallel assemblies of strips of the type shown in figure 5.4 may be superposed at various orientations to obtain a complete description of the two-dimensional luminosity function of an area. Three orientations are shown, but for high resolution many would be required.*

To accomplish this frequency selectivity, at least two mechanisms are imaginable. The simple field cells themselves could have specific tuning for spatial frequencies, produced via lateral inhibition, and a given complex field cell could integrate only across a strip of simple field cells selected for such tuning characteristics. Each of the complex cells integrating along this axis of orientation would then represent in its output the amplitude of a single component of the Fourier transform, along that spatial orientation, at some selected spatial frequency. This is shown schematically in figure 5.6.

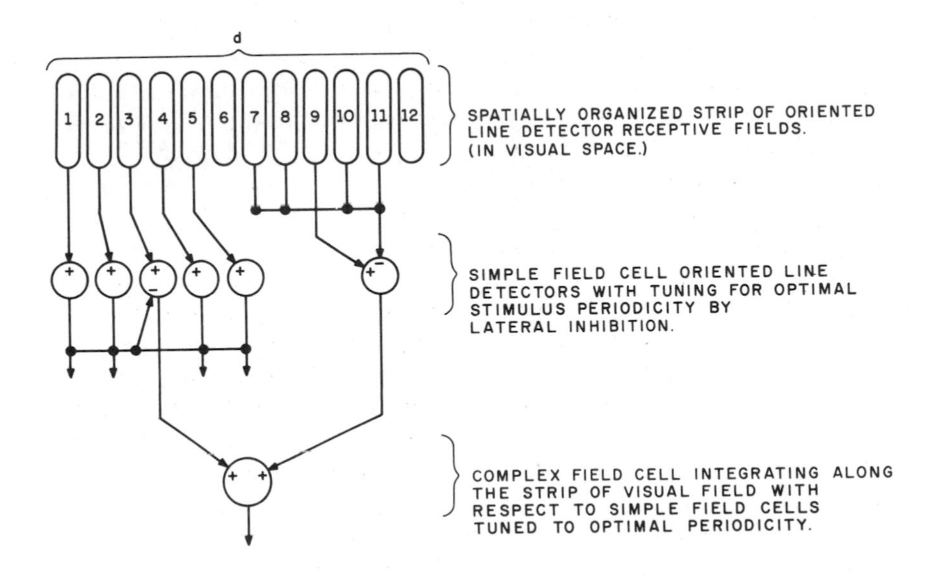

Figure 5.6: *A possible method of "tuning" complex field cells to specific spatial frequencies. This model relies on tuning of the simple field cells by lateral inhibition.*

Alternatively, complex field cells, integrating with respect to particular spatial frequencies, could select as inputs only those simple field elements which had particular spatial relationships between their receptive field spacings. In this case, multiple complex field cells would receive outputs from the untuned elements of a single strip of simple field cells, as in figure 5.7.

There is currently not enough evidence to decide which, if either, of these systems the complex field cells might employ (there are, of course, an indefinite number of more complex models), but there is evidence that they are arriving at this end result. The fact that the complex cells show translational invariance along their preferred receptive field orientations is the primary evidence that they are in fact integrating along such a strip of oriented, simple cell receptive fields. When these are explored with a single oriented line as a stimulus, there is no way to show spatial frequency tuning. However, if these cells do have such a tuning, it would be predicted that a stimulus consisting of two such lines with a variable spacing should

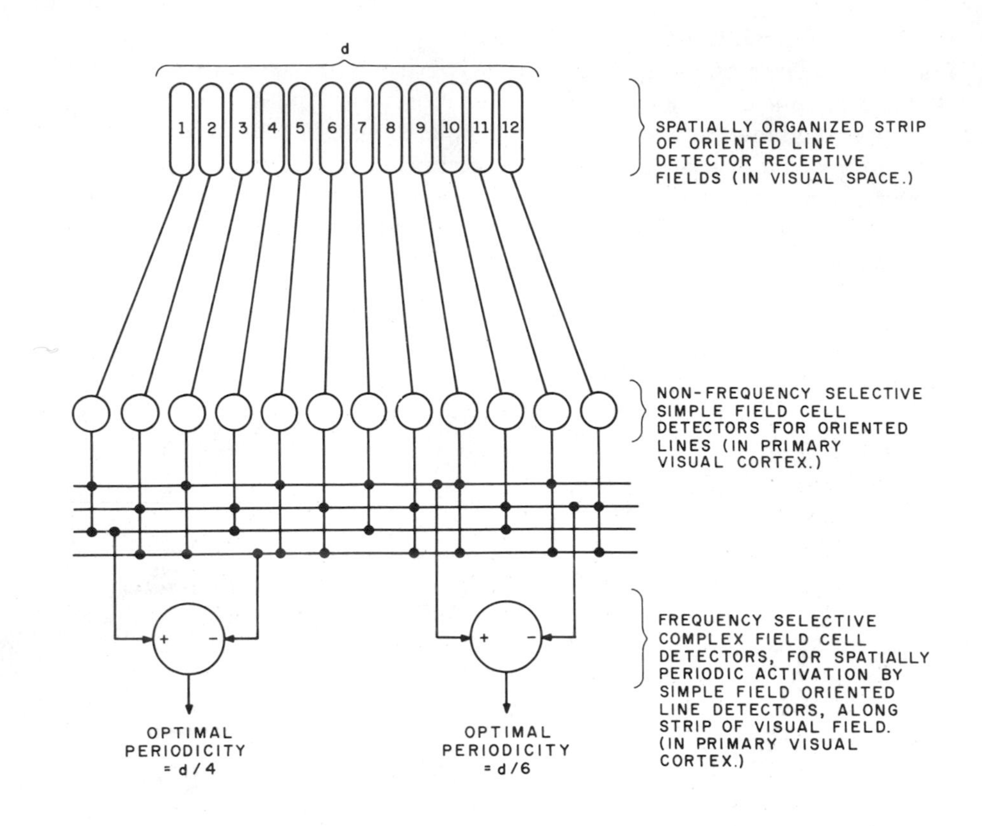

Figure 5.7: *Another possible method of "tuning" complex field cells to specific spatial frequencies. This model relies on selective convergence and inhibition at the level of the complex field cells.*

continue to show translational invariance, but that the degree of summation between the two should vary as a periodic function of the spacing; ie: it should show peaks at multiples of the preferred spatial frequency. (See figure 5.8.) The effect is identical with that of a tuned circuit showing resonance at multiples of the tuning frequency. This experiment has been performed, and it is found that complex field cells of V1 show the predicted minima and maxima in output with varying line spacing, even though their responses to either of the lines separately is uniform. Cells have been reported with as many as four peaks in the response. This result strongly suggests that the complex field cells are extracting spatial frequency information in the form required for a Fourier analysis of the area covered by their columns.

The phase information is available from the simple field elements too, but it is certainly premature to push the Fourier analysis too strictly as a model of visual cortex processing. There are a number of possible ways of representing spatial frequency information which would be useful in explaining the observed properties of visual object recognition. (Arrays of correlation integrals along the orientation strips would be just one such example.) The point is that the cortex seems to be encoding spatial frequency information, and whether the transform is a neat analogue of one of our standard mathematical transforms or not, the basic properties of a spatial frequency coded representation of visual images are very like those of the perceptual process.

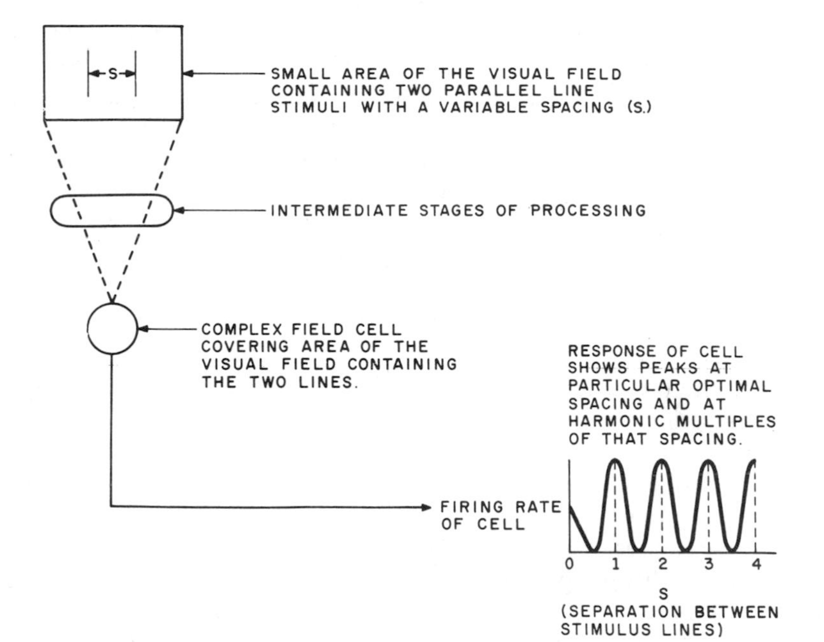

Figure 5.8: *An experiment which supports the hypothesis that complex field cells do have spatial frequency selectivity or "tuning." When two stimuli with a variable spacing are used, the cells' response is a periodic function of the separation. When only one stimulus is used, the cell only shows the normal selectivity for orientation.*

Note that such a statement is not in conflict with the "model building" hypothesis of infero-temporal cortical function or with our considerations of the brain's probable coding schemes. It only specifies that the code consists of some permutation of activity among a set of lines which represents, at least in part, spatial frequency values. If the basic operating code is in this format, the restrictions on the type of frequency transform become very relaxed. As mentioned earlier, one of the properties of Fourier transforms is that there are reverse transforms which yield the original spatial luminosity function. If the brain "recognizes" objects as their encoded transforms, there is no reason why a reverse transform is needed. All that is required is the inability to discriminate the encoded transforms. An irreversible transform, such as the power spectrum, would suffice. (Such a transform does not preserve phase information, but in this case phase is primarily information about an object's location.) It may well be that the line-sensitive cells of the ITC will turn out to show spatial frequency selectivity, in which case they would encode the frequency analysis "signature" of objects over large parts of the field. For reasons which will be discussed next, this would be a good code in which to describe the apparent interactions of this area with visual memory.

Some further properties of the visual perceptual process need to be mentioned. Principal among these is the group of phenomena called "constancy effects." It is clear that as the position of, say, a rectangular object varies in three-dimensional space, the two-dimensional retinal outline departs from a true rectangle and varies through all sorts of parallelograms (including trapezoids of different height/width ratios). Nonetheless, our perception is that of the undistorted rectangle. Our perceptual system "allows for" the angle of regard and presents us with a picture of the actual reality. This phenomenon is called "shape constancy." Similarly, we "see" objects as unchanged in size even though they may be presented through a wide range of actual sizes of the retinal image due to distance changes. This is "size constancy." To see how the retinal image size really changes, move objects towards your eye in a room illuminated by a 3 Hz strobe lamp. There are similar constancies for almost every perceptual dimension. To some extent, this can be explained by computation performed on the basis of information from depth perception and prior learning. However, it has been shown that a monkey can be given visual experience of an object through a peephole, from one angle of view only, and subsequently identify it from other angles, from which he has not previously seen it before. This sort of ability, plus the problems inherent in getting "matches" to memory for recognition of objects in whatever size or orientation they might occur, suggests that the brain must store or deal with encoded patterns in a form which permits these transformations.

Another property of perception, which has appeared under various names, is the ability of the system to "see" organized patterns in discontinuous objects. Figure 5.9 illustrates some examples. A contour-following explanation, such as that discussed in the section of active processes, is unable to account for the illusion. Empirically, a number of properties of the stimulus are found to be contributory; among these are proximity of the elements, similarity, continuity of line, degree of closure, etc. However, these findings have not suggested the underlying mechanism.

The reason for mentioning these perceptual phenomena is to point up the utility to perception of a spatial frequency-coded approach. Probably no other mechanism is as well suited to explain those properties of the system which were considered for years as being unexplainable by a straightforward interpretation. Further, because these characteristics describe the system's ability to perform recognition under conditions of several kinds of spatial translation, magnification, distortion, and "noise," they are the features that are most difficult to design in artificial systems.

In this regard, a spatial frequency-coded object representation would have the following advantages: first, translations of the object in spatial

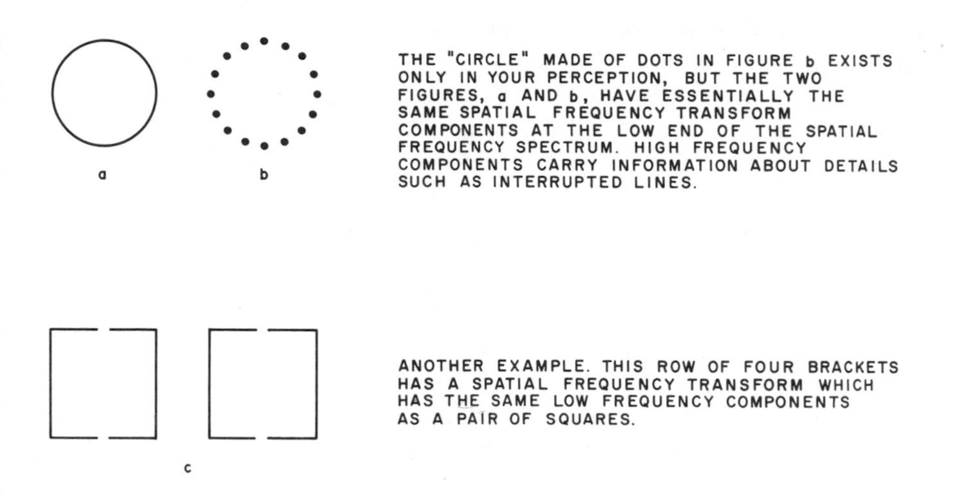

Figure 5.9: *An illustration of the fact that we tend to "see" a global object, whether or not its parts are really connected. This has many similarities to noise problems in pattern recognition. The explanation suggested in the text makes use of the fact that the interruptions in the basic form generate high frequency spatial frequency components, while the low frequency components of the interrupted and uninterrupted figures are similar.*

location do not affect its frequency signature; they only effect the phase relations of the components, which can be ignored. Thus, only one code would have to be stored for all the possible translations of the object. Size or scale *do* affect the distribution of power in the object's frequency spectrum. However, the *ratios* of the frequencies remain unchanged. Thus, the code itself could either be in terms of ratios rather than absolute frequencies or could be "read" by taking its ratio to different "test" frequencics. This latter possibility sounds less efficient, but there is evidence from certain kinds of brain disorders that suggest that such a mechanism, if it exists, may be functioning improperly. In these cases, there are perceptual distortions which appear as a kind of "zoom lens" phenomenon in which perceived sizes of objects shift erratically. Further, such a scanning model would preserve absolute frequency amplitudes for other tasks.

One could also postulate a model in which object recognition codes were stored as frequency ratios, and size perception was influenced by known size of the recognized object type. There is evidence for this possibility in the studies of constancy phenomena for an effect of prior knowledge of the stimulus. Object type recognition is required as a precondition for influencing size perception in this fashion. If the object's size were deduced entirely from recognition of its nature, size constancy would be already accomplished; comparing the data to the object's retinal size would be more appropriately used as a distance cue. Indeed, this type of influence is found among the many cues to depth, which occur at perceptual levels higher than those which rely upon binocular disparity.

A further advantage of the spatial frequency code for object representation, especially in the frequency ratio form, is that the process of comparison with stored codes is greatly simplified. This is essentially a matter of cross-correlation between the current code and the stored codes, in which a high correlation indicates a significant degree of similarity. Thus, the degree of cross-correlation could serve as an "index of fit" to a classification in storage (as in the model building process hypothesized for the ITC). What makes the spatial frequency code unique is that the cross-correlation function in the frequency domain is reduced to a simple process of multiplying the two frequency distributions. In terms of brain architecture, this is ideal, because it could be accomplished simultaneously, in parallel, for all frequencies and summed in real time.

The ability of the perceptual system to pull out patterns in discontinuous groups of objects is also handled easily in the frequency domain. The lower frequency components of objects encode most of the general outline information; the detail is encoded in higher frequency information. In fact, storing only the low frequencies would be a good way of establishing "class" codes or category templates which would match with current inputs despite differences in detail. The low frequency components of continuous and interrupted patterns (see figure 5.9) are virtually identical. Thus, there are reasons for the brain to pay special attention to the low frequency parts of codes, and there is perceptual evidence to suggest that it does. A unique use of the brain's different classes of simple field cells may also make it possible to encode the *distribution* of particular high frequency features in a low frequency template. This is detailed in the next section.

The ability of the perceptual process to recognize objects even after rotation under three dimensions is reminiscent of systems such as holographic imaging. The holographic analogy should not be overstressed (it has been in the literature), because there is no need for the brain to ever reconstruct and examine a spatially represented image. However, it does point up the fact that systems such as the hologram, which are based on frequency domain encoding, can carry information which permits relatively simple physical systems to extract information about the appearance of the object from any desired angle of regard. That is, the code carries within it easily accessed subcodes for the appearance of the object from a continuum of angles.

Another property of the hologram is that it loses only resolution, rather than parts of the image, when portions of the coded medium are destroyed. The hologram is also reminiscent of many experiments on cortical damage and memory. Because frequency coded information is distributed and location free, such encoding is appropriate to a system like the brain in which components tend to die.

Some kinds of orientation changes are difficult to handle with the model just outlined, if the task is to match a stored template for recognition, but it is in just those tasks that human perceptual abilities also fail (as for example, in inversion).

This brings us to the end of that part of the input system which can be identified as being the mechanism of higher perceptual processes. Beyond this point, the continuing process is multimodal and is more accurately described as "cognition"; the higher stages involve the construction of a world model, which includes elements of memory, and the formation of a sort of preverbal logic which underlies our "common sense" ideas about the nature and operation of the world around us. This is the sort of unconscious logic that we share with all the higher animals, and it is in many ways like a perceptual process. It is necessary to pause in the description of what is really a continuous process and consider the possible implementation of the perceptual models discussed thus far.

VII. SUMMARY MODEL OF THE HIGHER PERCEPTUAL PROCESSES

The advantages of the spatial frequency code are so obvious in the area of object recognition that the model outlined seems very reasonable. It lends itself to providing precisely those abilities which have always been difficult to build in artificial systems, and which conversely have always been the strong points of biological systems. Moreover, its chief disadvantages seem to be well complemented by other processes which are probably also employed by the nervous system. For the sake of discussion, let us assume that *all* of the hypotheses we have entertained are correct, and create a global model of the perceptual system.

Our model begins with feature extraction based on selective convergence; however, beyond the level of V1, this process would be applied only to more complex geometric forms in the case of stimuli of unique significance which justified "private channels" and their attendant hardware costs. This type of operation would be particularly suited to inherited (or in this case, hardwired) connections which serve species-relevant stimuli. The majority of the feature extractors are of a general nature and respond to simple elements, such as edges and corners and lines, which could be invoked in the service of either active or passive perceptual processes. Parallel to the feature extractor channels, there are motion sensitive channels which support subcortical orienting reflexes and, cortically, would provide another dimension of input to the tuning of the cortical extractors.

Active processes, which use tracking guided by edge and corner detectors that are sensitive to particular directions of motion, assemble a serial program of "eye position codes" generated by scanning the object. This program could be employed either as a code in itself or as a means to relate the detail codes obtained from the high acuity, central regions of vision. In either case, this mechanism could pick up the "low end" where spatial frequency encoding of the perceptual data is less efficient.

The principal "passive" process is a model building process in which the encoded location-free visual field content is matched against visual memory, and reorganizations of the receptive fields of earlier stages could be generated to test "hypotheses" represented by partial matches. If the principal code were a spatial frequency transform of some variety, receptive field reorganization could be a retuning of the elements to different frequency sensitivities, different orientations, or different bandwidths. In this process, the unique advantages of parallel architecture can be exploited to deal simultaneously with many features and objects.

It should be noted for the sake of clarity, we have discussed the two processes of geometric feature extraction and spatial frequency coding as different mechanisms; however, they could both be described as feature extraction processes by selective convergence. Thus, in the case of spatial frequency coding, the frequency band becomes the extracted feature. The real difference is simply a shift in our emphasis from geometric features in the early part of the system to spatial frequency features in the later parts of the system.

The degree to which the system is relying on these two kinds of information is unsettled. They may both be of equal importance at all levels. Spatial frequency would be useful for providing object invariance under translation (which is a global property of the visual field dealt with at higher levels), but it could also be useful in the analysis of texture, which is an important local property of the field available at early levels. There may indeed be two different spatial frequency mechanisms operating. The local process which tunes cells, from the RGC to the simple field cell to particular passbands, may operate to define texture or other fine details in terms of its frequency content. This process may account for the response of the system to grating type stimuli, and the poor low frequency response would then be seen as a result of the local nature of this sort of spatial frequency analysis.

This type of local spatial frequency process could be generated by the ratios of excitatory center and inhibitory-surround diameters in the RGC field. However, we noted that it was difficult to imagine a retinal process which could endow the RGC with a true Fourier analysis of a nonrepetitive waveform. If we adopt the notion that form recognition is based on a second type of spatial frequency analysis, derived from the strip intergration model discussed in the preceding section, we find that this process can explain frequency analysis of nonrepetitive stimuli. Further, this global process need not have as its target a set of neurons which look at particular frequency bands. Instead, we would expect that these frequencies would be targeted onto cells which were sensitive to permutations of a number of frequencies which represent objects. Thus, the frequency channels of the global object identification system would not necessarily be susceptible to the sort of threshold fatigue experiments that have indicated a lack of spatial frequency selective channels at low frequencies. This approach also agrees with the observation that texture is a visual feature that is certainly present in the "unanalyzed" visual experience which presumably arise from V1, and is not necessarily accompanied by the global object recognition which we have associated with the later parts of the visual perceptual apparatus.

A further interesting possibility arises in connection with the postulated strip integration method of spatial frequency analysis at the level of V1. The method described depends on integration along strips of oriented simple field line detectors, which can be shown to yield an approximation to the two-dimensional Fourier transform of the field. A potential problem arises if we wish to consider only low frequencies in order to facilitate classification irrespective of detail differences. This problem is that some important details are also high frequency phenomena. For example, corners and arcs have much the same low frequency content; however, corners require high frequency for discrimination by Fourier techniques. We know that corner detectors, edge detectors, arc detectors, etc, abound among the simple and complex field cells, and the existence of hypercomplex cells, which respond to these features in translation-invariant fashion, suggests that these hypercomplex cells are integrating along strips of such oriented corner extractors, etc. There is also indirect evidence for this type of organization from the perceptual results of certain kinds of cortical excitation, such as migraine, which produce the so-called ''fortification illusions'' consisting of nested patterns of features such as in figure 5.10.

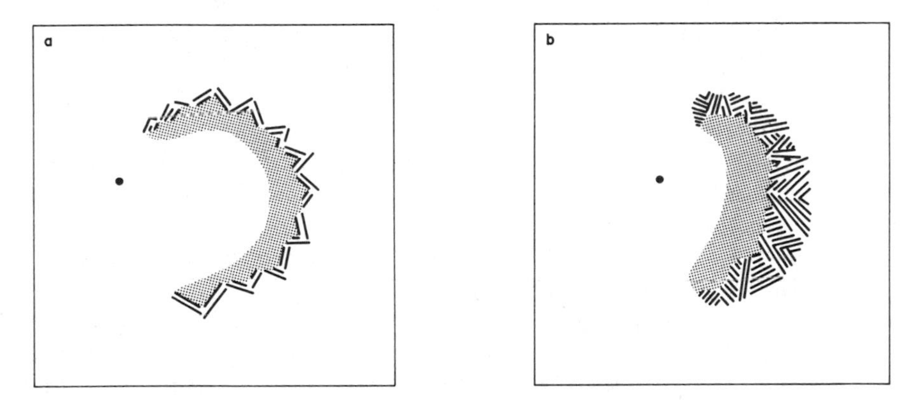

Figure 5.10: *The ''fortification illusions'' seen by persons suffering from migraine. The systems of short parallel lines at different angular orientations suggest that visual field cells activated in the wave of migraine process on the cortex have the perceptual properties expected of oriented line detectors (from ''The Fortification Illusions of Migrains'' by Whitman Richards. Copyright © 1971 by Scientific American Inc. all rights reserved.)*

Now suppose that such a hypothetical strip integrator for corner detectors has spatial frequency selectivity. The output of the collection of such cells would constitute a spatial frequency spectrum of corner features. It seems likely that such a spectrum could carry, in its low frequency components, information about the spatial distribution of common high frequency features. I have not worked out the mathematics of such a system, but it seems intuitively plausible. If anyone does work it out, I would be interested in hearing about it. With such a system it would seem possible to once again drop the high frequency components, which would only carry information about detail differences in distribution of corner features, and retain the low frequency information which would contain the essential items for classification purposes.

For the purpose of our model, let us expand the code to include several spatial frequency analysis: one for line detectors, one for corner detectors, one for edge detectors, etc. The total collection then would represent in spatial frequency form the spatial distribution of each of these types of features.

In order to produce a global synthesis of information from these local computational elements, the model requires the ITC stage to receive information at each processing point in the ITC from all parts of the visual field. While this condition is met in the real brain, the details of this addressing and the type of processing which is performed here permit several possible interpretations. For example, it is not clear whether each local unit of ITC processing addresses the preceding spatially mapped areas by location alone, or whether they are selective as to type of feature extractor addressed within each spatially mapped area. If the goal of this stage is to produce an encoded global object in a location-free format, it would seem most efficient to address the preceding elements by type, whether the postulated process was convergence of geometrically defined features or a convergence of selected spatial frequency extractors into a global expression of a frequency element.

Assuming that our model can, by one or a combination of these methods, produce an encoding of the content of the visual field which is essentially independent of the global spatial organization of the field (except as that information is explicitly retained), the problem of just how to assemble the features into perceptual objects remains. That is, which features belong together as objects? I have suggested that one approach to such a process could arise from the assembly of position codes generated by ocular or attention shifting mechanisms which spatially scan boundary elements. (One could even imagine the neural equivalent of the fact that integration around a closed curve sums to zero, but there is no evidence that the scanning process has anything like that sort of precision and regularity.)

One possibility which utilizes ocular position is suggested by certain facts of perception and by studies of the way in which people move their eyes or (attentional foci) when scanning objects. Examine figure 5.11. The area within the circle is divided into two fields by an irregular line. Although neither of the areas thus created is a meaningful object, our perceptual mechanism still attempts to make sense out of it in those terms. One tends to see one of the areas as the "figure" seen against the "background" of the other "partially covered" area. This sort of so-called figure ground relation is a common item in studies of perception. Such a figure can be drawn such that either area can be seen as the profile of a face. When this is done, one sees either one face or the other, and the remaining "face" is not perceived but seen only as "background." In the case of our completely random line, we see that the system continues to attempt an object organization of the picture, and we can switch with a conscious effort back and forth between the two possible object organizations of this simple visual field.

Which area is perceived as the "figure" is strongly influenced by which area currently contains your locus of visual attention. Although one of the two perceptual organizations can be deliberately *maintained* in the face of attention shifts, the simplest way of *generating* a figure-ground reversal is to switch your visual attention to the area included by the outline you desire to perceive as the figure. This fact, together with the pattern of eye and attention shifts which people employ while first visually investigating objects, suggests that one important process in object definition may be the "trapping" of the attention locus by boundaries. In other words, if the point of visual attention cannot be moved without encountering a particular type of boundary, it is probably within an object.

Figure 5.11: *The figure-ground phenomenon. Even when, as here, there is no real figure but only two irregular areas, the brain tends to interpret one area as the "background" against which the other area is seen as the "figure." The relationship can change, and this change can be accomplished by simply switching the point of attention and visual fixation from one area to the other.*

WHICH HALF OF THE CIRCLE CONTAINS THE PUDDLE OF SPILLED MILK ?

Note that I say a particular *type* of boundary, ie: from red to something else, from something at distance X to things at other distances, etc. This allows the system to handle objects with various sorts of internal subdivisions. In fact, we find that the more perceptual dimensions on which the boundary edges of an object differ from those of the background, the more easily discriminated the object becomes. A complete scan of all possible directions of escape from the boundary is not necessary. A few quick passes of the sort people usually make on an initial scan would suffice to suggest an inital hypothesis for the system which could then be refined through the dynamic model building and hypothesis testing process. In the real visual world, where objects typically differ strongly in several perceptual dimensions from their surrounds, the process is very quick. In cases where such differences are greatly reduced (as in the cow or in camouflage), the process may be very slow, because many equally good "hypotheses" must be tested.

The active process model provides a useful method for approaching the problem of object definition. What of the passive process? It is necessary to distinguish carefully between "analysis" and "recognition" of objects. Analysis refers to the segmentation of the visual field into objects. Recognition refers to the assignment of the object to a class for purposes of interaction with memory, logical, or emotional processes. Recognition by itself is an abstract experience without visual preceptual content (remember "blind sight"). However, recognition processes could form the basis of a passive process object analysis. Recall that the passive process appears to be most effective in situations requiring analysis of familiar objects. The process which is suggested for our model is one in which the encoded visual field information is scanned for partial matches to visual memory, and the hypotheses thus generated are used as feedback to retune the input feature filters to enhance selectivity for the features relevant to the suggested identification.

This "like begets like" operation has frequently been suggested as an explanatory principle in various purely perceptual phenomena, and it is certainly consistent with the facts of physical brain function so far as we know them. It does not matter in principle whether we are viewing the encoding process in terms of convergence of spatial frequency features or geometrically defined features. The essence of the analysis procedure in either case would be the progressive tuning of the feature extractors, or the elements which address the feature extractors, to pass information of the type relevant to the category of best partial match. This would be continued so long as such a process continued to improve the match to category.

Because we have hypothesized that match to category is a low frequency spatial frequency process, the global properties (ie: shape) of the object would be emphasized in the feedback retuning, and the perception of detail which was not a part of the match information would not be affected. Neither would the object's perceived location since the presumed recognition code is location-free. (Although the input code no doubt contains specific location information, this could have been separated from the code for the form and would not have to enter into the matching process.)

This type of object analysis is probably the sort that underlies the Rorschach, or "ink blot," test phenomenon in which people tend to analyze random visual fields into patterns (objects) which resemble familiar object categories. (The utility of such a test resides in the fact the hierarchy of the recognition search tree is an important personality variable.)

If the system selectively tuned itself to those spatial frequencies (and frequency ratios) which were heavily represented as features in the hypothesized match, it would tune out as "noise" the other spatial frequency bands containing primarily information arising in the hypothesized "surround." This may be the mechanism underlying the perception of the object "standing out" from the background when it is the subject of our attention. Again, such a process would be enhanced by maximal differences between the object and the background.

Such a dynamic tuning process offers an explanation of one apparent problem with the "Fourier-like" model of the perceptual mechanism. In such a process, the interaction between spatial frequencies arising from nearby objects should tend to distort them. Such distortions occur and are the subject of many illusions (see figure 5.12), but in general the process seems remarkably resistant to such effects. If a selected frequency spectrum were enhanced according to a predetermined template while its "competitors" were suppressed as noise, such interactions could be minimized.

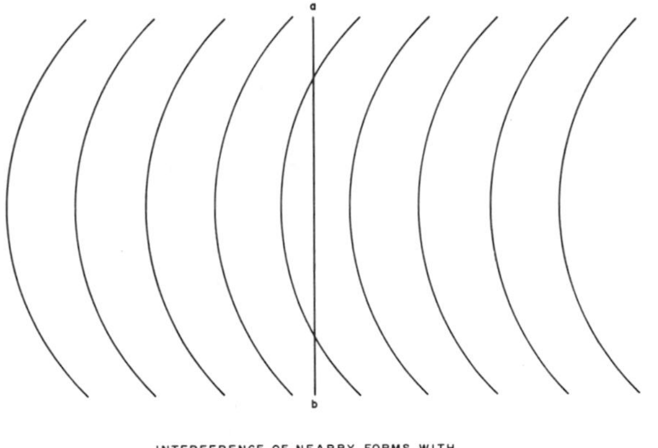

Figure 5.12: *Despite the brain's excellent abilities at keeping perceptions correct in the face of interference, there are small effects of nearby lines on the perception of a line. Such effects are predicted by interference of spatial frequency components under a transform hypothesis of figure perception.*

Clearly, in such a model one must be careful about the level of abstraction of the template pattern in memory. There are ways of dealing with the recognition of an object from any angle (for example the holographic-like process, which could be extended to objects encoded in the round with a search template consisting of a two-dimensional slice through the three-dimensional matrix of the code), but there are more difficult problems. For example, we recognize a standing dog and a dog curled up sleeping as examples of the visual category "dog," but they do not have the same shapes as abstract visual objects. Any codes for them would seem to be either so different as to make a match impossible, or so abstract as to make a match useless as a guide to object analysis. Are we to believe that we have templates, even fairly abstract ones, for dogs in all possible body positions? It seems more plausible that the code search is initially guided in such instances by analysis of important details (ears, eyes, teeth, tail) which could lead to a search of the category "animal," etc. Such details could be detected efficiently by the use of the small number of hardwired specific complex feature extractors which we postulated might co-exist with more general-purpose extractors as elements in the code. It is also likely that the analytic process, which is not yet part of the recognition machinery, would simply analyze the object as a collection of subparts. That is, the analysis would be in terms of legs, as objects for basic visual analysis by the tuning method, and only the recognition stage would then attempt to match to the higher level category of "dog." This actually occurs in some kinds of cortical damage, in which the parts are identifiable by the patient, but the more global "dog" level is not recognized.

In practice, of course, we would expect that the process of perception would involve an interaction between the active and passive processes, and the active process might well be coupled into the search procedure to facilitate the dynamic tuning of the passive process.

VIII. PERCEPTION IN ARTIFICIAL SYSTEMS

A black box summary of this model of the brain's perceptual process is shown in figure 5.13. Some of the general principles of processing seem applicable to artificial systems. In fact, many of them have already been employed in various forms. In most instances, these have been experimental situations in which the practicality of one or another of these processing principles was being explored. Thus, the several attempts at Fourier-type image analyses have not incorporated active processes (nor to my knowledge, dynamic tuning), while the attemps to build "perceptual maps" of the environment by exploration have generally not employed spatial frequency analysis.

Many of these principles suggested by brain function would seem to be useful in artifical systems, and the brain's apparent combination of them in ways that complement the weaknesses of one system with the strengths of another seems a reasonable approach in the artificial case as well. The simulation of such a perceptual system in software would be complex, but certainly not impossible in principle. Of course, the problem is once again efficient use of time. With the exception of parts of the active process, the mechanism is essentially a small number of serial steps performed in parallel on large arrays of data. Even in the case of the serially operating active process, the position code for the locus of attention could be applied in parallel to the content of all of the array's elements at that moment. To emulate this total perceptual process on a traditional computer would require an enormous number of serial steps, especially if we are thinking about three-dimensional Fourier transforms. Array processors are becoming commonplace today, and I imagine we will see them on a chip before long. Batteries of such chips would seem to be the minimum requirement for a good software emulation of brain function. Despite the very sophisticated nature of current efforts at perceptual processing software, it seems that a true breakthrough into real-time image recognition of the sort practiced by the brain is going to require hardware advances in the direction suggested by brain architecture: massive parallel processing. The use

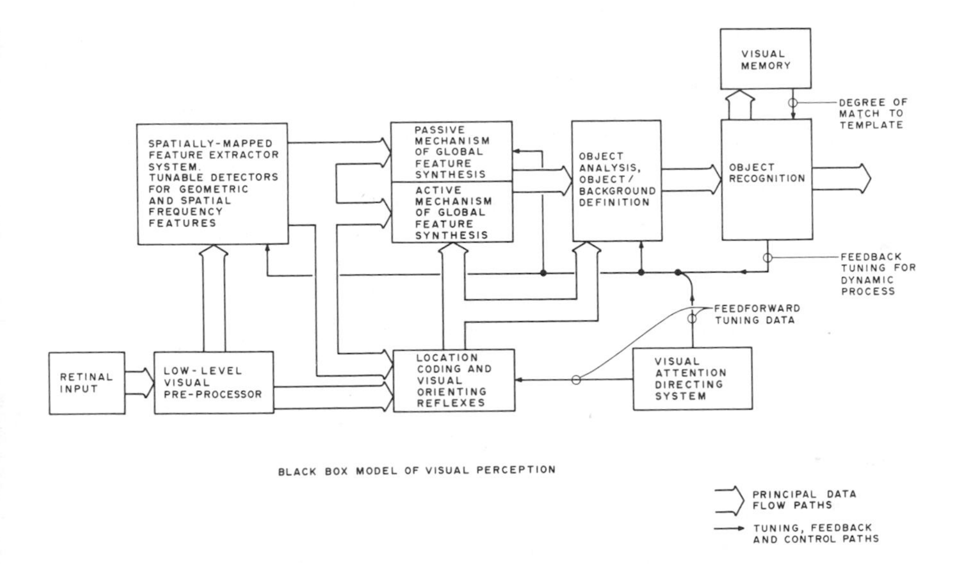

Figure 5.13: *A summary model of the perceptual system for vision, incorporating a number of the possible mechanisms discussed in the text. The role of feedback tuning paths and feedforward attention processes is shown.*

of feedback in the dynamic tuning of the input is a technique more commonly seen in analog than in digital devices, but its equivalent is seen in programming in iterative procedures. Its employment by the brain in a basically parallel, digital format to redefine the receptive fields of feature extractors seems immediately applicable to arrays of digital components.

All of the preceding discussion has ignored the two-dimensional nature of the brain's byte. The temporal dimension is not static, but is constantly changing as the information entering the system changes. As we have seen, these are principally "magnitude of stimulation" changes at the input, but by the level of higher processing, they reflect the current amplitude of a variety of spatially coded qualities. The system is not clocked, but is simply continually updated on each line. Therefore, there will be differences in the processing time when different kinds of information are made available to the higher centers. These differences can be demonstrated experimentally. However, in practice the longest times are on the order of a second, so that in terms of interaction with the world at any instant the temporal byte could be said to reflect the state of any analysis of the current sensory world at that moment. (I will return at a later point to the implications of the first derivative of this temporal byte.) As in the earlier discussion, there seems to be no obvious way to emulate the temporal byte aspect of perceptual function in a digital system, other than by encoding magnitude on a small bus representing each axon. This clearly becomes unwieldly in a system of any size. Perhaps it will ultimately be necessary to follow the brain's lead in intermixing analog and digital techniques at the component level.

IX. WHERE THE PERCEPTION GOES

We have reached that level of the brain's input system at which it is no longer possible to continue in a "straight line" from input to output. Before considering some of the branches which the information processing paths take from here, let us examine the higher perceptual processor as a black box to see what kinds of inputs and outputs it has. If the model I have developed from the visual system is approximately correct for the other modalities, we have defined a black box system which is replicated in more or less complex form for the other modalities. All of these will have inputs from different sensory preprocessor systems, but as I have suggested, they are most likely concerned with moving their information towards a common coding scheme. In addition to this obvious input of sensory information, the higher perceptual processors will have other inputs from the more central parts of the system serving to direct and assist in the analysis of the sensory information.

In the visual system in particular, I have mentioned inputs from motor systems that provide feedback on eye position. Analogous functions exist in at least some of the other sensory systems. An input from frontal cortex is well known and is probably the source of our postulated "attention" input which serves to direct the active process in a straightforward manner, and probably can play a goal-related role in the dynamic organization of the passive process as well. A final source of input are the data from visual memory which may be the basis for improving the search strategy in the dynamic phase of passive perceptual analysis.

On the output side, it is important to note that not only do the outputs of the process feed a variety of other processes, but also that different sorts of outputs leave the perceptual system at several levels. The sensory preprocessors themselves apparently have outputs which can be tapped by higher levels of the system for simple information such as total luminous flux. It seems that at each stage of progressively more refined analysis of the input, the output of that stage is available to all parts of the system as well as to the next stage of perceptual processing. This is eminently reasonable, because it is always possible that a subsequent level of analysis may not be able to further refine the representation of the input. It also appears that the organization of the system is such that the output of any given stage of perceptual analysis does not contain the information in the preceding stages, but only the additional code which it wishes to add to the data. Thus, the primary cortex provides us with data on the detailed nature of the visual field and its spatial framework without analysis of the patterns into objects, whereas the higher levels provide us with the additional experience of object organization, which is apparently independent of our subjective receipt of the code from the primary visual cortex. Our total visual experience appears to be a montage of the outputs of several levels of analysis, although we are accustomed to think of it as a unified experience. In ordinary life, we usually experience only the absence of the higher levels of analytic output in which we fail to recognize what we see. In unusual circumstances, the reverse appears possible as well; we can recognize what we cannot see.

These various levels of output from our black box are routed to a variety of other systems within the brain. These systems draw on the various levels of output to different degrees according to their needs, and they are essentially autonomous parallel functions. Among those of importance are the process of recording in memory what we have seen (memory function), the process of organizing the data from all of the sensory perceptual processors and from past experience into a logical world model (cognitive function), the process of responding emotionally to what we perceive (a division of the motivational function), and an assortment of higher rational functions which range from verbal description to complex planning of future action. All of these functions which utilize the input are in a continual state of interaction with one another, so that there ceases to be any fixed sequence of processes representing the flow from input to output except in the very simplest cases. I will proceed now to the examination of a number of these central processes.

6 The Logical Functions

I. THE PROBLEM

In this chapter, I will explore the operation of some of the "higher processes" in the central machinery of the brain; these functions include the construction of a perceptual model of the current state of the world, some applications of symbolic language, and the organization of certain logical processes.

A word needs to be said concerning the nature of the evidence and some of the underlying assumptions that guide such an exploration. In the first place, there is a technical difficulty: for data on higher functions in man, we are dependent almost exclusively on cases of disease and injury. That is, we have to rely upon clinical case histories in which such niceties as before and after control tests are not necessarily available, and which have been reported by medically rather than by scientifically-oriented workers. Thus, definitions tend to vary from one report to another, and essential points for deciding theoretical issues are frequently omitted or unavailable. Further, the types of damage that occur are rarely restricted to the anatomical area or specific function that we wish to investigate, which enormously complicates interpretation of the results.

However, all of these difficulties must take second place in comparison to the problems posed by the complexity of the functions we are trying to investigate: we do not know what we are looking for, and we tend to see things in terms of our preconceptions. If you knew nothing of the internal workings of a general-purpose digital computer, but were accustomed to giving it complex numerical problems and receiving answers, how would you respond when asked what the basic functional composition of the machine might be? You would very likely say that the machine possessed an "arithmetic" function; you would hypothesize that there were some areas of the machine devoted to multiplication and others to addition, and so on. Although these inferences might be correct, given the existence of parallel floating-point processors and hardwired arithmetic elements, they do not really characterize the operation of a typical general-purpose central processor.

In the same way, we tend to approach the brain by looking for a set of preconceived black boxes which contain the functions we expect to find. These preconceived functions are built up out of experiences derived either subjectively, from our own internal processing, or objectively, by observing the acts of others. These are experiences with the overall functioning of the total machine, and just as different important logical "functions" may share the same hardware elements in a computer, the execution of different "functions" in the brain may be distributed over many systems whose roles are transparent in the final output. Nonetheless, destruction of one of these hardware-defined operations will impair the operationally-defined output function, which leads us to suppose that we have found the area concerned exclusively with executing our preconceived functional "unit" of brain operation. When this is the case, other functions, which require the same hardware, will be impaired as well, but this deficit may never be systematically examined. For example, a lesion in a particular part of the cortex reliably impairs both arithmetic calculation and the ability to describe the relative spatial positions of objects. What "function" has been disrupted? This problem becomes acute in research on higher functions, because the "obvious" symptoms are the patient's "problem," as defined by his inability to operate in a customary fashion. Such symptoms of dysfunction become the focus of the clinical investigation; however, these symptoms may not be related in any obvious way to the real "hardware" function of the damaged region, and a truly revealing method of evaluating some deficit may not exist.

For some artificial intelligence purposes, this lack may not be an important problem. One approach to the problem of artificial intelligence is concerned with modeling the global functional characteristics of human thought. This approach (like the analogous approach of cognitive psychologists to the mind) holds that the actual hardware processes are irrelevant, so long as the desired input/output relations exist in the model. Similarly, cognitive psychologists are frequently concerned with the analysis of global functional processes in human thought and feel that what is of interest is an adequate functional description of mental activity. Any correspondence between the terms of their models and the actual organization of the hardware of the brain is viewed as coincidental and uninteresting. The psychologists are essentially seeking a description of the mind's software. This point of view is certainly defensible if one's goal is the description of mental processes rather than brain function. In the present context, this approach to *what* the mind does is analogous to the design of intelligent software systems which are hardware independent.

On the other hand, the modeling of these global intellectual processes on traditional computers has proven to be very difficult; it tends to take forever to run even the simplest sorts of simulations. One immediately suspects that we are again faced with the necessity, for purely practical reasons, of moving to machines with a more brainlike, parallel structure. Given this requirement, it becomes of great interest to know just what sorts of functional divisions the brain finds advantageous in *hardware* designed specifically to perform the operationally-defined functions which we recognize as reasonable or even experimentally demonstrable components of intellectual processes.

In considering the higher functions of the brain, my primary goal will be to discover, as best I can, what sort of neural processing units exist, and in what way they are employed by the brain to produce various, seemingly unitary, intellectual processes. My principal concern will be with the number, type, and relationship of such major processing units, and the degree and order of their employment in different operations. This is a formidable task in itself, and I will have little to say regarding the detailed neural operation of these processing units.

The point of departure will be the functions of the parietal lobe of the cortex, and the junctional areas between this region and the temporal and occipital lobes. (See figure 6.1.) There appear to be two levels of operation in this system: one can be described as the level of our immediate experience in which a unified model of external reality is synthesized out of the products of the various senses' perceptual processes. The second level, which appears to occur principally in the human, is the use of these same neural mechanisms to manipulate symbolic information.

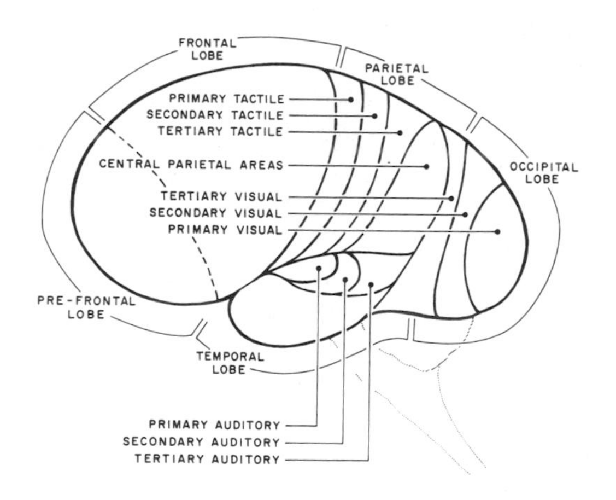

Figure 6.1: *The major cortical lobes, showing the relationship of the central parietal area to the primary, secondary, and tertiary sensory areas for hearing, vision, and touch.*

According to this view, the fundamental "logic" of the relations among physical objects in the world is learned by the brain and employed by its apparatus in dealing with the physical world. When, subsequently, this same apparatus is employed for the manipulation of symbolic data, the everyday logic of "common sense" relationships among physical objects is already available to be employed on a more abstract level of code. This "natural" level of operation seems to cover most of the logical thought of human beings when they are not engaged in the application of formal logical systems (which clearly require additional learning of new rules).

I will begin, therefore, with the brain's methods of synthesizing its model of the external world.

II. THE MULTIMODAL ANALYSIS OF THE EXTERNAL WORLD

In our subjective experiences the world is presented to us as a set of objects that occupy space and have a variety of properties and relationships which result in sensory experiences of all sorts. Objects can be seen, felt, smelled, and tripped over. There is a clear relation between our visual perceptions of their form and location, and our tactile and kinesthetic perceptions when we touch them. If we felt them to be shaped or placed differently than we saw them, we would be astonished.

We have no subjective sense of the process by which this cohesive impression of the object as a physical unified ''thing'' is built up out of the variety of our multiple sensory inputs (and our past experiences). This multimodal preconscious synthesis is presented to us as an accomplished fact, much like the processes of perceptual organization within the particular modalities which have already been discussed.

The nature of multimodal sensory synthesis is illustrated by a comparison of normal and brain-damaged experience. Imagine being blindfolded and given an object to identify by touch. Suppose that you could feel it clearly in your hand, note its shape, and easily identify it by touch among several other objects presented to you. Suppose that I then removed your blindfold and asked you to identify it by sight, and that among the objects which you saw before you, as clearly distinct *visual* shapes, you could find nothing that seemed to be related to the shape you had just felt. Assuming that you had not been tricked and the target object removed, you would be mystified at this lack of relation between tactile and visual shape. This is just the sort of loss of trans-modal recognition processes that is experienced by people with certain types of parietal lobe damage. This kind of correspondence or equation between the properties of objects perceived in different sensory modalities is the first step in the construction of our cohesive experience of the environment.

The flow of information in perceptual systems includes the distribution of the ''finished product'' of each stage of analysis to other brain functions and the transfer of these data ''downstream'' to further stages of perceptual processing. If we examine the layout of the cortical areas in figure 6.1, it is apparent that this ''downstream'' information flow begins in the primary cortical receptive areas for vision, hearing, and touch, proceeds ''inwards'' to the secondary perceptual cortex of each modality and converges on a small region in the central part of the parietal lobe. In fact, it appears that, here at least, form follows function: disparate modalities' inputs are brought together for synthesis in that small central region and are also put ''on the data bus'' for the rest of the brain.

The nature of this processing sequence becomes more apparent after examining the reports of human patients with damage to these various cortical processing stages. Taking vision as the example, and beginning at the primary visual cortex, we find, as expected, that damage here produces blindness. If the damage is restricted to part of the primary visual cortex, blind areas are found within the corresponding portions of the visual field. Moving to the secondary visual areas, symptoms are seen of a sort that are understandable in terms of the mechanisms of visual object perception which have already been examined in some detail. Patients with damage at this level have no loss of conscious visual sensation, but they do show various deficits in interpretation or comprehension of the visual scene. They may fail to recognize objects, or fail to group visual experiences into objects at all. Depending on the location and extent of damage, they may recognize some features of objects, but be unable to identify the object except by employing verbal logic to make reasonable guesses. A patient might remark, "eyes, legs, jaws . . . it must be an animal!" After damage in related areas, they may fail to recognize an object even though it has been adequately differentiated from its background.

As lesions occur closer to parietal cortex areas, the more elementary perceptual processes remain intact, while deficits appear in complex constructs. In one type of deficit, the patient may be able to see and recognize objects, but only one at a time. He will be able to identify or describe one of a group of objects without difficulty, but as soon as his attention is directed to a nearby object, the previous one undergoes perceptual disintegration and becomes part of the general meaningless jumble that is the rest of his visual field. (It is interesting to note, that in relation to the active process hypothesis, this condition is usually accompanied by disorders of involuntary eye movement.)

Other disorders of perceived relative location are seen in cases in which a patient attempting to draw an object may reproduce the parts of an animal with fair accuracy, but puts them together in no particular order. Another related effect is an apparent interference with location coding mechanisms which can result from irritative damage in these cortical regions. With this type of disorder, patterns may be seen in improper extent and location, so that the victim will perceive, appropriately, a woman in a patterned dress, but in addition will see the pattern as extending over the face of the figure as well. We can probably interpret this type of effect as confirmation of the hypothesis that the location and identification functions of visual perception are complementary but somewhat separate processes. In this case, there is improper activity in an area involving location-related functions, so that the properly identified pattern is coded with an improper spatial extent and, therefore, is entered into the perceptual experience in that way.

In all of these cases, there is no damage either to general intellectual processes or to other modalities' perceptual processes. A patient who is unable to recognize a common object visually will remark, "Of course! It's a key," when allowed to handle it. In many cases, there is an apparent inability to encode *visual* memory, but there is no apparent interference with memory processes in general. Such a patient may be able to copy a picture of an object accurately, but will be unable to draw it from memory. In these instances, there is usually a profound impairment of the ability to perform such visual memory functions as recognizing faces.

These clinical effects seem quite compatible with the more detailed, neural models of perceptual mechanics, derived from animal experiments, which I have already discussed. (This is encouraging, because I will have to turn shortly to processes which are not easily investigated.) To round out this clinical picture of the secondary cortical perceptual areas, I should mention the analogous deficits which are seen in cases of damage to the secondary areas for the senses of touch and hearing. Such effects support the idea that the type of sensory or perceptual analysis occurring there is fundamentally of the same sort seen in the detailed analysis of the visual system.

In patients with damage to the primary tactile areas, there is a basic loss of the sense of touch in the corresponding region of the body, which seems entirely analogous to the blindness that results from damage to the primary visual cortex. As the site of damage moves further into the parietal lobe, there is less impairment of basic sensations of touch, but the disability is manifested in impairment of successively higher levels of perception.

In the secondary tactile areas, one finds damage which leaves the tactile thresholds intact, but produces patients who state that while they can feel an object, they can't tell what it is. They may be able to describe adequately the tactile sensations, but cannot seem to assemble them in a meaningful way into a perception of an object. These are the "inverse" of those with damage in the visual areas: the patient may feel the key and describe its temperature, texture, bumps, and curves, but he can only come up with, "Of course! It's a key," when he is finally allowed to see it.

There is also an interesting failure of the "body scheme" which can result from damage here: the patient loses the ability to understand the position of his body parts or to recognize them as parts of himself. He is able to identify an object such as an arm or a leg correctly, but may be amazed to discover that it is his. He may even refuse to believe it and refuse to accept the limb as his own. Related deficits in dealing with location in tactile space include anomalies such as an intact ability to recognize individual objects by touch, while still being unable to arrange them into desired patterns without visual assistance.

The primary auditory cortex is located on the upper surface of the temporal lobe. (See figure 6.1). As mentioned earlier, damage here does not result in total deafness, but in loss of more advanced auditory analysis. Thus, the primary cortex of the auditory system is somewhat like the secondary cortex of the tactile and visual systems. There is one fundamental difference in processing between the auditory system and the visual or tactile systems which may relate to this difference in functional level: the auditory system is primarily a serial input system, whereas the visual and tactile systems are primarily parallel. To be sure, the auditory system is concerned with the localization of sound in space, and the tactile system usually "explores'' objects with the fingers in a sequential process. Nonetheless, the important features of the visual and tactile worlds are simultaneous groupings of stimuli, while the important features of the auditory world are usually temporally serial groupings of stimuli. Therefore the auditory cortex is a sequence analyzer. In order to recognize ''objects'' consisiting of temporal sequences of features, these features (individual complex sounds) must be stored briefly; this process may require that the features be categorized earlier in the analytic system than is the case, for example, in vision.

One of the most important examples of these serially presented features of the auditory world in humans is the set of speech sounds known as ''phonemes.'' These are the elements of the spoken words which, when assembled, constitute auditory ''objects'' which can be recognized in the same way that the visual system recognizes visual objects. The phonemes, then, have the same standing in the auditory system that features, as components of objects, do in the visual and tactile systems. (The phonemes are clearly learned features, for they differ from one language to another. The sounds which are important in word identification in one language are not the same as in another. That is why you perceive speakers of a language unrelated to your own as speaking in a very rapid babble; you cannot tell where the words start and stop. It is entirely likely that a large number of visual features are learned too, although this is more difficult to establish.)

Damage to the primary auditory cortex impairs the recognition of these phonemes. It also gives a variety of other impairments of fine auditory discrimination; as we would expect, these deficits influence perception of the temporal order of stimuli. Persons with this type of damage are usually unable to understand spoken speech, although there is no particular impairment of hearing. The speech simply sounds like a foreign tongue to them. It should be stressed however that there is no loss of intellectual understanding of the meanings of words, only of the ability to discriminate them as perceptual objects. Reading of familiar words remains intact although there may be problems with difficult or unfamiliar words, which probably reflect the practice of dealing with them by ''sounding them out'' to oneself.

When the damage is localized in the secondary auditory areas, the ability to ''hear'' words correctly is retained, as evidenced by the patient's ability to repeat them, even though he still may not be able to understand their meaning. This deficit seems analogous to the visual case in which the object is seen and can be drawn, but has no ''meaning'' for the patient. The sensory input is correctly analyzed, but not recognized. Such a patient may be able to copy written passages correctly, but cannot copy from dictation (or, interestingly, write compostion).

In cases of partial damage, the general meaning of words is better preserved than are specific meanings. It is possible that this reflects damage to a process of the sort which I discussed in terms of the ''dynamic tuning'' of the passive perceptual process. The initial match can still be made to a general category, but the subsequent refinement process is not adequate.

It appears then that, although there may be some differences in the cortical level at which particular sorts of analyses occur, the auditory system follows essentially the same pattern as the visual or tactile systems in the perceptual integration of information reaching it. As in the other systems, this process proceeds through several stages which correspond in a general way to anatomical progression across the cortical surface.

Each of these sequential stages of analysis uses only those aspects of the external object which can be detected through one modality. Many objects in the real world have properties ranging from simple ones like shape or texture, through complex ones like membership in a category, which can yield information to more than one modality. It is clear that in an advanced brain, some means must be available for the same properties, as analyzed by the different modalities, to be linked with one another, both in memory and in subjective experience, if they arise from the analysis of the same multimodal object in the external world. This type of linkage seems to occur at the junction areas between the cortical regions involved in analysis of the various modalities, and in particular to involve certain specific intercortical fiber systems which interconnect equivalent parts of the cortical sensory systems.

On the surface of the temporal lobe below the auditory centers, is an area which has been implicated in the storage and retrieval processes of memory. (Because damage which produces deficits in speech memory might here be attributable either to memory processes *per se* or to problems with categorization of words as perceptual objects, it is difficult to infer which functions are involved: the patient can understand words, but cannot correctly repeat phrases, and "searches for words" when trying to express himself.) This lower temporal area also abuts the infero-temporal cortex, and damage here also seems to lead to an inability to associate the words with their visual images, even though the word may be understood in all but the visual sense. Such patients, for example, can draw from a copy, but not from verbal direction. It would appear that the connection between the verbal perceptual object and the visual perceptual object is broken by damage in this region.

Similar effects are found at the junction between secondary visual and secondary tactile cortical regions, where damage destroys the ability to transfer object recognition from one modality to the other. Something felt cannot be identified later by sight, and vice versa, even though it may be possible for the person to recognize a previously known object through either modality. That is, the basic perceptual abilities of both modalities are intact, but there seems to be no communication between them. This is a different situation from that in which the perceptual ability is lost in one modality but retained in another. It appears that these junctional regions are involved in *establishing equivalences* between object level perceptual operations in different modalities. This kind of operation is necessary in preparation for the final stage of model building in which the multimodal objects are incorporated into a spatial world.

An interesting phylogenetic point is worth mentioning: the ability to *transfer* information from one modality to another is apparently well developed in only the highest brains. Humans have no difficulty viewing a set of objects with the eyes only and then picking the one they want by touch only, or *vice versa*. Apes can also perform well on these tasks. Monkeys, on the other hand, have great difficulty and can only do so under special circumstances where the objects are of extreme relevance to the animal. Lower mammals cannot do it at all, except in the case of extremely simple stimuli (such as equating a pulsing sound with a pulsing touch).

One such intermodal equivalence in perception is the equation of words (auditory objects) with objects, seen or felt. It seems that the ability to use symbols may be unique to humans and, to some extent, apes, because of their well-developed ability for intermodal object recognition in general.

The physical basis of this ability appears (not surprisingly) to be the degree of interconnection between the cortical secondary sensory areas of the senses involved. In man, there are large, well-developed fiber bundles which interconnect the various higher level sensory-processing areas around the occipital, parietal, and temporal lobes. Among lower animals, only the apes have similar, well-developed connections. Other mammals have these connections in only a meager and poorly differentiated form. When these connections are severed in humans, the ability to perform intermodal recognition between senses processed in the disconnected areas is lost; ie: if the auditory connections are involved, the ability to name objects upon sight, touch, etc is destroyed.

Correlation between touch and vision is every bit as much a learned phenomenon as is this relation between word (auditory object) and visual or tactile object. A great deal of time is spent in infancy learning the relationships between touch and vision. Such relationships also help us to coordinate our visual perceptual space and our tactile or kinesthetic perceptual spaces. This is a necessary ingredient in the construction of a multimodal perceptual object, because such an object should be in the same position relative to us in all its representations in all of the modalities by which it may be perceived.

III. THE INTEGRATED WORLD MODEL

Once the identification of equivalencies between objects as perceived by the various sensory modalities has been accomplished, the way is cleared for further perceptual analysis to proceed in a nonmodality-specific fashion. It can operate upon any perceptual information obtained about the object through the perceptual apparatus of any of the individual modalities, and the results of such processing can be added in parallel to a single set of codes for object qualities which are now all part of a common set representing the unified object, regardless of the modality of origin.

What sort of perceptual analysis might occur beyond the point where the codes from different modalities are equated? Clearly, the multimodal objects are already recognized as such, and their relation to the organism in space is known; an additional fundamental perceptual quality is the relation of an object to an external spatial framework and the objects contained in them. If a human brain sustains damage to the central area of the parietal cortex, where the process of perceptual analysis seems to converge, a variety of deficits are observed in the ability to recognize or employ the spatial relationships among objects. These deficits are not confined to a perception of the object in any given modality, but appear to operate on a multimodal representation of a unified perceptual object. Because damage here does not affect the ability to perceive the object as such, or even the perception of its location relative to the observer, we can plausibly infer that "spatial relationship" is another abstract perceptual quality which is present or absent as an independent in-parallel addition to the set of neural elements whose activity encodes the object. Although there is scant evidence to test such a conclusion, it does seem likely that the same mechanisms of neural processing that support the extraction of other perceptual qualities at lower levels of analysis are also employed here.

At this point, it is important to consider just what is implied in the concept of spatial relationship. It is a different thing altogether from the straightforward property of position or location of the object relative to the observer; in fact, it is a property of relationship between two objects in the external perceptual space. More importantly, "spatial relationship" implies an analysis of the position of the object in relation to some external frame of reference, a "perceptual space," within which the observer may move (thereby changing the object's position in relation to him) without the object(s) moving. The relationship is not a property of the objects, but of a "world model" which contains them as features.

This type of perceptual analysis is imperative to all higher processes involving rational interaction with the environment. The simple identification of objects may provide information about what type of action is to be taken, but without information on the relationship of objects in space no logical course of action can be undertaken. It is a level of perceptual analysis which is well developed only in the higher brains. In simpler brains, the typical response to a situation in which a visible goal object (eg: food) is behind a fence and out of reach, is to strain against the fence. A higher mammal will show an appreciation of the spatial relationships involved by going around the fence.

I don't mean to imply that *only* a comprehension of spatial relationships is involved in this simple sort of reasoning, but this understanding is clearly an important component of such intelligent behavior. To simply move towards an object, as in the case of straining against the fence, does not require a central model of the spatial relations of the physical world; instead, a simpler model is used in which the goal is encoded as a direction relative to the organism. To go away from the goal and thus go around the obstacle, implies a conception of the surrounding space, and the relationship of both the objects and the organism in it.

Most of the simpler robotic devices that have some ability to maneuver in the environment and to home on stimuli such as power sources have very simplistic algorithms for dealing with obstructions. Usually these involve a trial-and-error process of repeated backing and sideways motion until forward progress is again possible, whereupon the device continues to home in on the salient stimulus by feedback processes. This sort of operation is entirely analogous to the operation of simpler organisms. One reason it is so difficult to design a more advanced rational response into these systems is because a sophisticated model of perceived spatial relationships is a prerequisite to more advanced behavior; this model can come only from an advanced stage of input analysis after object definition has occurred. (A different approach has also been used, one in which a robot "maps" its environment to define the limits of obstacles by trial and error, but this "learning" paradigm is of little use in the generalized novel environment.)

It is hardly thought of it as a "reasoning" process when we go around the fence. The solution is obvious; we "see" it immediately. The very terminology conveys the intimate relation between this type of intellectual feat and the underlying perceptual processes. However, the fact that this solution is so easy that it does not require the intervention of verbal symbolic processes does not mean that it is not a reasoning process; one of the points which I wish to develop is that "seeing" the solution to a problem in spatial relationships forms the fundamental substrate out of which *all* static logical relationships proceed.

Look again at figure 6.1. The central area of the parietal cortex, which is bounded by the secondary perceptual cortex of the three major spatial modalities, is ideally suited to receive the required location-analytic and object-analytic information in preprocessed form. Damage to this area in humans profoundly interferes with their ability to understand spatial relationships and renders them incapable of solving problems, such as running around the fence. They lose the distinctions between directions and even are confused when they have to decide between vertical and horizontal. They understand words such as "above" and "below" or "in front of" as sentence elements, but can attach no meaning or significance to them. They have no problems with object recognition or location, but they cannot operate in a spatial framework or arrange objects in required relations. They can reach for and pick up objects; this is a simpler feedback function, like trying to run through the fence, but they cannot relate objects to one another in any meaningful way. They experience difficulty dressing themselves or telling time from the relative positions of hands on the clock face. They find it hard to write or copy figures, because they tend to confuse mirror images, go in wrong directions, and are unable to decipher the relationship of parts of an unfamiliar figure.

These patients also have a variety of profound disturbances of higher intellectual functions, but I will return to these. For the moment, I want to concentrate on their deficits in spatial relationships to make the point that destruction of this area produces a fundamental disruption of the final integration of the perceptual world.

If we accept this general picture of the development of the world model, several principles can be identified which may be relevant to the design of artificial systems. The addition of successive stages of analysis as abstract qualities added to the perceptual code in parallel with prior results has already been stressed. It could be schematized as a series of hierarchically organized processors employing a common bus structure as in figure 6.2. What I would like to point to now is the process of employing these codes in the construction of a unified model of perceptual space. To the raw visual or tactile code of features, two lines of analysis are added: one for identification and one for location. What should be noted here is that location in this sense is *not* equivalent to a perception of spatial relations. It is an encoding of the location of the object, at that moment, in that modality, in relation to the observer. This is a much more primitive analysis, which is at once necessary to the construction of the more refined unified model, and also immediately employable by the system in direct actions relating to the object. These actions only require information on its position relative to the observer. In simpler brain systems, it is all that is used.

The development of relationship as an abstract perceptual property is probably made possible by first identifying and employing object recognition across different modalities to generate an equivalence between the object's location codes. In the human infant, this is done through a learning process by simultaneous visual and tactile exploration of objects, which results in perceptions which can be related to the preceding motor output commands that define movement in space for the organism. (In artificial systems, one presumes the perceptual codes for location could have these equivalencies with space-defining movement commands defined in an *a priori* fashion.) The next stage is the definition of a spatial framework which is independent of, and contains, the perceiving system. That is, there must exist a code which maps the observer onto a surrounding space which is absolute in the sense that its referent is not the organism, but some external frame, such as the earth. This permits the object location code to be translated into a location code in a coordinate system which can remain unchanged as the organism moves. The only change resulting from movement is the position code for the organism itself. Thus, relative position codes of object and organism are already available for direct interaction, and the relative positions of external objects are fixed in a code which need not be recomputed for every motion of the organism.

This set of "absolute" location codes is available for further analysis of rational action and constitutes a map of the world in which the system finds itself, but it need not be consulted in order to achieve control of immediate interactions with objects. The fact that this spatial map includes the organism as a spatially-defined object is of extreme significance: when the level of sophistication necessary to map one's own movements onto an absolute space is achieved, a code for observer location can be dealt with in the same fashion as the code for any other object. The manipulation of this code, in logical processes, is equivalent to having a self concept which enables one to deal with oneself as an abstract quantity. This opens up enormous advances in the rational planning of future operations. Considering that the full-blown perceptual code constitutes a large share of the

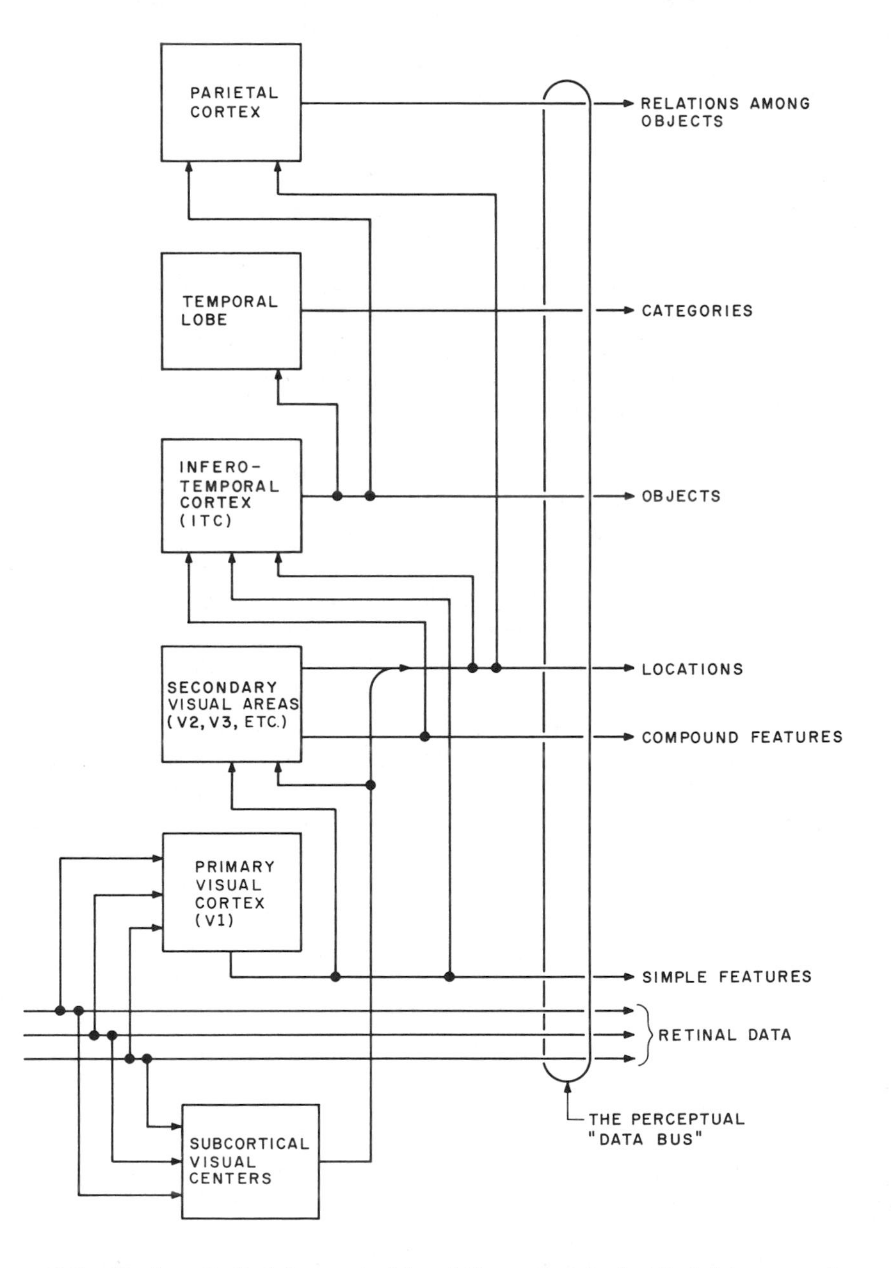

Figure 6.2: *The hypothetical "perceptual bus," illustrating the fact that data processing proceeds in the perceptual function (and in the brain generally) by the addition of new lines to the existing data lines rather than by a series of transformations of the data on the input lines.*

content of our conscious experience, it is not unlikely that this process of mapping the observer into the world is a major component of self-awareness as we experience it. That is, the self is an object with a location and other attributes rather than simply an experiencer.

In artificial systems, it might be simpler to deduce an absolute spatial location and enter it immediately into the perceputal code for the object. However, the brain's ordering of these processes seems to offer advantages of economy. Because the brain seemingly adds object identification, as an abstract quality code, to the already coded perceptual properties of the object, this multimodal quality of "identity" can serve to represent the object in processes which do not require its elementary perceptual attributes as data. (These attributes have already been the "data" for the object identification stage.) For most processes involving the object's location in the spatial framework, its identity alone is a sufficient processing token. Waiting until this level of abstraction to introduce the spatial relation code presumably saves a great deal of code over entering it earlier in the system, where it is of little use. It seems apparent that the brain itself is encountering a processing burden by this level of analysis. It is difficult to attend to the spatial relationships of more than a few objects at the same time. We cope with this problem by shifting the "focus of attention" to the objects of immediate interest and by relegating others to less finely differentiated "regions" within the perceptual space. The attention shifting process will be the subject of much later discussion, but here we see one of its useful properties. It seems to be a mechanism for putting our available processing elements to work primarily on the data which are most urgent. In any event, it seems clear that the central parietal cortex in its dealings with the problem of spatial relationships needs all the help it can get, and one approach may be to reduce the amount of code for analysis by only adding spatial relationship to the final "token" code for identified multimodal objects.

IV. FROM STATIC SPATIAL RELATIONSHIP TO STATIC LOGICAL RELATIONSHIP

The brain's practice of making additions to the perceptual code whenever a new perceptual quality is abstracted, rather than carrying the entire encoded world through a series of sequential stages of analysis, means that a subsequent analytic stage need not be concerned with what the total code represents, or even if it is all present, so long as it contains the necessary information for the operation of that stage's analysis. Thus, when the perceptual processors reach a point at which spatial and temporal relationships of objects are extracted as perceptual qualities, relationship in time and space appear as abstract qualities of objects (or probably more accurately as local properties of the perceptual field). These can be, but need not be, added to the perception of the objects. In theory, this also implies that the neural activity which encodes these qualities must be directly perceivable in the abstract, without requiring perception of any other properties of the objects. In the normal course of perceptual events, the other codes for the perceptual object would be present because they serve as the input data upon which the processors of spatio-temporal relationship act in the generation of their contribution to the total code. Conceivably these processors could be activated in some other fashion to generate an abstract perception of relationship without actual visualization of the objects. (Remember "blind sight.")

The development of this abstract quality of "relationship" paves the way for the development of symbolic logical operations. From our previous discussion, it is apparent that in animals such as man, with sufficient cortical space devoted to complex auditory analysis, words can form auditory "objects." More importantly, with sufficient development of parietal lobe mechanisms, as is the case in most primates, these auditory "objects" can be equated, by learning, with visual and tactile objects in essentially the same way that the equivalence between the visual and tactile objects is established. (I am still trying to avoid the problem of where the memory is and how the processors access it, but functionally that is irrelevant for the moment.)

What distinguishes this particular intermodal equivalence is the fact that the verbal object, the word, does not have a clearly defined spatial location, so it can be equated through learning with *any* visual/tactile object. The sound source, of course, does have a spatial location, but that location does not differentiate the different verbal objects. When we are learning the meaning of words as infants, the verbal object is regularly associated with particular visual/tactile objects, while the sound source is associated spatially with *all* verbal objects. Therefore discriminations among verbal "objects" in the organism's spatial world model is dependent on their association with particular spatial objects of the visual/tactile modalities, in much the same way as the nonverbal auditory objects which are the sounds actually produced by the physical objects. The word comes to be entered into the perceptual world model as an auditory modality equivalent of the object to which it refers by a process which is not, in principle, different from that by which the visual and tactile attributes of the object are equated. All that is required is that the occurrence of the verbal auditory object be frequently associated with the spatially-defined visual/tactile object. Of course, this is exactly the way in which we do learn names of objects as infants. The verbal objects also come to be identified as attributes of the speaker but only in the general sense of undifferentiated "sound," because the source does not identify the word in any unique way.

We are dealing with an early stage of word learning, where words refer to concrete objects. "Mommy" is comprehensible as an object, "Johnny's Mommy" causes confusion because "mommy" in that sense is a more abstract entity. The point to be made here is that once an equivalence is formed between the visual/tactile spatial object and the verbal object, the perceptual code for the verbal object can be used alone to encode for the object, just as its visual or tactile codes can be used alone. Seeing the key, feeling the key, and hearing "the key" can all serve as ways of accessing the representation of a complete multimodal object.

When we look at the properties that the central parietal cortex is abstracting in its analysis of the spatial relations of objects in the physical world, we encounter such notions as direction, position, symmetry or asymmetry and grouping. We experience these as perceptual qualities of our world; naming them comes later. Remember, however, that these perceptual qualities are parallel additions to the perceptual code that are not required for the more basic perceptual experience, and more importantly here, apparently do not require it. For this reason, relationship as a perceived quality can be evoked by verbal objects which have learned equivalence with physical objects, even though these verbal objects, as perceptual code, do not consist of the same full lower-level perceptual experience that, when analyzed, also gives rise to the object perception in the visual or tactile modalities.

At the primitive level at which we have considered the learning of verbal-physical object identities, this would not seem to be a very great advance. The ability to evoke the immediate perceptual experience of the spatial relationship "above/below" by the words "sky/earth" in the absence of an actual view of the sky and earth hardly seems like advanced reasoning capability. It is, however, the crucial first step in abstract symbolic logic. It is beyond the scope of this book to give a detailed account of the steps in the verbal learning process by which we acquire names for ever more general and abstract categories of objects, and ultimately give names to actions and even concepts. The basic process is one of abstraction, categorization, and learning, for which I shall try to account in general in a later section. What should be noted here is that once one has learned to equate a verbal object with an abstract category, the verbal object that represents the category is equated with the properties of things in that category, which may include many kinds of abstract spatial relations, just as "above and below" are associated with the very concrete object names, "sky and earth."

At this point, the perception of abstract relationships, in the absence of the rest of the perceptual code for a concrete object, is capable of being evoked by verbal objects, names, which may in turn evoke rather abstract perceptual qualities. It is clearly a small step to deliberate verbal representation of spatial relations and subsequently to naming the relations themselves.

Let me emphasize that I am not yet talking about a reasoning process, which is a dynamic, sequential process that requires a good deal of additional machinery. What we are looking at is a hypothesis for the development, by means of a perceptual processor evolved for the purpose of representing the physical world, of the ability to experience the perception of static logical relations in the abstract under the control of symbolic "objects."

I don't know how this is achieved on the detailed neuronal level. Fortunately, our artificial systems can make use of readily available logic gate functions which we can supply in finished form. What I think may be of some interest in this scheme, from the standpoint of artificial systems, is the means of data representation. The idea of building the brain's "logical relations" processor into its perceptual processor has some intriguing points. It enables one literally to "see" certain logical relations in the world directly as part of the process of building the world model. Because the world model includes code for representing the position and other attributes of the observer, the brain can use it directly to work out schemes of action. The brain's apparent method of parallel addition of code of the perceptual input with successive stages of analysis is what makes it all possible. In effect, this amounts to a scheme of variable byte width, with each successive stage of analysis adding code on new lines of a very wide bus. Abstract properties can be accessed independently by "masking out" the other unwanted portions of the perceptual byte or can be taken in association with the output of the preceding stages which give rise to them. The "masking" process is under the active control of the "attention" function which we have encountered before and will be explained later.

A moment's reflection will convince you that the logical properties of spatial relation, order, direction, symmetry, grouping, etc are directly equivalent to relations to which we give different names when they are associated with nonspatial concepts. Thus, symmetry and asymmetry are the basis of our notions of equality and inequality; relative position is equivalent to ordinal relation and so on. We recognize this equivalence between spatial logic and arithmetic logic in a formal way in analytic geometry. Both in personal and historical development, the logical relations of algebra and arithmetic are first associated with very concrete object relations in space. Our direct perceptual appreciation of spatial grouping is appealed to in order to teach us arithmetic, "three apples, take away two apples...." Our very names for operations indicate their conceptual origins; consider "squaring" a number. Multiplication was historically the additive grouping of spatial areas long before it is thought of as an abstract operation.

Number itself is not part of this process, but prior to it. Quantity like object identity appears to be a more primitive perceptual analysis than spatial relationship. For a limited amount of quantity (about seven to twelve items depending on the quality of your brain), you can perceive numbers directly. More than that becomes "many." You don't have to count four items to see how many there are, but you have to count fifteen (unless you have an unusually wide-byte brain). This direct perceptual impression of quantity (as a quality!) is independent of the spatial relations of the objects. However, the convenient symbolic representation of large numbers in terms of a base such as ten relies on spatial relations. Thus, the fact that you see 401 as larger than 389 depends on your assigning different weights to different symbolic numbers on the basis of spatial position.

The assignment of symbols as names for quantities is similar to the assignment of words as symbolic names for objects, actions, and relations.

Therefore, it is not surprising to find that although the correspondence between object and word is at a more primitive level of perceptual analysis than spatial relation, the relations between words (ie: grammar) is based on the spatial and temporal relations among the objects and actions represented. Again, both the individual and historical orders of development of grammar seem to show a transition from the concrete to the abstract and general; the concrete original is tied to spatial object relations. The basic relationships of grammar are ways of associating words which tell us the logic by which the individual components are related. It allows us to determine spatial and temporal order of events, to know that the action was transacted *from* one object *to* another, or that certain words are to be *grouped* together as *related* to a common idea. Thus, although different grammars have very different rules, all these rules are ways of indicating the basic concepts of order in time and space, direction, grouping and position that are most fundamentally perceptions of spatial relation. This is not to say that these relations may not later by applied to abstract sentences that have little or no obvious spatio-temporal reference, but only that the major logical relations of the grammar are derived from the natural logic of directly perceivable spatial relations. The role of prepositions, possessive forms, tense, subject-object distinction and the like are particularly clear. Entities such as adjective and adverbial phrases are ways of grouping, a spatial logic relation, which indicates that the whole collection is to be taken as a related set.

With these concepts in mind, let us now return to the study of the effects of damage to the human parietal lobe. In the earlier discussion of the effects of damage to its central area, I focused deliberately on its role as the high end of a system for perceptual analysis of the spatial model of the physical world. Having seen how it thus fits into the hierarchical sequence of perceptual processing and having briefly reviewed the fundamental relation between the natural logic of spatial relation which it processes and the logical relations of symbolic objects, such as numbers and words, we are in a position to understand some of the other results of damage to this central parietal area.

Early observers of the results of such damage were impressed by the patients' loss of a number of functions of "higher intellect." In particular, a number of syndromes were defined with names such as "alexia," the inability to read, "agraphia," the inability to write, "asymbolia," the inability to use symbols, and the like. Of particular note was a loss of the ability to think in words or at least to express such thoughts in them. With such impressive symptoms to explore, disorders of perceived spatial relations received little attention.

Because these profound intellectual deficits were not related to the loss of perceptual ability required for the reading of letters, or the hearing of words, nor vocal abilities for forming words, it was naturally assumed that the seat of the various intellectual faculties had been found. It is of course true that such patients may be profoundly impaired on standard tests of intellectual function, but it is unlikely that the "seat of intellect" or anything remotely that complex is represented in this one area. What is much more likely is that we are observing the effects of the loss of some function which is an important link in the chain of symbolic logic, but which is in itself only one aspect of the larger process of rational thought.

It appears plausible to suppose that what is really occurring in these cases is a loss of the ability to extract the perceptual quality of spatial and temporal relation from the encoded model of either the external or the symbolic world. I have already remarked on the impairment of such patients in this regard. On the basis of the relation just outlined between verbal symbolic objects and physical spatial objects, and the relation between the deep structure of algebraic or grammatical relations and spatio-temporal relations, the following hypothesis can be made.

One necessary component of the total process of symbolic thinking is the ability to analyze the static logical relationships between symbolic objects, that is, to analyze the relationships implied between symbols by algebraic statements of relationship or between words by grammatical indicators of relationship. It is hypothesized that this kind of operation is actually a perceptual process of the same sort that is involved in the analysis of relations between physical objects of the perceptual world model, and that the same machinery is employed in either case. I have just outlined a process by which words or symbols associated with objects or their relationships can come to activate portions of the perceptual code for those objects or relationships. It is a simple extention of the idea to suppose that once evoked, those portions of the perceptual code would be processed by the machinery of the parietal cortex in the same way that it processes them when they are evoked by stimulation through the senses from physical objects. Further, the outputs of this machinery are encoding the relations existing among its inputs in a similar fashion, and words or symbols may come to be associated with these codes for abstract and complex relations by a subsequent process of learned association.

According to this view, the fundamental code of the brain for relationships is the perceptual code which is generated by its machinery when those relationships appear among physical objects at the input. If, in later development, these same codes are generated by other processes, the processes which operate upon it neither know, nor care.

In Chapter 9, I will discuss the difference between coding of information at the verbal level and coding at the "semantic level." This semantic level represents the deep structure or sub-verbal representation of things in the brain, and it is my hypothesis here that the code of the semantic level is the perceptual code.

Suppose that I am confronted with "A contains B and B contains C." This might refer to geometry, set theory, or a stack of boxes. If I observe a stack of labeled boxes, I immediately "see" that physically, "A contains C." I can also, however, appreciate the "contains" relationship as an abstract quality of space. I can use the word "contains" to signify that quality. That's an easy example, of course, because there is a rather compelling similarity between my notion of "contains" in the real world and my notion of ''contains'' in set theory. Most of our everyday logical operations, however, whether grammatical or algebraic, do have rather nice analogies to physical object relations. (This is probably simply a matter of preference in words, for all that the above hypothesis really requires is that there exist *some* relation between a symbolic logical entity and some combination of real world relationships. Trial and error will serve to construct the proper relation between the code and the symbolic objects or relations. This is the process of learning to think in abstract terms in early childhood.)

Of course, once I have worked out some particular abstract symbolic relationship (or accepted it on authority), I can bypass the whole reasoning process and install it as a learned sequence of verbal responses, but that is getting ahead of the story. We are concerned here with real symbolic logical processes. One supposes that complex symbolic operations may be several stages removed, so that one learns later in life to understand certain special symbolic operators by association with whole series of simpler symbolic processes which are more naturally related to physical relationships. It is certainly no secret that our patterns of symbolic thinking tend to follow the lines of ''common sense,'' which is to say the perceived relations of the physical world. Advanced symbolic logical structures which go counter to our everyday perceptions of the world are notoriously difficult to understand. We can, however, follow a series of simple relations which we can ''see'' to a complex one which we cannot ''see.''

This hypothesis, of the relation between symbolic verbal and mathematical abilities and the underlying process of extracting spatial and temporal relations as abstract perceptual qualities, essentially follows the lines suggested by the great Russian neuropsychologist Luria on the basis of his studies of the intellectual deficits of thousands of patients suffering brain damage in these central parietal areas. He observed that, in such cases, the supposed loss of ''intellectual functions'' could be much more accurately characterized as a loss of ability to comprehend some rather particular sorts of symbolic relations.

Luria found that the meanings of individual words or phrases of many sorts is not lost. Interestingly, these patients apparently retain an understanding of even very abstract meanings. Words such as ''causation'' and ''capitalism'' are understood. What is lost is the ability to logically relate the parts of a sentence into a meaning. Concepts of order of transaction escape them; they cannot see any difference between ''The sun lights the earth'' and ''The earth lights the sun.'' They understand each word, but the implications of the word order are lost to them. Although these patients show a kind of word memory impairment (they have trouble providing the names of visual objects), it is a different sort of verbal memory problem than that seen in patients with temporal lobe damage. The defect is more nearly a loss of connection between object and symbol than an inability to remember hearing a word; it cannot account for the loss of understanding of complex sentences because the length of a phrase is not what determines its comprehensibility. Very short phrases are completely unintelligible to them if they contain words that can only be understood in spatial terms, such as prepositions like "above" or "behind." Logical relations of asymmetry such as "John is bigger than Sam but smaller then George," are impossible for them.

The type of "aphasia" or speech deficit that occurs in these cases has long been recognized as an inability to formulate verbal thoughts rather than a deficit in actual speech mechanics. In the context of our hypothesis, it is seen specifically as an inability to employ the relational processor to generate symbolic expressions.

Yet another type of intellectual impairment which is seen following damage to these areas is a so-called "acalculia," or inability to calculate. This is demonstrably not due to the inability to perceive or recognize numerals or arithmetic symbols. The patients appear to have an inability to combine them in meaningful ways, or even to understand the conception of the operations. They do not, however, lose the basic concept of number. Just as in the case of patients who lose grammatical construction without losing the meaning of words, these patients can demonstrate a recognition of numerals and an understanding of magnitude.

They cannot, however, generally comprehend compound numbers. They see 401 as less than 389 because most of the individual digits in each place are smaller; they cannot understand the "tens" and "hundreds" categories based on ordering of the digits. The best they can manage with concepts like multiplication and division and substraction is a vague "getting larger" or "getting smaller." There is thus a loss of the system of mathematical relationships and numerical category, but no loss of number as a concept.

The interesting thing is that there appears to be no real loss of abstract thought or basic intellectual competence in such cases. If one makes tests that allow for their peculiar difficulties with certain kinds of verbal operations, these persons will show excellent ability to perform such tests as choosing antonyms of words with abstract meanings, selecting the correct choice in words with cause-effect relations, etc. Despite the impairment of specific kinds of grammatical and algebraic operations (and, of course, the inability to perceive spatial relationships), they seem to suffer only from the impairment of these specific symbolic logical operations. The dynamic process of organized intellectual activity is not lost. Aside from the initial difficulty they have in getting a grasp of the conditions of the problem, they seem to retain the general process of reasoning. If allowed to simply indicate the point where certain kinds of logical operations of the sort we have discussed would occur, they are capable of producing a correct plan for the solution of a problem, even though they are incapable of executing the plan due to the "missing" operations. In other words, they can demonstrate that they retain the general form of reasoning.

In summary, the higher functions of the parietal lobe can best be seen as a set of logical operations which can be employed in the service of symbolic (grammatical or algebraic) manipulations, or physical (spatial object) manipulations, but which comprise only one particular link in the set of operations which constitute rational thought. It appears likely that this relational processor has its origins in the perceptual apparatus as a feature extractor for the features of spatial relationship, and that its subsequent employment in symbolic processes is made possible by the development of learned equivalences between verbal symbolic objects and real objects or categories of objects, and between physical spatial relations and symbolic relations. Parallel processing and encoding of perceptual features enables these relations to be obtained as abstract qualities, devoid of other perceptual baggage.

This discussion has necessarily centered on the role of cortical mechanisms, because these comprise the bulk of the information on higher functions in humans. It would certainly be a mistake to suppose that the operation of this machinery is different from that of other brain systems in that a constant interplay of cortical and subcortical systems is required to achieve the end result. Here as elsewhere, it is probably the case that the subcortical structures define a direction or problem and the cortical structures elaborate and respond with an analysis, whereupon the cycle repeats, and moves to other areas as required. There is some evidence from animal experiments that a part of the limbic system, called the *hippocampus*, may be involved in the total process of performing the actual spatial mapping of the environment and recognizing the organism's position in it.

The material I have covered here could be broadly defined as the analysis of exteroceptive information, that is, the information which enters the organism from the external environment. I have followed that thread all the way from the receptors to the highest levels of analysis of symbolic representation of that information. What must concern us now is the process by which rational action is generated on the basis (in part) of that analysis. In order to do that, it will be necessary to follow the action of several other large-scale analytic systems. To begin this process, I return to an examination of higher motor system functions.

V. THE SYNTHESIS OF ACTIONS

When I examined in an earlier chapter the lower portions of the motor output system, I mentioned that the basal ganglia and the cerebellum both received massive inputs from the cortex. These were set aside at that time as being primarily concerned with the synthesis of the plan of action, in distinction to the execution of specific movements. The time has now come to trace the processes involved in the synthesis of these plans of action. In this section, I shall be concerned with the role of cortical structures in the synthesis of the "order of movements" that constitutes a behavioral sequence. I shall assume for the moment that both the definition of the overall goal of the behavior and the evaluation of its outcome are being attended to elsewhere, and defer discussion of these functions to a still later chapter.

The preceding chapter considered the tactile, or touch-related, functions of the somato-motor cortex, in which it acts as the primary sensory cortex for the tactile sense. It is also the primary sensory cortex for the analysis of kinesthetic sensory information. I have already discussed the kinesthetic inputs in relation to subcortical levels, where they are used to provide feedback information for the mechanisms of the lower motor centers. In the current stage, I will be concerned with their use in providing information about global states of body positions to the brain's activity planning functions. This global analysis of body state is a high-level perceptual function, similar in most respects to the global perceptual analysis of tactile inputs, and is handled by similar cortical apparatus.

Just as there are perceptual mechanisms for the recognition of object-level organizations of tactile inputs, so there are perceptual mechanisms for the recognition of high-level constructs of kinesthetic inputs. Because these inputs tell us about the state of our body's configurations in space, rather than about the external world, "object analysis" is not quite an appropriate term; however, we have few words in the language for describing different body configurations. I shall have to make do by speaking of analysis of "posture," and by thinking of "postures" in kinesthetic perception as being somewhat equivalent to "objects" in other sensory modalities.

The process of kinesthetic perceptual analysis has not been sufficiently studied, but there is no evidence to indicate that its principles differ radically from those of other modalities, which have been examined in greater detail. The secondary cortical perceptual areas for the kinesthetic sense seem to be more or less coextensive, anatomically, with the corresponding areas for the analysis of tactile information. Damage in these areas produces, in addition to the disorders of tactile perception which I have already reviewed, a variety of disorders of motor functions which seem best explained as a loss of the provision for kinesthetic feedback.

With damage to the kinesthetic secondary perceptual areas, there is a "paralysis" which superficially resembles that seen in damage to the motor system proper. In this case, however, the muscular contractions retain their potential power, and the defect appears to be a loss of guidance to the motor outputs. The muscles show no selectivity in their movements, but rather a diffuse simultaneous activation of flexors and extensors which results in little or no movement. With less severe damage, the movements occur, but they are gross and undifferentiated. I am referring here primarily to the skilled "voluntary" movements of the hands and similar high-level motor functions that require the participation of cortical levels of kinesthetic analysis. Writing, finger movement, and similar functions are severely affected.

That these disorders are problems of feedback control can be demonstrated by the fact that substantial improvement may be obtained by permitting visual feedback and guidance. This is not an adequate substitute in cases where quick, skilled voluntary movements must be rapidly controlled, but visual analysis of movement does play some role in normal operation, particularly in regulating feedback control of the gross movement of the body with respect to the environment. Vestibular feedback is also of importance.

The effects of damage to cortical motor processes in general are proportional to the degree to which the species' behavioral repetoire is composed of planned or "voluntary" movements as opposed to sterotyped or reflexive responses. At the level of the rat, the majority of behavior is capable of being processed by the basal ganglia on the basis of "homing" stimuli from environmental goal objects, and cortical damage produces little effect on the general pattern of movements. In man, on the other hand, planned behaviors occupy a large portion of the normal ongoing behavior, and cortical damage is particularly disabling.

In the review of the lower motor mechanisms, we saw the role that tactile (and kinesthetic) inputs play in directing the fine corrective movements that are the principle province of the motor portion of the somato-motor cortex. What I must attend to now is the similar but more general interaction of more highly analyzed postural and sensory analyses with the more globally defined motor patterns generated by the "pre-motor" portions of the frontal lobe which lie forward of the somato-motor region. (See Figure 6.1.) There is a continual interplay of feedback between these cortical premotor regions that generate the motor plan of action, and the sensory processes which they employ in order to refine it. This interaction exists both in the form of direct intercortical connections and in the projections of cortical areas upon the basal ganglia and the thalamus. In the generation of these motor plans, there appears to be a sequential proccss of continual development and elaboration which proceeds from the most global definitions of the desired actions to the most concrete statement of required outputs into the basal ganglia, motor cortex, and cerebellum.

Let us assume that we begin with a goal of behavior, in the form of some desired state of the world, which is to be achieved through our actions. This goal state could be defined in terms of the postural, visual, tactile, vestibular, and auditory conditions which would be obtained at the conclusion of the action sequence. Initially, however, the desired state of the world is probably much less well defined. It may include only a few important abstract features of this picture, which consist of high-level perceptual constructs, devoid of "content." That is, we are hypothesizing that some as yet undiscovered processing unit responsible for setting goals, can produce a state of activity in the higher levels of the set of parallel lines which encode the outputs of the perceptual analytic processes. Such activity could, of course, exist without the accompanying lower level perceptual codes which would be generated in the course of its production through sensory channels.

The synthesis of action appears to be a continual recursive elaboration of this "goal state," together with an extrapolation of the movements required to achieve the state. This procedure may range from very simple cases in which planning and execution are carried out in real time, to complex situations in which an action is built up in advance via symbolic processes; the most complex type of "action" may consist entirely of symbol manipulation which is never overtly expressed in motion. This last stage is called "rational thought." I will begin with the simplest case.

Suppose that the goal is to get to the other side of the room. Assuming that this is some sort of planned or "voluntary" behavior and not simply a feedback-controlled process that is both activated and guided by stimuli from a goal object, it will be expressed in the form of some conceptual representation of the desired end. We might better say "perceptual" representation, because the goal is given to the relevant circuits as a perceptual code (discussed in the previous chapters). This may be a high-level fragment of the full code, which carries only an abstract perceptual experience, such as position rather than a full blown "picture" of the world.

There is a general hierarchy of organizational level of action in the pre-motor areas of the frontal lobe (figure 6.1) which ranges from the most global to most specific motor operations as we pass from the more forward parts of the pre-motor region towards the somato-motor cortex. This motor sequence is the inverse of the perceptual system's "layout" of specific feature analysis to abstract object analysis as one moves across the surface of the parietal lobe. In figure 6.3, the process of synthesis of movements is shown schematically in relation to both the time frame of execution and the level of organization of the action. As indicated, the transition from the most global level of action description to the most concrete corresponds to the transition from the forward to the more posterior regions of the pre-motor cortex.

The highest level of figure 6.3 corresponds to the "desired action" which intervenes between the desired and the current state of the world.

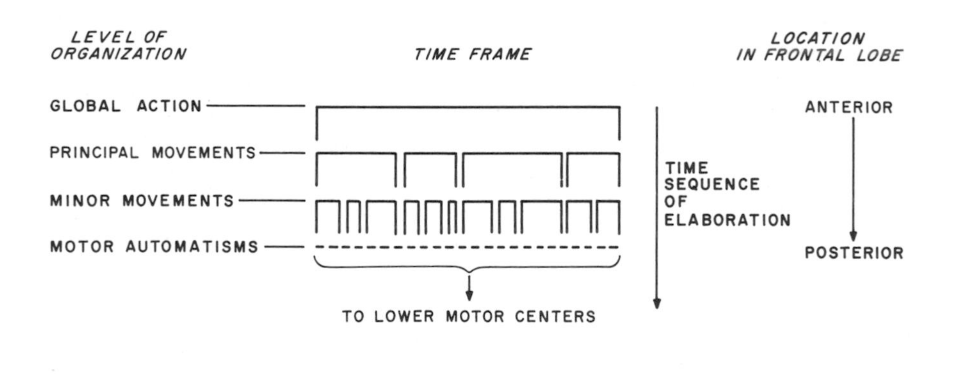

Figure 6.3: *A schematic representation of the brain's approach to the synthesis of action sequences. The time sequence of elaboration of the action runs vertically from global statement downward to specific detail. Although the action sequence is to be performed as a readout in time, indicated by the horizontal arrow, the brain can employ its fully parallel architecture to generate the consecutive parts of the sequence simultaneously (see text).*

The problem facing the apparatus is how to specify the intervening set of brief, specific actions labeled "automatisms." These are so labeled to indicate that they represent the level of specificity at which they could be issued to the motor automaton of the lower brain centers for execution guided by local feedback control without further concern on the part of the action-planning system.

Figure 6.4 outlines the general features of a process which could construct the action sequence; three levels of action planning are shown. Each consists of action sequences that are at once more detailed and also of shorter duration than the plan of the preceding level. The highest level (1) consists of a single action, A. This is the global statement of the action for which a specific final movement sequence is desired. This action covers the operations needed to intervene between the current perceptual world state (WS1) and the world state represented by the desired outcome (WS5). At the second level of planning, this action is shown broken down into two subunits, B and C, which intervene between (WS1) and (WS3), and between (WS3) and (WS5′) respectively. Level 3 carries the process a step further.

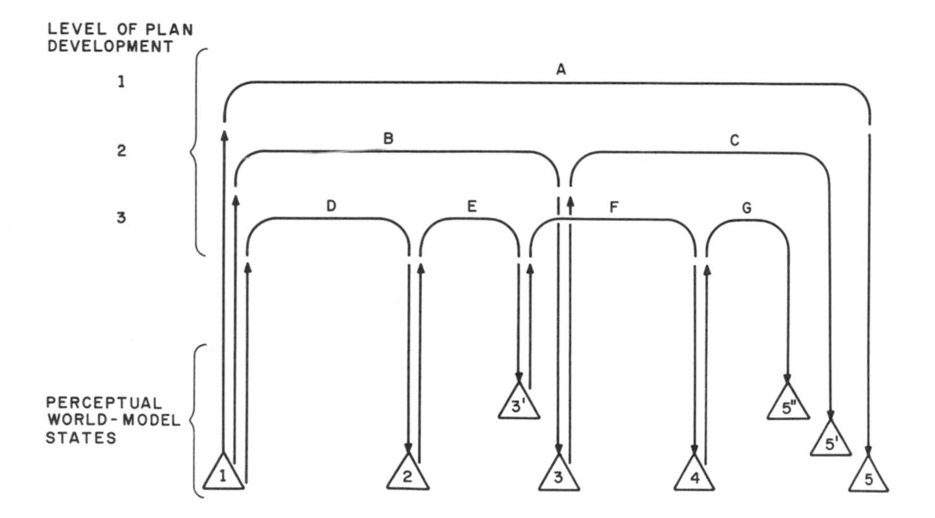

Figure 6.4: *The development of an action plan. The triangles represent states of the perceptual world model elaborated by parietal areas, with number 1 representing the current state and number 5 representing the desired end-state. A comparison of 1 and 5 may lead to the attempted action (B) through learned expectations about the ways in which the world model transforms under various actions. This process applied to 3 and 5 produces (C) and 5′. If 5′ is similar enough to 5, the next stage of resolution is begun, operating on (B) and (C) simultaneously. If 5″ is sufficiently similar to 5, the next level of resolution is entered, and the process proceeds. This kind of parallel processing is possible because the brain addresses problems that have a variety of possible solutions that result in a range of final states which are a "good enough" fit to the desired state (see text). When the brain is required to solve formal logical or mathematical problems, this system is not possible, and the brain becomes a poor processor in comparison to a traditional serial computer.*

The following sequence is hypothesized to take place: WS5 is compared with WS1, and a difference statement is generated. This difference statement can be used to select the action scheme B. B in turn has associated with it (through prior experience) a set of expected differences which it will produce in perceptual space if successfully completed, and that set is used to generate WS3 from WS1. The difference statement generated from the comparison of WS3 to WS5 in turn can be used to select action scheme C, which will also have a set of expected alterations of perceptual space that, operating on WS3, will produce something close to WS5 (shown here as WS5'). This procedure has brought us down one level of analysis. The problem has been resolved into two sub-actions, and an intermediary desired state, WS3, has been generated.

Because the brain is a fully parallel processor, it can now set to work on the resolution of B and C simultaneously. By an extension of the preceding process, B is resolved into the more fine-grained action scheme D and E, and the intermediary world state, WS2. At the same time, C is resolved into F and G, and WS4 is produced. The process continues until the required degree of resolution has been obtained. At each level of resolution the differences between the final extrapolated result and the desired features of WS5 are diminished.

Because each of the successive resolutions is producing more finely detailed action schemes, the intervening world states must also be more finely detailed (or at least they must contain some fine details; they can now omit the more abstract analyses that are required for generating more global levels of action). Thus, the difference statement, generated by a comparison of WS1 and WS3, will consist of gross differences. As WS1 and WS2 are more nearly alike, their difference will be smaller. Unless the features specified in WS2 are finer than those employed in WS3, the difference information will be able to encode fewer potential action schemes for the next level of analysis. It appears that we need global abstract world features for WS3, and finer, more "textured" features for WS2. This may be accomplished automatically by the nature of the interconnections between the perceptual and motor areas, as we shall see shortly.

Such is the overall scheme of the required processing. The details of the process are, of course, largely obscure. There are some points worth noting. What has been schematized here as "World States" (WSs) are subsets of the perceptual code. This includes *both* exteroceptive information, such as position of objects in the environment relative to the organism, and interoceptive information about kinesthetic postures. When I referred to particular difference statements being used to "select" code for particular action schemes or the selection of an action scheme addressing an "associated" code for the anticipated difference result, what I was talking about in real brain terms was the result of a learning process which goes on in early infancy through childhood. This kind of "motor learning" is the process of forming learned equivalences between the execution of different actions and the changes which they produce in the kinesthetic and spatial perceptual worlds. This is a process which occurs along with the previously mentioned development of equivalences between, say, tactile and visual objects in what we call "perceptual learning." I should perhaps emphasize again that this kind of learning is never to be confused with the common usage of the word "learning" which conjures up an image of memorizing and recalling verbal data, or some similar process. The underlying physiology may be similar, but here we are not talking about a process in which conscious recall plays any role. Learning in this sense is the "Pavlovian reflex" type in which simple neural circuits form functional links on the basis of temporal contiguity of activation. Thus, when execution of a particular action repeatedly produces a particular set of differential changes in the spatial and kinesthetic world, the activity of the perceptual apparatus evoked by these sensory input changes subsequently acquires the ability to be similarly evoked by activity of the circuits which encode execution of that action scheme.

The actual site of the change in the neural circuitry might be at the synapses of the interconnections which form massive communication routes between the motor scheme generators of the pre-motor cortex in the frontal lobe, and the perceptual cortex of the parietal, occipital and temporal lobes (figure 6.5). However, this might only be an entry point for such learned information. There is evidence (which I discuss in the section on memory) that memory of spatial and temporal patterns and relationships is handled separately from, and in a more localized fashion than, memory of the object recognition sort.

The learning of "expectations" of how sensory inputs will be transformed as a result of movements may be a reasonable process to emulate in artificial systems. Some mechanism for encoding the relation between a movement and its expected spatio-kinesthetic consequences is necessary for the kind of action-synthesizing operation described here, but encoding it *a priori* would be quite a task. In the long run it might be simpler to let the machine "learn" by experimentation; this would have the same benefit as it has for us: the basic brain circuits could be put into various realizations of the physical body mechanisms without having to respecify all the connections for every different set of arm lengths, etc. It would also ensure a "fine tuned" system that would adapt to minor changes automatically.

Of course, it is unlikely that we would want to go so far as to emulate the brain in the site of incorporation of such learning. More likely, some sort of table lookup scheme could be employed in which the motor action to sensory-transformation relations could be entered by the machine itself on the basis of experience. Allowing it to modify the table, when experience failed to confirm a prediction, would give it the ability to adapt constructively to injury or wear, just as we do.

The general scheme of organization of the sensory motor processor of the brain is also of potential interest. I mentioned that global perceptual world features were appropriate for earlier levels of action scheme generation, whereas more finely grained features were appropriate for later stages of detailed action component synthesis. Just this sort of transition of level of abstraction seems to be incorporated into the wiring of the interconnections of the frontal and posterior lobes of the cortex. The interconnections of the lowest level cortical sensory perceptual areas are made with the frontal pre-motor areas nearest the somato-motor primary cortex, those of the perceptual cortex furthest "downstream" in the perceptual process. The extractors of the most global features are made with the most anterior portions of the pre-motor cortex, which are concerned with global organization of action schemes. (These connections are among those which I described earlier as branching off at each level of the perceptual process.) The general scheme of this organization is shown in figure 6.5, which provides an overall view of information flow in the perceptual/action processor. Information enters through sensory channels and travels to the left through various phases of feature extraction. The final analysis of the perceptual world state is available at the far left for the use of the goal decision systems which I have not yet discussed. Once the goal is set, the desired global action scheme enters the motor synthesizing sequence at the left and is moved successively to the right as each step of resolution is completed. At each stage along the way, the interconnections of the two systems provide for the operation of the process at the requisite level of detail.

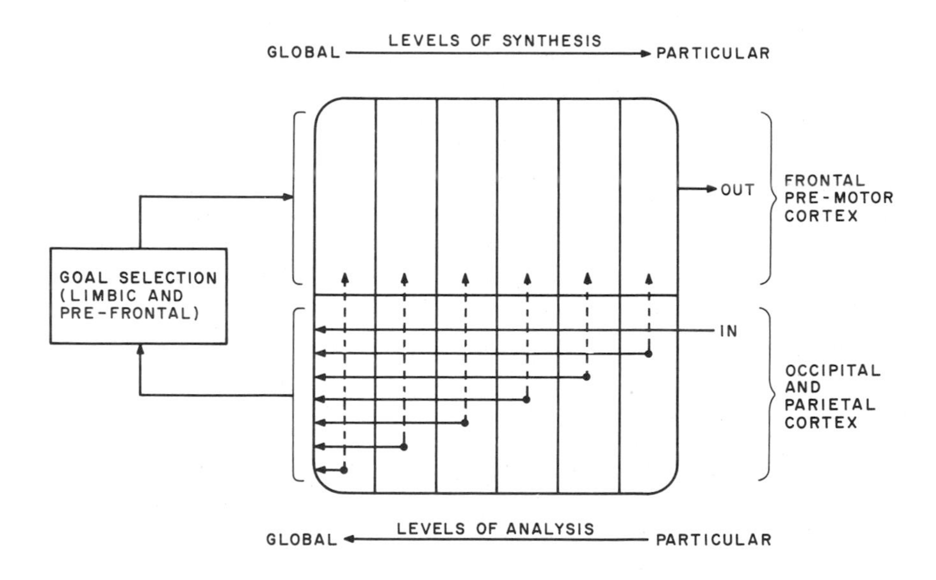

Figure 6.5: *The interconnections of the parietal and frontal lobes are shown here in a schematic form, which emphasizes the fact that the analysis of the most detailed level of input is made available directly to the areas synthesizing the most detailed level of action, while the analysis of the most global aspects of the input is made available directly to the areas synthesizing the most global levels of action (compare figure 6.2).*

VI. EXECUTION OF THE PLAN

Note that the brain has the option of either having the lowest level of steps in figure 6.3 output to the motor apparatus as they are produced or of withholding execution. This has importance in that there are some excellent reasons for retaining the "virtual action" as an "as-yet-unexecuted" plan. I shall deal with this shortly. First, however, I must account for the fact that this is a parallel coding of a scheme for action, generated more or less simultaneously in all of its elements, while action itself is the serial appearance of these elements in time.

The problem is how to obtain the desired sequential execution. When some of the operations of the lower motor system were examined I specified two major portions that were responsible for the actual direction of movement: the cerebellum and the basal ganglia. Kornhuber, who is responsible for this theory of the division of labor among these systems, suggests that these systems are presented with data from the cortex in a parallel format which specifies a sequence of operations. Such a parallel coded sequence could be developed by the model of voluntary action synthesis which we outlined above. According to Kornhuber, it is a handshaking interaction between these portions of the lower motor system and the cortex that establishes the actual physical sequencing of the action components. Let us first look at the cerebellar case.

When I discussed the operation of the cerebellar cortex as a tapped delay line, it was noted that this allowed it to act as a parallel-to-serial decoder for elementary sequences. I further noted that if the final element of such a sequence could be employed to initiate the next elementary sequence, long chains of action could be built up. What I am now explicitly suggesting is that the final output "word" of the cortical voluntary action synthesizing mechanism is composed of "bits" which represent just such elemental cerebellar movement sequences. That is, the interaction between the pre-motor and sensory cortex is synthesizing the global action by ultimately specifying it in terms of a sequence of elemental movements. These are presented to the cerebellum, where each of the elements of this code is used to trigger the action of a granule cell ("G cell"), which sets off a timed chain of activity in yet lower centers. The crucial addition is the hypothesis that the final action of such a delay line is the enabling of the next "bit" of the parallel action code into its respective "G cell" delay line. Presumably, the full parallel action code should specify not only which cerebellar lines are selected, but also their sequence, so that the returning trigger signal will be "switched" to the element to be enabled next. This scheme is illustrated in figure 6.6.

After examining the nature of the connections between the cortex and cerebellum, figure 6.7, it becomes clear that the cerebellar outputs return to the cortex via the thalamus. These connections are thought to constitute the physical basis of the feedback-based sequencing system for the cerebellar component of the output code. However, the portion of the action sequence executed through the cerebellar circuitry comprises only part of the output.

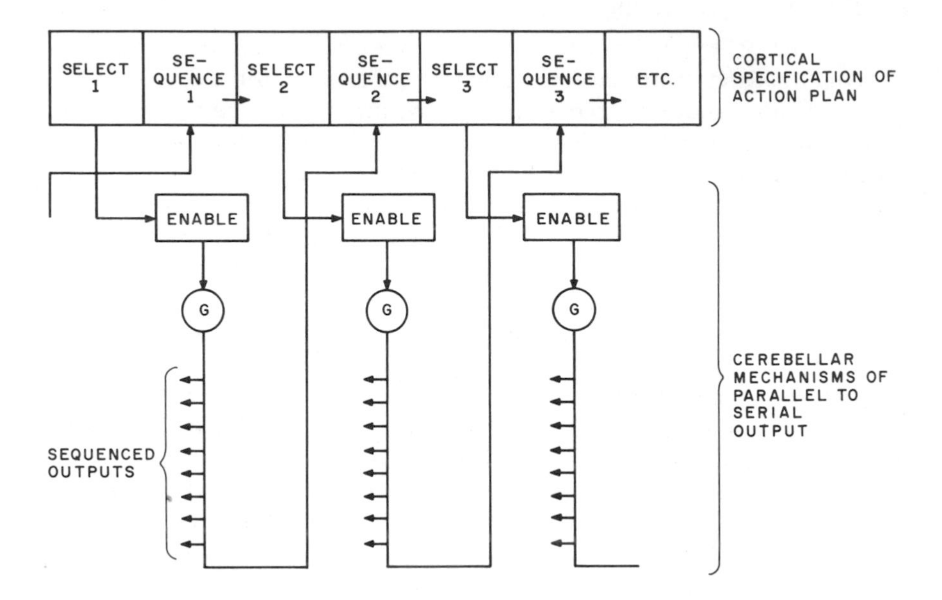

Figure 6.6: *A highly schematic model of the mechanism for sequencing the elemental outputs of the part of the action plan executed through the cerebellum (redrawn from Kornhuber).*

Another portion is executed through the motor system of the basal ganglia. Sequencing here is more complicated than in the cerebellar case, because it is based on input feedback contingencies rather than on timed sequences of execution; ie: the cerebellar circuitry is thought to handle ballistic movements according to preset timing circuitry which runs to completion. The basal ganglia apparatus, on the other hand, is thought to execute movements under a feedback control, which determines when the movements are complete.

This indicates that the action code to the basal ganglia must specify both the movement itself and the sensory input conditions which will constitute its completion. This necessity probably accounts for the fact that the basal ganglia have a massive input from sensory as well as motor portions of the cortical apparatus. The motor sequence specified is initiated under control of the basal ganglia; the resulting conditions of the perceptual world state are compared with the desired state, and continuation or correction of the movement is produced, until the desired state is achieved. This operation requires that the code for the desired world state be modified to take into account moment-to-moment local effects, such as the orientation of the organism when it effects the comparison. In fact, the basal ganglia have important inputs from subcortical centers such as the vestibular system and have been shown to be important in operations involving tasks such as comparing visual input with verbally defined categories, such as "vertical," in the presence of body tilt. The outputs of the basal ganglia are directed to the lower motor center, as noted earlier, but there is another important output from these structures, via the thalamus, to the pre-motor region of the cortex. These connections are diagrammed in figure 6.8. Presumably, the closed loop thus established between the pre-motor regions and the basal ganglia functions to achieve the desired sequencing by indicating the completion of the current segment of the action scheme. In this case, it seems likely that the next action component is specified to the basal ganglia in terms of a code for the next input state to be achieved (relative to a particular limb, etc) instead of as a code selecting a particular series of fixed movements as in the cerebellar case, but the sequencing principle is the same.

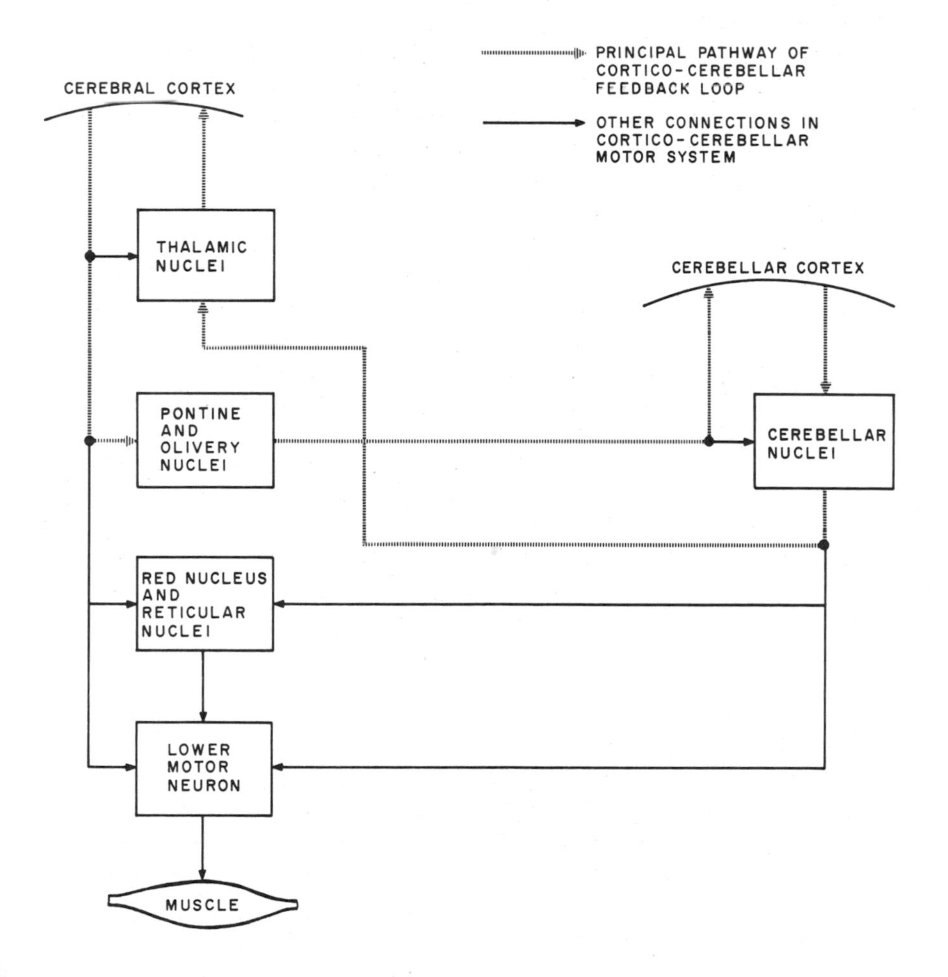

Figure 6.7: *The major connections of the areas involved in the control of cerebellar output. The hatched lines indicate the cortico-cerebellar feedback loops involved in the sequencing operation shown in figure 6.6.*

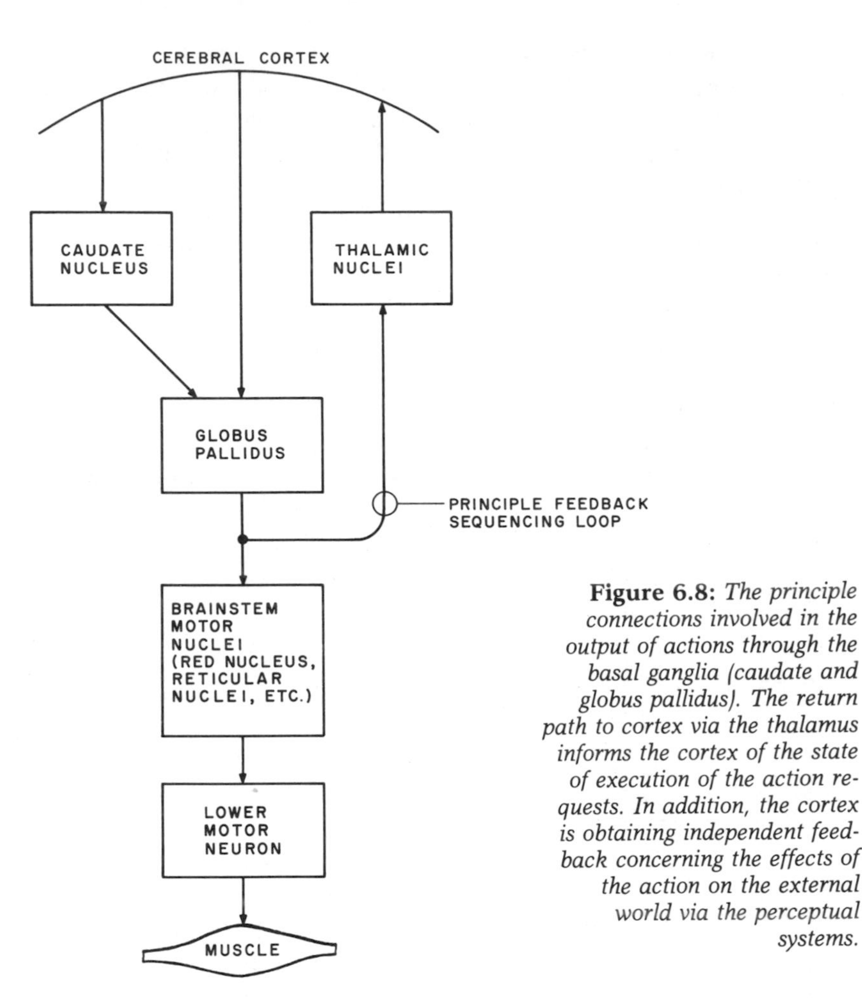

Figure 6.8: *The principle connections involved in the output of actions through the basal ganglia (caudate and globus pallidus). The return path to cortex via the thalamus informs the cortex of the state of execution of the action requests. In addition, the cortex is obtaining independent feedback concerning the effects of the action on the external world via the perceptual systems.*

There is some similarity here between the hypothesized action of the basal ganglia in its feedback-controlled pursuit of a target state specified by cortex, and the hypothesized action of the cortical system in its use of real and transformed "world state" models to synthesize action schemes. The differences should be pointed out: the basal ganglia system applies real-time corrections to maintain a behavioral trajectory which continually diminishes the difference between the specified arm position and the actual current position. It is essentially a direct "approach to goal" system capable of handling the problem of moving all or part of the organism through space towards some objective. It is a more primitive output processor which requires and uses only the lower level spatial perceptions of direction and distance relative to the organism. The cortical mechanism is an evolutionarily later elaboration of this kind of process which has the capability of employing spatial relations among objects in a perceived model of external space. Additionally, the cortical system need not function as a real-time, correction-only system, steered by a goal object or position, but can employ plan-ahead algorithms which test the desirability of predicted resultant intermediate states.

In simple organisms whose behavior is dominated by the basal ganglia, the desirability of a goal object is determined by primitive, wired-in mechanisms for recognition of food, etc, and stimuli from the object fill the roles of defining the guidance criteria (maximize that smell) and of triggering the performance of the behavior. Such behavioral systems are called "stimulus bound." On the other hand, the cortical system can provide much greater flexibility in the choice of actions and their defining sensory inputs and in the choice of timing and initiation criteria for the behaviors.

The hypothesis of feedback to the cortex, from cerebellar and basal ganglia systems, accounts in general for the conversion of the parallel scheme of action into time sequences of movement through the interfacing of the pre-motor cortical areas with these motor executive units. As pointed out earlier, the existence of yet lower-level mechanisms for dealing with local perturbations, and the existence of preprogrammed "automatisms" which can be called as units, allows even the smallest elements of the cortical voluntary action scheme to be of a fairly global nature. Presumably, if the nature of the task requires a detailed control of fine movement in its execution, there is nothing to prevent the cortical mechanism from extending its level of analysis to whatever extent is required. It is probably just in such cases that the fine tactile feedback and control capabilities of the somatomotor cortex are called into play in the service of voluntary action. There is also evidence to suggest that learned motor sequences are handled in detail by the cortex as they are first being worked out and are subsequently relegated to the status of subcortical "automatic" processes. The utility of such a process is apparent, but the details of the transfer from cortical to subcortical control are obscure.

The view of synthesis of voluntary action presented thus far seems adequate to explain simple action sequences. We can understand at least in principle if not in detail how the interaction between pre-motor and sensory cortices can generate parallel coded action sequences for getting across the room, and how these can be sequentially executed by the motor apparatus. But how can we account for the generation of more complex operations? What about the production of "reasonable" behavior?

In the first place, by generating a sequence of actions in its entirety before execution, rather than simply continually minimizing an error signal as in simpler systems, the cortical apparatus produces a projection of future actions and their associated sensory world states, which can be examined before execution. For example, a projected term in the action sequence may produce a transformation of the organism's location in the spatial model which indicates an undersirable proximity to something dangerous. This is available for inspection by other systems (yet to be discussed) which can "veto" the proposed sequence, and a new synthesis can be initiated. This kind of ability produces behavior with a rational appearance, especially when the process involves long and complicated action schemes. The organism is anticipating at least the immediate future consequences of behavior and modifying the behavior accordingly.

VII. SYMBOLIC ACTION SCHEMES AS VERBAL REASONING

In advanced models such as man, a further development is possible. The same sort of associative learning, or "Pavlovian conditioning" process, that allows the neural activity decoding verbal perceptual objects (words) to activate the neural circuitry encoding the multimodal qualities of an object in perceptual space can also enable words to be associated with the activation of neural processes representing action sequences. That is, actions can have names. The immediate advantage is that actions, as well as world states, can be manipulated in a much more compactly encoded form so that complex sequences of greater length are possible. This stage represents the achievement of at least a concrete form of symbolic reasoning. It differs from what was available on the basis of the symbolic capabilities of the perceptual cortex alone in that a process is available for the synthesis of a symbolically specified goal by means of symbolic steps representing actions to be taken. The perceptual cortex alone does not have the capability to develop a process, even though it has the capability to analyze and express logical relationships.

Humans, of course, go far beyond this concrete level of symbolic reasoning which simply allows for a more compact perceptual coding of both actions and world states. We develop all sorts of formal logical rules and define words for abstractions which can be combined by means of these rules. This process, which is arduously learned, involves the same basic mechanisms; it simply requires experience (or instruction) which provides for the association of words with abstract perceptual properties that are fitted by whole classes of objects and the acquisition, by practice, of conditioned associations between cortical representations of symbolic action sequences.

In principle, there is nothing different between the kind of motor learning which enables us to run off sequences of skilled movements without reasoning them out, and the kind of learned symbolic manipulation which enables us to run off logical sequences without reasoning them out. If you recall the first time you had to figure out how "AND" works in bit masking when programming a computer, you will probably remember that it took you a moment to "see" what was required. With the development through practice of skilled manipulations of this sort, the goal of masking certain bits now "automatically" invokes the neural activity associated with the symbolic manipulation of "AND". The process of reasoning it through can then be short circuited. We build up in the course of our lives all manner of conditioned associations of action sequences, both symbolic and actual, which enable us to avoid the need to reason out problems anew on each occurrence.

Seen in this light, high-level symbolic logical operations are not different from the generation of a rational scheme of motor behavior. The same basic physical cortical operations underlie both, and the development of the symbolic mode rests principally on the conditioned (learned) connection of perceptual objects and action sequences with perceptual verbal objects and later visual symbolic objects. Again, the major advantage of the human brain in this regard seems to lie in its enhanced capacity for correlations between different modalities. Your rational behavior would seem much less impressive in comparison with that of other animals, if you were prevented from using symbols and had to think your actions through in actual images when confronted with a novel situation.

Speech is the most obvious motor output of symbolic activity. The perception of words and the association of them with various concrete meanings is paralleled on the output end by the association of the results of symbol processing with the generation of appropriate motor sequences for the production of words. When the rational process has produced some result, its symbolic description can be converted into the appropriate motions of tongue, lips, etc that will produce the associated sounds. Speech is a skilled motor activity which is learned when we first begin to compare the effects of our infantile noises on our ears with the effects of adults' words. We seem to learn by trail and error how to organize motor movements that will mimic sounds heard. Initially, the expression of our symbolic operations in words requires many trips through the rational process of action synthesis and perceptual extrapolation to get the grammar and vocabulary correct. Later, such expression becomes much more automatic, through conditioning, so that we only have to think out the plan of what we are going to say in difficult cases, and the mechanics of making the sounds are almost always at the automatic level.

In this view, the identification of logical symbolic thought with the basic mechanisms of "rational" motor behavior is therefore total, and it is no surprise to discover experimentally that the same brain regions are involved in both, or that damage to these regions leads to similar deficits in both symbolic and nonsymbolic voluntary behavior. Damage to pre-motor areas in human beings results in a variety of disturbances of the execution of voluntary and particularly symbolic behaviors. As a general rule, damage to the more forward portions produces problems in initiating global action schemes, while damage to regions nearer the primary motor cortex tends to interrupt the execution of specific elements of behavior sequences. There is no very specific mapping of the body musculature onto the pre-motor areas as there is in the primary motor cortex, but there is a general tendency for operations involving the head, and particularly the musculature of vocalization, to be most involved with lower regions and operations involving the limbs and trunk to be associated with upper regions of the pre-motor areas. This is also the general ordering of the detailed mapping of the body onto the primary motor cortex.

As a result, complex motor action schemes for speech generation tend to be most affected by damage to an area towards the lower front of the pre-motor region. It was once thought that this area was uniquely specialized for the operation of speech production as a separate function, but it is now generally held that its operation is not different in kind from the rest of the pre-motor regions. It has extensive interconnections of the nature indicated in figure 6.5 with the portions of the secondary auditory cortex involved in the recognition of speech phonemes as perceptual features, and it seems to participate strongly in the synthesis of motor schemes involving the formation of phonemes by the musculature of the tongue, lips, and throat. It can best be conceptualized, therefore, as that part of the action synthesizing pre-motor cortex which is well connected for functioning in the synthesis of the motor action schemes involved in word formation. Damage to this part of the cortex produces an inability to speak, or an impaired ability in cases of subtotal damage, which has a characteristic symtomatology. This problem has been extensively studied and can serve as an illustration of pre-motor function.

Whereas dysfunction of the portions of the temporal and parietal lobes involved with word recognition may lead to inability to find the desired word, and damage to the portions involved with kinesthetic perception from the organs of speech may lead to a slurred speech and difficulty of pronouncing the desired sounds, lesions in the pre-motor areas concerned with speech tend to interfere with the dynamic organization of the speech. Patients suffering from this kind of damage typically are capable of forming individual sounds of any sort. Where they have difficulty is in the elaboration of an ordered sequence of sounds constituting a meaningful expression. The major difficulties appear to be with the ordering of the sounds and with the transition from one sound to another. They may speak words out of order, or be unable to move from one word to the next. This latter problem can either take the form of an inability to initiate the next sequential word, or an inability to terminate the current one, leading to repetitious sounds. These difficulties are most marked, as we might expect, with speech which is not a well-learned or habitual pattern and requires the synthesis of motor action rather than reliance on learned associations for its generation.

If the damage is somewhat forward of this classic speech area, the symptoms are more global. In this case, there is not so much an interference with the execution of the speech act as with the initiation of it. That is, the disturbances of ordering and transition just described may not be present if the speech can be started at all. The patient, however, has great difficulty doing this and simply stands mute. In terms of the hypothetical model of action synthesis which we are evaluating, this would seem to correspond to the loss of the first echelon of the process. The initial statement of the action is not implemented and so the subsequent stages cannot proceed to refine it into executable elements.

Before passing to other types of disturbances arising from damage to the pre-motor region, I should mention the involvement of written speech. It is not readily apparent why written speech is also interfered with by lesions which affect auditory and verbal processes. In fact, damage in the auditory word-recognition areas also tends to affect reading, and damage in the pre-motor speech areas also tends to affect writing. It appears that this arises from the fact that we learn to read and write by a conditioning process which associates the visual symbol or the motor output of the hand with the previously learned sound symbol. That this is the explanation is dramatically illustrated by the fact that reading and writing are only affected in patients whose languages are alphabetically written so that the construction of the symbol encodes its pronounciation; eg: the Chinese written language consists of pictorial ideograms, and reading or writing are unaffected by this lesion, while speech is still impaired. Apparently, Westerners learn to mediate reading and writing by internally "pronouncing" the word to themselves. When this fails, Westerners lose the ability to recognize or formulate the written symbol.

Other voluntary motor acts are affected by lesions of the pre-motor area in a manner consistent with our analysis of the generation of action schemes. Gross, well-learned, movements are least affected, and novel activities requiring motor control are most affected. The disturbances again take the form of disruptions of the sequential organization of motor acts. The sequential scheme is lost so that the movement appears as a general attempt at the desired operation without integration into a smooth sequence of related operations. Depending on the lesion site, fragmented motor sequences may appear, but may be out of order, or show repetition or inability to proceed to the next part of the sequence.

All of these observations are in accord with the hypothesis that the premotor cortex is operating to synthesize the sequential details of movement required to achieve a dynamic temporal solution to the problem of the design of voluntary movement. What remains is to examine the kinds of intellectual deficit which result from damage to this area in humans and to seek evidence for the hypothesis that the same machinery, used in the same way, is responsible for the dynamic organization of rational problem solving and symbolic thought.

Unlike the patient with a parietal or temporal lobe problem, pre-motor cases have no difficulty with the immediate meanings of words or phrases, calculations, or static spatial relationships or grammatical constructs. On the other hand, they have problems following or synthesizing complex lines of thought. They have great difficulty deciphering the meaning of passages with complex relations and contexts, and especially have difficulty changing to a new interpretation once they have mastered a previous one. This preservation characteristic is pronounced in all sorts of problems, including ones that they can master initially. Thus, they have little trouble with static spatial ralations and can follow instructions to place symbols in certain relations, but once having done so they find it very difficult to switch to a new sequence involving a slightly different relation. They are unable to devise workable schemes for the solution of problems, even when they understand the relationship involved and the objective desired. In other words, their principle intellectual deficit is in the generation of a reasonable plan for solving a problem. They comprehend the goal desired, and they can appreciate the logical relations of the relevant elements of the solution. Where they have difficulty is in the specification of a set of sequential steps for converting the given situation to the desired solution.

All of these difficulties are reminiscent of the kinds of problems that this damage produces in the execution of voluntary motor movements and of speech. It is difficult not to agree with Luria's contention that the underlying problem with the intellectual process in these patients is a disturbance of the sequential design of the "internal speech" of symbolic representation, and consequently of the synthesis of symbolic manipulations required for the attainment of symbolically defined ends. Reasoning is thus seen as fundamentally the same process as the design of complex goal-directed motor behavior, and in fact supported by the same hardware.

This analysis of the brain's procedure for the synthesis of a rational solution for physical or symbolic goal achievement is by no means complete. I have outlined the general procedure, but not the details. The details are unknown. It is true that the cortex seems to have the same kind of local cellular organization into small functional columns in all of its parts. (It is tempting though to think of the synthesis process as one of feature *insertion* and to imagine the perceptual cortex being run in reverse!)

Our model also fails to account in detail for the manifold complexities of verbal reasoning and syntactical interpretation. What we see is an approach to the basic mechanism of the hardware. Any number of advanced "programs" for its employment are undoubtedly built up through learning and instruction processes and must be taken into account in describing the final operation of the machine. An analysis of these functional programs is more properly the province of cognitive psychology in the case of the brain and artificial intelligence programs in the case of the machine.

What is of relevance here, however, is some insight into the architecture of the brain's "processor" for these types of operations. The general plan of attack schematized in figures 6.3 thru 6.5 will not seem novel to anyone who has worked with computer programs designed to accomplish the same ends. All of the basic concepts of recursion are there. The fact remains, however, that no one has been able to write such a program, capable of running in a reasonable time, while dealing with complex and general problems involving large data bases. The brain does it with slow components to boot. The answer seems fairly obvious. On a conventionally designed machine, the program handles one operation at a time; the program counter cannot be in two places at once. In the case of the cortex there is no such restriction. As the solution moves through successive stages of resolution, the number of yet finer steps requiring attention increases. The cortical processors respond to this by working on all of them at once. The process yields to simultaneous solution in parts, and the brain excels at that, just as it does in the perceptual case with its simultaneous attention to all the points on the retina. This only fails in the case of formal mathematical and logical problems where the next stage of a sequential solution is an unknown quantity until the proper procedure is found, but it is here that our performance falls apart in comparison with any serial computer.

VIII. EXACT VERSUS SUFFICIENT PROCEDURES

This may be a good place to explicitly mention one feature of the solution-synthesizing process discussed in this section: the scheme is strictly geared towards problems which yield to a ''sufficiently good'' solution. There are an indefinite number of ways to solve most of the practical problems of daily life which the brain faces; some solutions will be better than others. A number will be ''good enough,'' as long as it has some criterion for ''good enough''; it knows when to quit and implement the scheme. This is not true of formal logical and mathematical problems. There is a right answer to finding the factors of a quadratic equation, and there is no class of solutions which are ''good enough.'' That is why the process of figures 6.3 thru 6.5 cannot be meaningfully implemented in parallel on these types of problems. For example, there is no way of knowing whether or not WS3 is going to be a satisfactory intermediate state until the operation is complete. If WS5 is a completed book that you are trying to write, there are any number of sufficient approximations to it. If WS5 is an unknown result of a formal logical process, there are no ''sufficient'' approximations and the unknown goal cannot be used to generate intermediate states. Even if we have an idea of what we are seeking in such a case, the intermediate states still may be right or wrong, and we don't know until we get to the end.

This is a very important distinction which explains why the brain can employ its parallel architecture to excel in solving the kinds of problems for which it evolved, while performing poorly on the problems at which serially structured computers excel. If we want to build machines that can deal with the types of problems that brains handle, one of the first things that must be realized is that these are problems that do not have a solution, but a class of sufficient solutions. Once this is grasped, it becomes apparent that the parallel-processing approach gives a vast advantage and that the solution to emulating the brain's problem-solving techniques lies in adapting our hardware to the problem rather than in greater sophistication of our programs.

It also becomes clear that new difficulties are introduced. How, for example, do we define a ''sufficiently good'' solution? What are the criteria for recognizing solutions? Even more importantly, a ''rational'' machine is not one which simply performs logical operations. Our understanding of this word includes our estimation of the reasonableness of the goal as well as the accuracy with which it is attained. In the pursuit of all of this we enter an area where hardware has yet to tread, and software has only begun. I turn now to such topics as motivation, emotion, attention, and the specification of goals.

7 The Goal-Defining Systems

I. MOTIVATION

There are several fundamental differences between the initiation and the implementation of rational behavior. Among these is a function which psychologists refer to as "motivation" when it occurs in organisms. We may think of it as the means of defining the goals of behavior, although other functions are also connected with this concept.

It is clear that however logically synthesized a behavioral procedure is, and however exquisitely precise its analyses of the world model are, they do not fit our idea of a "rational" behavior unless they are directed towards some "reasonable" end. The definition of these "reasonable" ends is so deeply embedded in our thought processes that we usually are not even aware of them. Our problem generally seems not to decide what we want to accomplish, but how to accomplish it. Nonetheless, this is not a trivial problem, and the brain devotes a substantial amount of machinery to it.

The term "motivation" covers such concepts as emotions, drives, wants, needs, desires and the like. In most standard computer situations, these goals are analogous to the purpose of a program, which is in turn implicit in its operation. They are rarely defined in terms of computer hardware. This means that one of the major differences between brains and present-day computers is that brains determine their own behavioral directions or objectives, whether by pre-wired reflex or by learned processes, whereas computers are built to passively accept whatever purpose is inherent in the current program. There is no fundamental reason why a robot brain could not be built to operate like a natural brain in this regard.

Many psychological terms are used in various contexts, so let us review some that have been used to describe various kinds of motivation, and what motivations do for an organism. If we start with the notion that motivations are that general class of cerebral events which determines the objectives of behavior, a wide variety of things immediately comes to mind; almost everything we do has some objective or purpose. However, there are certain broad categories into which these objectives can be grouped, and these categories can be further subdivided as well. This can be done in such a way that there is a rough correspondence between the categories, and some of the functional systems of the brain.

First, our motivations can be divided into the two categories of "seeking pleasure" and "avoiding pain." Immediately one thinks of many situations in which our behavior seems to have as its object things that may bring us pain, but these are probably best considered as instances where the avoidance of duty or the failure to behave in a commendable manner would bring the greater displeasure. These types of motivation could also be seen as the pursuit of higher goals such as self-esteem, and the avoidance of pains such as guilt, instead of the more obvious direct results of such behaviors. (This type of behavior involves learned motivational processes, and here the concern is with simpler types for purposes of illustrating the basic principles.) This division into pleasure-seeking and pain-avoiding motivational states reflects one of the brain's most fundamental operating principles. Brains evolved as control systems for organisms that have clearly defined bioligical needs to seek certain things and to avoid others. All things with brains, however simple, seek to eat and to avoid being eaten.

To this end, the brain has evolved two major functional systems, operating in synergy, which transform information on bodily needs and environmental events into levels of activation and choices among behavioral options throughout the rest of the brain. It is an oversimplification to describe these strictly as pleasure-seeking and pain-avoiding systems, but such a description has considerable validity. In my laboratory, we are devoting most of our attention to studying the operation of these systems, in particular their goal directing functions, in the belief that this will reveal some of the most basic foundations of behavior, however far removed that behavior may be from the simple purposes around which these systems were initially developed. The most basic portions of these goal-directing systems are located largely in the regions of the brain called the limbic system, and in the hypothalamus and mesencephalon. Most of their detailed operations are of marginal interest here, because a robot system or a rational computer will probably have very different motivational requirements than a biological organism. I shall concentrate instead on their functional roles and on their interactions with the other brain systems.

One of the subdivisions of the motivational complement of all organisms is a rather loosely defined set of operating states, often called "drives." In general, this term refers to states which we recognize by such names as hunger, thirst, sex drive, suffocation, pain (in the more specific sense of a particular bodily sensation), and a host of similar, familiar terms. Mostly these refer to physiological need states of the organism and define the most obvious sorts of goals which might direct and energize operations in the rest of the biological brain. Two general classes of these drives have emerged which are differentiated by their mode of operation, and which have some applicability to robot systems as well. These are "homeostatic" and "non-homeostatic" drive systems. Homeostatic systems, of which hunger is the customary example, operate in such a way as to maintain a proper level of some important quantity or state in the organism. In the case of hunger, there is a mechanism which monitors the level of energy reserves in the body's chemistry by means of altered states of activity in certain sensors in the brain when these levels fall critically low. The activation of this system changes the operational state of a variety of other systems so as to cause the organism to engage in behaviors which result in the acquisition and ingestion of food. Stimuli resulting from food intake in turn cancel the operation of the hunger system, and the organism's behavior returns to other goals. The whole process could be likened to the operation of a thermostat in maintaining temperature. In practice, its precise functioning is extremely complex due to the need for hysteresis in its operating curve, the need for interactive monitoring of multiple parameters, and so forth, but in its overall operation, it is basically a feedback mechanism.

Non-homeostatic mechanisms do not operate under the control of such cyclically recurring needs and do not function to maintain a particular control level of a quantity or process. However, in other ways they are similar. An example is the response to pain. Here the motivational system operates only when and if certain external stimuli happen to occur. If the organism is unfortunate enough to encounter a painful set of circumstances, such as blundering into a patch of thorn bushes, receptor systems which detect damage to the skin immediately activate motivational mechanisms which assign the highest priority to behaviors which will remove the organism from the painful situation. This non-homeostatic motivational state will persist until the noxious environmental stimuli are escaped or eliminated. Both types of drive mechanisms serve to direct behavior towards situations which will meet the organism's immediate needs.

II. EMOTIONS

It is obvious that efficient operation requires more than simply being programmed to consume food, if we blunder into it when hungry. The organism must also respond to non-food stimuli which can signal the availability or direction and location of the food. Certainly, when hungry, we find the smell of food or the sight of food, or even the sight of a restaurant "pleasurable," even though the pleasurable taste of food is not the real object of the hunger drive. The joy produced by the taste of food serves as a stimulus to continue eating, but the object of the hunger drive state is repletion of energy levels, not the joy of eating. Nonetheless, the pleasure associated with stimuli which signal food or the fear associated with stimuli that signal pain serve to both energize and direct our behavior.

Because we have assigned this function to motivational states, we must consider these emotions to be part of "motivation" as well. Emotions are recognized as powerful "motivations" in our daily experience, and they are intimately related to drive states in governing our behavior. It is important to distinguish between the operation of the brain's physical emotional systems with their effects on behavior, and the subjective experiences we usually call "emotion" which depend upon the activation of these systems. Let us first consider the relationship of emotions to drive states and brain operation.

In general, the brain's emotional systems deal with stimuli which *anticipate* the actual objectives of drive states. That is, the brain responds to stimuli which normally occur prior to contact with the actual goal object. In the homeostatic case, the emotional systems are activated by the drive state operating at the time, so that a whole host of stimuli associated with the object of the current drive state is selected for special response. When, in the case of pleasure-seeking systems, these stimuli are detected, behavior patterns are initiated which lead the organism nearer to them, and hence (hopefully) to the actual objective of the drive state. The contrary holds true in the case of the pain avoidance process: stimuli that signal a situation which can lead to painful stimuli, but are not in themselves painful, still lead to the energizing and directing of appropriate pain-avoiding behavior. Looking down over the edge of a tall building's roof is not painful, but it signals potential pain and it is a stimulus which activates an emotional system (fear) that directs us away from the dangerous situation just as actual pain would.

In the case of homeostatic drive systems, the ability of the goal-relevant stimuli to activate emotion is dependent on the operation of the drive state. The smell of food is not pleasurable when you are already stuffed. In the case of the non-homeostatic drives, detection of the goal-relevant stimuli usually leads to operation of the motivational system as readily as does the drive stimulus itself. Thus, it is not required that we first feel pain in order for us to feel fear of the pain-related situation and to act on such fear.

The actual goal-related stimuli which serve to activate the emotional systems may be simple and "hard-wired," or they may be complex and effective only after learning processes associate them with the basic object of the drive state. Through experience, we can expand the range of stimuli which we can use to energize motivational systems; this serves to make our goal-seeking behavior much more efficient. The learning process is the basic "Pavlovian" mechanism of learning by repeated temporal contiguity. The primary goal stimuli (eg: nutrition, pain) activate the emotional mechanisms directly. Stimuli which reliably precede these basic stimuli can acquire, by conditioning, the ability to activate these same emotional mechanisms and provide motivational activation and guidance.

When operating, the motivational system not only energizes and directs behavior, but also appears in our subjective experience as "emotions" or "desires." The operation of the hunger drive state mechanism is perceived subjectively as the feeling of "being hungry," or "desire for food." In these circumstances the sight of food activates emotional systems whose operation is subjectively experienced as the "feeling of joy." The activation of emotional systems which detect danger-signaling stimuli is experienced subjectively as the feeling of "being afraid."

It should be noted that making this distinction eliminates a problem posed by the old question of whether or not a machine could have emotions. If by "having emotion," it is meant having a mechanism which detects certain types of situations and calls for certain categories of response, the answer is yes, and it could function just as usefully and efficiently as yours does. If, on the other hand, it is meant can a machine have a *subjective* experience of fear when this mechanism is operating, the question is probably unanswerable. Strictly speaking, we cannot even say whether or not another person has subjective emotional experiences: we can only know that he engages in emotional behaviors in appropriate circumstances, including behaviors such as saying, "I am afraid." With respect to the construction of a robot device, the question of subjective experience is not the essential point; what is important is to recognize that the brain processes which give rise to subjective, emotional states do not have these experiences as their goal: their purpose is to provide certain useful types of information processing relevant to increasing the power and efficiency of our behavioral responses to the environment. These processes remain as essential to efficient action in a robot brain as they are in an organic brain, regardless of whether of not their operation is perceived as a subjective experience.

III. MOTIVATIONAL PROCESSING IN GOAL DETECTION

Let us turn now to the motivational system's actual mechanisms of operation. To a limited extent, motivation is analogous to an interrupt system in a computer, and it has a priority structure too. Motivational systems are more complex than an interrupt structure because they do not simply detect specified conditions and turn control over to specific programs of action. Instead, they enable *classes* of activity and define *types* of relevant goal stimuli. The actual activity undertaken within the operation of a particular motivational state may vary widely; it is only the desired end result which is determined. The behavior that leads you to the food when hungry may be anything from reaching into the refrigerator to looking for a job, or even going hunting.

This method of reaching the goal is a heuristic rather than an algorithmic process; this is one of the most important distinctions between a brain and most ordinary computers operations. In an algorithmic process, the goal is assumed, and a sequence of steps is executed which, if executed correctly, is known to lead to the goal. In a heuristic process, the potential results of different paths of action are compared with the goal, and those which seem to lead to states closer to the goal state are executed.

The process is then continued and may include backtracking when it is discovered that a promising course led to a blind alley. This heuristic approach has been implemented occasionally in software when there are no known algorithms.

The chess playing programs are good examples, as there is no algorithm for winning a chess game; however, simple heuristic systems were implemented in hardware even in the earliest robot systems. In one of these systems, the goal was recharging the battery. The "drive state" was initiated by a relay that opened when the voltage fell below a prescribed level, and its action was to place the forward motor drive under control of a directional photocell. The goal-related stimulus was a light mounted over the battery charger. When the light was in line with the device's forward direction of motion, the photocell amplifier was activated (emotional response) and the machine rolled forward (motivated behavior). When the device was not lined up, it turned at random but did not move forward. It was never guaranteed that the sequence of behavior was correct (consider the consequences of a mirror), but within the behavioral potential of the device, it was a best guess.

In conjunction with the reward system, the basal ganglia provides an example of this sort of heuristic operation at a relatively simple level in the brain and illustrates the action of the motivational system in selecting classes of response. Recall that the kind of goal-seeking behavior which can be handled directly by motor and sensory systems at this level is dependent upon the ability of a reward system to recognize a rewarding sensory input, and then to enable the continuation of whatever sort of behavior led to the rewarding input. This is the simplest possible sort of goal-seeking mechanism. It will eventually lead the organism to the goal even if the behaviors are generated at random; ie: if the organism moves at random except for the continuation of the behaviors that increase rewarding sensory input and the discontinuance of those that lead to decreases in rewarding input, it must ultimately approach the source of the rewarding sensory stimuli. Of course, in any advanced brain, the behavior selected for trial is not randomly selected, and I will discuss that fact momentarily. The role here of the motivational system is the same regardless of the manner in which the behaviors are selected for trial; ie: motivation selects those classes of sensory stimuli which will be able to activate the reward system.

If the organism is low in energy reserves, it must respond to food-related stimuli rather than to extraneous information related to some other (potentially important) class of goals which is not relevant at the moment.

This defines one aspect of the operation of the motivational system in a heuristic device; it must gate the reward-detector to respond to that class of stimuli which is relevant to the goals of the drive state in question. For example, in order to give the simple robot two motivational options one needs a system to switch between the photocell and something else, say a microphone, to activate the forward locomotion circuit. Then a speaker could be used to define the location of some other goal (lubrication?), and which goal was approached would depend on whether the locomotion circuit was activated by the photocell or the microphone. The major difference in the brain is that entire categories of complex stimuli are "enabled" into the reward circuitry by the drive state, rather than just a single simple stimulus. Moreover, these range from straightforward stimuli, which are probably hard-wired into the motivational circuit from birth (the taste of food for example) to the most complex stimuli which are learned in later life. Like most other systems of the brain, this one has a hierarchical representation at various levels, and more complex stimuli are dealt within more advanced structures, while simpler ones are handled at anatomically (and evolutionarily) lower levels. The very highest levels deal with anticipated stimuli extrapolated from projected future actions.

It is easy to see how a system which detects a need state and initiates a drive operation might gate some simple and specific stimuli into the reward system, but it is not so clear how complex learned stimuli are to be dealt with by a system that must specify classes of events. Research has revealed that the structures of the limbic system are important to our emotional functions as well as to our reward system, and these regions seem to be necessary for activating the reward system when the goal-relevant stimuli are complex and learned. The limbic brain is also in close connection with the hypothalamus and other lower centers which function in the detection of our homeostatic need states. Further, the limbic system receives a wealth of projections from cortical areas which are involved with the higher levels of perceptual feature extraction.

In particular, the limbic structures which lie beneath the cortex of the forward portion of the temporal lobe, such as the "amygdala," receive massive projections from the visual and other perceptual areas of the cortex. Interestingly, these projections do not come from regions which are concerned primarily with object categorization or other already highly abstract perceptual functions. Rather, they originate at the earlier stages such as the secondary visual cortex, prior to the infero-temporal cortex. This suggests that while the perceptual input is being assembled and analyzed in ways relevant to a description of the surrounding space, by way of the processes which we have already followed, the limbic system may be pursuing some other feature-recognition process involving this same data and is attending to different sorts of important relations in the incoming information. There is no essential reason why the "feature" in this process must be related to a perceptual construct of the type we were dealing with in building the spatial world model. In other words, although the process described was presented in terms of the elements of visual object recognition, it would be equally applicable to what might be termed "emotional perception." That is, the emotional relevance of a visual stimulus could be considered to be a feature to be extracted just as much as its intellectual identification. Perceptual studies in humans suggest that emotional and intellectual perceptions of stimuli are not only separate, but that emotional perceptions may be more rapid.

Consider how an emotional "feature extractor" might be constructed. Let us say that as a start we already have a small number of lines which encode the elementary features of some object in the visual field regardless of location, etc. Now suppose that the output of such elements converges, among other places, on elements of the limbic system, and that the convergence is selected so that the "feature" which is extracted by the target neuron in the temporal cortex or limbic system is "relevance" to some drive state rather than membership in a geometric category or other "intellectual" property of the stimulus. Geometrically related objects are not necessarily members of the same category in respect to their association with system goals. (Exactly how these connections are determined is not important; they might be hard-wired or established through a later process of learning.) We would then have a group of limbic system neurons which could encode the presence of stimuli relevant to a particular drive state. Such limbic system elements, each driven by a multitude of perceptual objects, could easily be inhibited or enabled by inputs from the motivational system; thus, motivationally-defined classes of drive-relevant stimuli could activate the reward system. Presumably, the rational for diverging the perceptual input stream into separate analyzers at an early level of processing is that cognition deals primarily with similarities of spatial form among stimuli, while emotion does not. That is, different stimuli relevant to a particular motivational state need not be similar in appearance. The essential thing here is to see that the formal problem of decoding stimuli for emotional content is the same, in terms of information processing, as the problem of decoding them for logical purposes; some of the same perceptual circuitry can serve both functions, because it is really only after the level of feature identification that the convergence patterns need be different.

The general flow of information through this system is shown in figure 7.1. Partially analyzed data from the perceptual areas of cortex are routed to the limbic system structures where the object categorization process is continued to extract the occurrence of objects related to some drive state of the organism. That is, the classes are determined by their emotional features rather than by their physical similarities. At the same time, the drive state portion of the motivational system is decoding other internal data to determine if the organism has current needs. If such a condition is recognized, the motivational system can enable the appropriate class of emotional feature extractors to output to the reward system. This relationship is symbolized here by the ANDing of the drive state system's output with that of our hypothetical emotional feature extractor. Some classes of input, such as those appropriate for eliciting fear and fleeing, are non-homeostatic and are applied to the reward system as non-maskable interrupts without being enabled by the motivational system.

If structures of the human limbic system which receive the inputs from the perceptual cortex are stimulated electrically, various emotional responses, either positive or negative, are reported. Similar responses can be produced in animals, too. Damage to the mechanisms which detect the need states for the organism will cause it to ignore stimuli relevant to the need; an animal may starve to death in the presence of ample food. On the other hand, damage to the connections between the reward system and the motor output system appears to be responsible for a phenomenon called "sensory neglect," in which animals simply cease to attend behaviorally to events in their environment. They will not even orient to novel stimuli. They must be force fed to survive and generally seem unable to initiate a response to any sensory stimulus. In terms of the present model, this would result from the fact that no external stimulus would be able to serve as a means of disinhibiting the motor output system.

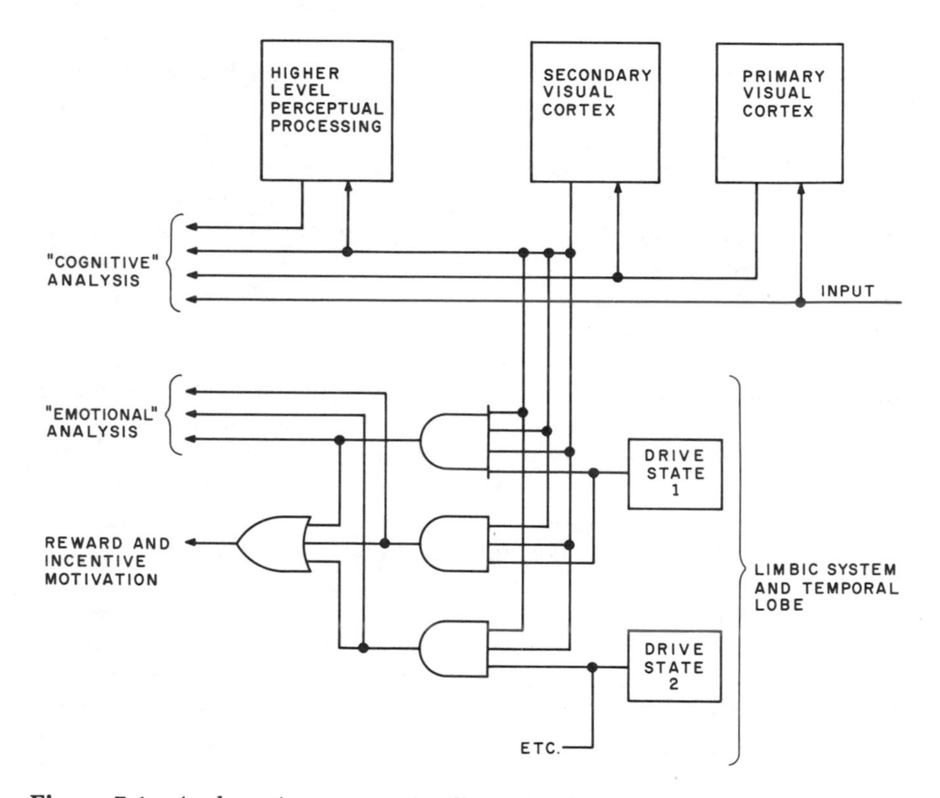

Figure 7.1: *A schematic representation illustrating the way in which the same input data might be analysed for "emotional features" as well as for "logical features." The two modes of analysis need not respond to the same features in the input, nor group them in the same fashion. The emotional analysis is shown as being under the control of the "drive states" that recognize organismic needs.*

Exactly how this emotional perceptual apparatus is connected is not yet well understood, but it appears that whatever the precise nature of its operation, it will not be very different in principle from the model I have just described. In terms of behavior, its general functional operation is established. For potential applications to robotics, the present model will adequately summarize these facts. The general scheme is presented in figure 7.2. In this diagram, the data flow from the receptors follows two main routes. After preliminary analysis in the sensory cortex, the data are

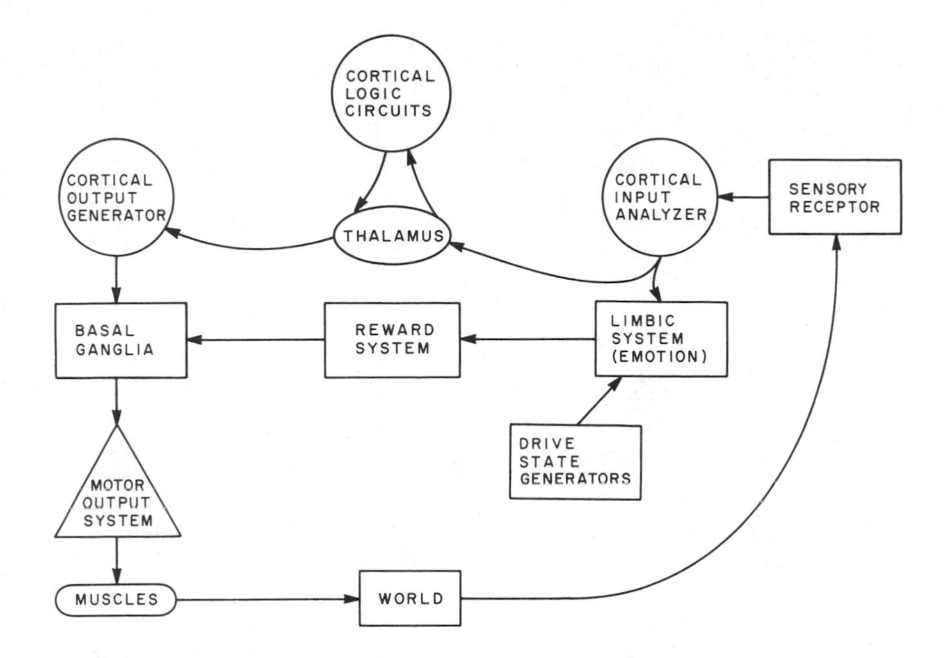

Figure 7.2: *The general plan of information flow through the system in regard to the control of output by the goal-directing system. The input is analyzed for the generation of logical action schemes in the upper portion of the circuit, while the lower portion evaluates the same inputs for their emotional relevance. The emotional relevance depends on the current needs of the organism as reflected in the activity of the drive-state generators. Actions resulting in a positive change in the emotional analysis of the world activiate the reward system which permits the perpetuation of the action generated by the logical analysis of the situation. Not shown are connections that enable the logical portion of the system to employ knowledge of the drive-state of the system in generating goals for directing the logical synthesis of action.*

available both to the limbic system for motivation-relevant feature extraction processes and to the other areas of the cortex for logical analysis. The information which is processed in the limbic system can activate the reward mechanism if (1) the information decodes to features relevant to a drive state and (2) the limbic system elements which decode it are gated onto the reward system bus by activity of the appropriate drive state mechanism. When these conditions are met, the behavioral strategies developed by logical analysis of the sensory data can continue to be translated into motor patterns by the basal ganglia and other portions of the output system. (I should mention that there seems to be a large component of the reward value of positive stimuli which is due to the rate of increase, or derivative, of the decoded stimulus rather than its absolute value; this feature seems to improve system response characteristics.) In the case of escape and avoidance behaviors, the *reduction* of activation of certain stimulus elements serves to activate the reward mechanism, possibly through release of inhibitory elements. As mentioned earlier, in the case of such non-homeostatic behaviors, the situation is reminiscent of the difference between maskable and non-maskable interrupts. Stimuli relevant to non-homeostatic drive states are "non-maskable" in the sense that the drive state does not have to be activated in order for them to have an effect. What seems to occur here is that the decoding of these stimuli serves to initiate a drive state, thus energizing the organism's behavior, and enables detectors for the subsequent reduction of these stimuli by behavior which thus activates the reward mechanism.

IV. OTHER FUNCTIONS OF MOTIVATIONAL SYSTEMS

What I have described accounts for the most basic functions of motivational systems—their elementary goal-directing functions. Several other simple functions remain to be mentioned before considering their role in complex logical processes: these are the energizing or activating properties, the cue or stimulus properties, and the property of reinforcement of learning. To take them in order, from the above discussion it is clear that there will be an increase in general activity level after an encounter with stimuli associated with a goal object: directed locomotion towards a goal arises when the motor system is gated open. In psychological jargon this is referred to as "incentive" activation and is clearly a result of the conjunction of drive state and relevant stimulus. In addition, there is an increased activity level resulting from the operation of a drive state generator in the absence of any relevant stimuli; this appears to be a general increase in levels of behavioral motor output, irrespective of any specific relation of the activity to the drive state. Many different kinds of drives will increase an animal's general activity as measured by wheel running, cage jiggling and the like. This phenomenon is called "non-specific" activation, and its utility to the organism seems apparent. If a drive state develops, the organism has a statistically greater chance of encountering relevant stimuli if it moves around in the environment, even when such movement is random and undirected. (Notice how people tend to pace when highly motivated.) The same argument would apply to a robot system.

The "cue" property of motivational states refers to the fact that the operation of motivational mechanisms generally gives rise to internal sensations which may themselves be considered as sensory stimuli capable of entering into the information base on which the logical functions operate to produce behavioral strategies. Thus, the organism "knows" when it is hungry, and this internal sensory data can be employed in conjunction with sensory input to recall information from memory and generate behaviors for trial. These are much more likely to lead to successful results than are random behaviors. This makes the system much more efficient than one which, like the simple robot, simply gates the continuation of successful behaviors. This is because successful behaviors (ie: those that lead to encounters with goal relevant stimuli) are much more likely to occur with this advanced scheme. For example, if "hunger drive" status is used in conjunction with information about current location to form a "memory address," any information about known encounters with food sources in the vicinity can be accessed and used to generate behaviors with greater heuristic value than random exploration. I shall consider this process further in conjunction with the operation of the frontal cortex.

Finally, the role of motivational systems in the reinforcement of learning must be considered. You will recall that in the first chapter, I mentioned two kinds of learning processes, one of which was "classical (Pavlovian) conditioning" which depended only on the temporal contiguity of a stimulus which was initially neutral and a stimulus which naturally evoked some behavior. We saw that repetitions of this pairing led to the ability of the initially neutral preceding stimulus to evoke the behavior normally evoked by the following second stimulus. It appears that this type of conditioning may be a fundamental property of neurons. A distinguishing feature of this type of learning is that only the pairing, not the consequences of the response, is important. The second type of learning process, called "operant conditioning" is entirely dependent on the consequences. Behaviors which the organism emits, and which are followed by activation of the reward system, tend to occur more frequently. Those which fail to generate activation of the reward system will tend to occur less frequently in the future. This type of learning is enormously useful in improving the range and efficiency of the organism's behavioral repertoire, because it can modify existing behavioral chains by the insertion or deletion of elements or build up entirely new behavioral sequences out of accidental successes. This occurs even if the reward system is driven directly by electrical brain stimulation, in which all the circuitry is bypassed which normally decodes the environmental and drive-state stimuli required to activate the reward system. It appears that the reward system must be involved in this process of operant conditioning. There is mounting evidence that the basal ganglia may participate in important ways as well. It is possible, although not definitely established, that a process of classical conditioning operating between the inputs to the basal ganglia could account for the phenomenon of operant conditioning. This theoretical model will be presented as the approach easiest to model in a robot brain.

From information given in earlier chapters, it is clear that a successful behavior involves two major events at the basal ganglia: first, the activation of a pattern of neural firing in the basal ganglia by its cortical and thalamic inputs, and second, a sustaining input from the reward system if the resulting behavioral output produces a more favorable (ie: rewarding) stimulus situation with regard to goal-related stimuli. Any number of possible motor activation patterns may be sent to the basal ganglia by the cortical and thalamic mechanisms as a result of analysis of current events and memories of similar situations in the past. Some of these will be more successful than others in inducing firing patterns in the basal ganglia, given the "self-quenching" action of the inhibitory elements in that structure. One function of this self-inhibitory mechanism may be to insure that only one of these competing response patterns will emerge to direct the motor mechanisms. That is, the strongest input will most easily withstand the inhibitory action, and the inhibitory action which that input's own successful operation initiates will further undermine the attempts of competing firing patterns to seize the upper hand. Think of a bistable flip-flop circuit when power comes on. Only one transistor winds up in the on state, and the other is completely off. In the same fashion, the mutually inhibitory actions of firing patterns in the basal ganglia would tend to insure that there was only one winner.

Just as in the case of the flip-flop circuit, whichever one of the possible firing patterns "wins" may be determined by very minute differences in the ability of the competing activation patterns to fire their target neurons first. Furthermore, a variety of other factors (eg: how recently a particular neuron has fired) will cause the usual "winner" to lose some of the time. Thus, the patterns that emerge as behavior will have a ranking of probability, but it will never be guaranteed that one will always be the winner, only that it is the most likely to win. This is very different from most computer approaches in which it is assumed that you know the best way and do it that way always.

In developing new behaviors to suit an unknown and changing environment, it is important to be able to experiment a little. Now, if we could somehow alter even slightly the sensitivity of the target neurons in favor of one or another of the competing patterns, so that the pattern could get a small jump on the competition, the probability of that pattern of behavior emerging as the victor would be increased. If this could be done in a way which was contingent on the behavior being "successful" (ie: activating the reward system), when by chance it was the winner, we would have all the features of operant conditioning. That is, whenever a behavior was successful in producing a rewarding (ie: drive relevant) stimulus situation, the probability of that particular behavior pattern emerging as the winner in the motor output competition would be increased, and the behavior would become more probable. It appears that this happens, and that at least with respect to complex, feedback-controlled, environmentally-oriented behaviors, it does involve the basal ganglia.

The process is called "reinforcement," and at the cellular level, we don't know how it works. A reasonable guess would be that it is essentially a process of Pavlovian conditioning of the cells of the basal ganglia by sequential activation: first by the cortical and then by the reward system's inputs. We know from studies of simple nervous systems that this classical Pavlovian conditioning mechanism, based on simple temporal contiguity of inputs, can occur within a single neuron when two inputs are activated sequentially. The operant conditioning process may then be the result of classical conditioning of certain cells, such as those of the basal ganglia, by inputs from the reward system immediately after they are successfully fired by input activity from the cortical action scheme generators or perceptual analyzers. The resulting small increase in the ability of the cortical inputs to fire those particular target neurons would increase the likelihood of that pattern emerging as the dominant output. The distribution of these two input systems to the cells of the basal ganglia is consistent with this idea. The cortical inputs carrying activation patterns to generate specific outputs are restricted to selected cells. The reward system's input, relevant to any just completed action, is diffuse and widespread.

In the case of a robot system modeling this kind of action, one might envision a system in which the processors which decoded the equivalent of the cortical inputs into the equivalent of the basal ganglia outputs to the lower motor system would apply a numerical weighting to the various inputs received, and would increment the weighting if the "reward" input were active within a short period thereafter. What would be important would be (1) the provision of some mechanism—perhaps a random number table—for making the predominance hierarchy probabilistic rather than absolute, and (2) provision for changing the weighting according to the success of the behavior when tried. With these two features, within the limits of its behavioral capabilities, the robot would achieve the flexibility and adaptiveness that characterize brains; this achievement also includes the ability to make mistakes, but it seems to be a price that evolution has found acceptable.

I have presented this model of operant conditioning as though it were occurring exclusively in the basal ganglia. While that illustrates the general principle involved, and would suffice for simple behaviors, it is clear that in the real brain, more is involved. The projections of the reward system extend into large portions of the cortex, and other things than the probability of output patterns can be modified by similar learned processes. For example, the weight attached to particular stimulus features in generating the output patterns in the first place, or the weight attached to particular items from memory in generating these action patterns, can all be altered by the reinforcement process. Such things more likely occur at cortical levels than in the basal ganglia, and their details are obscure. However, there is no reason why the principles described here could not be applied to such functional processes in an advanced robot. Once the principle of operant conditioning is employed, in theory it is possible to achieve in the machine any degree of fine tuning of any process.

It should be noted, by the way, that the more the machine relies on operant conditioning, rather than hard-wiring or unmodifiable software to generate its response, the longer training period it will need, and the longer "infancy" it will have in which it will need a "mother" to keep its major errors from being disasters. In organic brains, the period of helpless infancy is directly proportional to the flexibility and adaptiveness of the adult brain for just this reason.

V. MOTIVATIONAL CONTROL OF FOREBRAIN FUNCTION

An important brain function, that is closely related to the concept of motivation in respect to energizing and directing behavior is loosely referred to as "arousal." There are two general types of arousal systems. The most fundamental is called a "general" or "tonic" arousal system, which refers to the fact that this system sets the general level of activity in the nervous system over long periods of time; ie: it sets the "tone" of the system. The most important part of this tonic system is a group of diffuse nuclei forming a long column in the central part of the spinal cord, medulla, pons, and mesencephalon. These neurons are collectively referred to as the "reticular formation." One of the great early discoveries in brain function was the finding that the forebrain does not simply run of its own accord. It requires constant drive from the reticular formation to keep it processing. In this sense, the reticular formation acts like the brain's on/off switch. However, whereas most computers are either on or off, the brain is capable of operating at various levels of activation. Intuitively, you know the difference between feeling highly alert and excited, and feeling awake but inattentive and relaxed. These states reflect different levels of operation of the forebrain under control of the reticular formation. An imperfect analogy could be made with a computer having a variable clock speed, but it is not strictly speed of processing that is affected. The general modulating inputs from the reticular formation are diffusely connected through the forebrain and serve to bias the cells there towards or away from firing level. As we saw earlier, this not only has implications for the rate of activity in these cells, but also influences the nature of the processing they perform, because the "analog" and "digital" factors in the cell's input are interactive. (Recall that the biasing level of a feature extractor may determine the degree to which the input must resemble the optimal input in order to cause an output.) Thus, not only level of activity and speed of processing, but also the nature of processing is affected, in ways appropriate to the degree of attention required by a particular situation.

At the two extremes of reticular formation control of arousal we have high excitement on the one hand and unconsciousness on the other. (Do not confuse unconsciousness with sleep. Sleep is a separate *active* process that reflects a different type of organization of forebrain activity. I shall not discuss sleep, because this state may not be relevant to robot brains, and nobody knows what it's good for in humans either.) Part of the utility of such an activating system in a robot might be to conserve power when the environment was not demanding much attention. The power saving function is not inconsiderable in your own case; the brain draws about 25 watts, which is a healthy fraction of your available energy. A more important point, however, is that the raised arousal levels provide a substrate out of which particular patterns of forebrain activity can be carved by selective inhibition. This selective process is part of the function called "attention" and is handled by the frontal cortex.

Let us look first at the sources of general arousal. The reticular system is activated by three major types of input. One, which I have already discussed in part, is an input from the mechanisms which detect and regulate specific motivational states. We mentioned a general, non-specific component to the arousal produced by motivational states. This appears to function through activation of the reticular formation by the motivational systems, or in some cases directly by the internal stimuli that activate those systems.

A second source of control over reticular activation comes from the sensory systems. This occurs in a rather unique way. As the axons carrying sensory information to those areas which analyze its contents pass by the reticular formation, they give off branches which contact cells in the lateral part of the reticular formation. (See figure 7.3.) These cells in turn activate cells in the central portion of the reticular formation which give off long axons that carry activation-regulating "control" impulses to the rest of the brain. What is interesting here is that there is little attempt to keep the sensory input lines separate. Inputs from receptors for touch, sound, light, and all other senses, all synapse on the same neurons. There is no clear qualitative component to the reticular formation's input data. What it is responding to is the total amount of sensory activity in the receptors, or in other words, the general level of environmental "noise." Like all sensory systems, this one responds most strongly to things that change. Thus, abrupt changes in the level of activity at any sensory receptor will serve to activate the reticular formation and arouse the rest of the nervous system. Strange though it may seem, it is not the conscious perception of the sensory input which is arousing. In fact, if the reticular formation is damaged, and the rest of the sensory apparatus is left intact, the organism will not be aroused by the most intense stimulation.

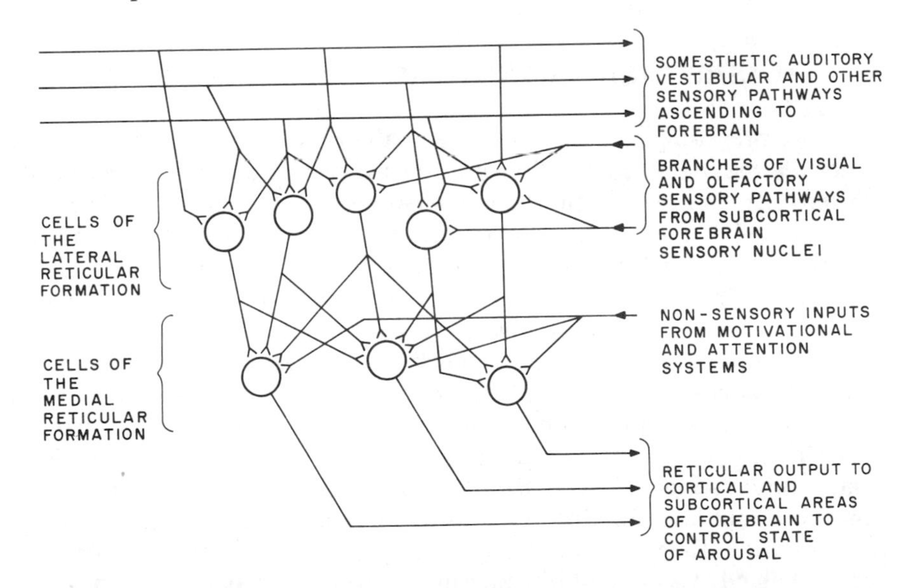

Figure 7.3: *The basic connections of the "reticular activating system." Three classes of inputs can result in activation of the cortex. These are the branches from the external sensory inputs entering the lateral reticular formation, the inputs from motivational systems in response of organismic need states, and the inputs from "attentional" systems that allow higher cortical processes to regulate the level of nonspecific cortical arousal.*

The function of this mechanism is clear; if there is much changing activity in the environment, the organism probably needs to be alert and attending to it. The reticular formation is very old, evolutionarily, and this mechanism probably subserves one of the earliest modes of response to the environment which is still useful even in the most advanced brains.

If the process were to stop here, we would have an organism that could respond adequately to external events or deprivation states, but which would not be spontaneously alert in the absence of a need state or an external stimulus which required attention. This is approximately the behavior pattern of creatures such as amphibians and reptiles, which have little or no cortex. In the case of animals with a well-developed cortex, we see the development of a third set of inputs to the reticular formation, which come from the cortex. This establishes a positive feedback loop between the reticular formation and the cortex which enables the cortical-reticular system to "lock on" and sustain its own activity. This important development allows for control of forebrain arousal levels by the results of the forebrain's own internal processing, and the advanced brain is freed from total dependence on need states or environmental input for its maintained activation. This is clearly advantageous when a level of cerebral sophistication has been reached which permits current action on the basis of anticipated future events. With this feedback loop, the more advanced brains achieve the capability for maintained conscious thought and regulation of their own operation.

VI. GOAL DIRECTION AND THE HIGHER FUNCTIONS

In the preceding sections, I reviewed the process by which the lower portions of the brain's motivation and reward system can act to give a goal-directed character to the input/output relations of the organism. At the level of the elementary reflex, the input stimulus alone carries all necessary information about the desirability of action, and the nature of the action itself does not require guidance that involves evaluation of results in relation to goals. At the level of operation I have just considered, the principle advance over reflex operation is that the response to a stimulus is made contingent on the relevance of the stimulus to the organism's need states, and the form of the response is controlled by a mechanism which attempts to maximize the desirability of the current stimulus situation. Such a level of behavioral organization is a great advance over the simply reflexive organization of the spinal cord and lower brainstem, and its implementation requires subcortical portions of the telencephalon. This degree of organization supports the level of behavior seen in sub-mammalian forms such as reptiles, and it is clearly a feedback process (otherwise known as "stimulus-bound" behavior).

With provision for the goal-signaling stimuli to be learned by association with the wired-in primary stimuli, and for chains of action to be built up through the operant conditioning process, this sort of stimulus-bound behavior is capable of quite advanced functioning in a machine with complex input processing and output synthesizing capabilities. Even human brains rely on this sort of functioning for routine processing. When the relevant stimuli and the appropriate motor responses have been learned, you can drive an automobile without any higher level conscious activity, until you come to a point where a novel decision is required. It is entirely likely that in such cases we are proceeding in a stimulus-bound, feedback-controlled mode of operation, which does not differ in principle from that of a lizard approaching food.

However, in Chapter 6 we have already seen that certain cortical portions of the input and output systems have achieved an advanced capacity for anlaysis of events and synthesis of behaviors which includes symbolic processing. It is obvious that this approach can be better employed than simply providing elaborate analyses in the service of simple-minded feedback-controlled response to current stimulus inputs.

In particular, the evolutionary development of these capabilities has presented the brain with the opportunity to develop a more advanced, goal-directing system. In the prefrontal cortex, we find an advanced form of the motivational-reward processor suitable for controlling the activities of the cortical analytic and synthetic functions efficiently in the service of the primary motivational system. This cortical branch of the motivation system uses the power of these analytic systems very effectively in its own operations. The development of this system has stressed a variety of optimizing techniques which provided the initial evolutionary advantage driving the system's development. I shall examine these optimizing techniques first. Following this, I will suggest a possible interpretation of the way in which the interaction of the system with the other advanced cortical systems can provide an understanding of the physical basis for the "insightful" aspect of human problem solving.

In effect, the input and output systems, at each of their levels of operation, are like a multitude of parallel processors which can function independently with fixed programs which determine their functions and interactions. Greater efficiency can be obtained if these systems can be coordinated by a supervisory system which has access to the system goals. This can occur both by the use of feedforward reprogramming and by the use of look-ahead procedures which anticipate the consequences of actions.

The fundamental control structure of the machine is such that the operation of all of its parts, even the most advanced, is ultimately directed by the primary motivational systems. To give flexibility and scope to its operations the fully developed motivational system must be capable of interpreting these primary goals into sophisticated plans of operation that free the system from a state of primitive stimulus-bound behavior. To this end, the motivational system itself has developed a number of levels of elaboration. We have already seen that the emotional mechanisms of the limbic system permits a limited feedforward action such that the system responds to stimuli which occur prior to the goal, as though they were the primary goal stimuli. These secondary stimuli may be learned through a conditioning process, or may be genetically specified or "hard-wired" into the system. Up to this point, the motivational system may be thought of as including the primary goal-defining machinery of the lower brainstem, mesencephalon, and diencephalon, and in addition that portion of the limbic system of the telencephalon which functions in the processing of emotional responding.

Some parts of the limbic system are in fact portions of the primitive cortex (called paleocortex), which is distinct from the more complexly layered cortex (called neocortex) which comprises all of the other cortical systems that have been examined. There is a portion of the neocortex, however, which functions in close relationship to the limbic system and other lower portions of the motivational and reward systems. This is the prefrontal cortex, which lies forward of the pre-motor cortex in the frontal lobe of the brain. It probably interprets the goals defined by the lower motivational system into advanced plans of action for the rest of the system, and it may be thought of as the highest level of the motivational system. It is also the most recently developed portion of the brain, and in its role as "supervisor" of the higher functions for the primary goal-defining portions of the motivational system, it is responsible for much of the rational aspect of human behavior.

There are two distinct aspects to its operation: feedforward and look-ahead planning. The feedforward mechanisms provide for the preparatory programming of the other functional systems on the basis of the system's goals; an example would be the process of attention. This process does not plan for the future in the sense of anticipating consequences and taking preparatory actions, but the effects are similar. True look-ahead capability probably develops out of an initial feedforward function of the prefrontal cortex. The feedforward processes all operate in real time on the basis of presently available information; they do not require the capability of operating in future time. They have an effect which gives the appearance of foresight and planning, because they can employ present-time predictors of future states to improve the quality of the goal-directed behavior by adding to, or modifying the goal value of, the current stimuli directing behavior. They do this on the basis of demonstrated utility of such operations in the past.

For example, an emotional response is a feedforward mechanism and not a look-ahead process, because the system is not actually developing, and responding to, a predicted model of future situations. It has simply identified a number of indicators which are found to improve performance if they are treated as goal stimuli in current time. In psychological research, this process has a number of names such as secondary motivation, which refers to the ability of such goal-associated stimuli to act as motivators for behavior, and secondary reinforcement, which refers to the ability of these stimuli to act like primary reinforcers or "need satisfiers" in their ability to direct behavior, provide incentive activation, and even support the learning of new responses when they act as rewards. (George Pugh has termed these stimuli "secondary values" in his fascinating analysis of the relations between human behavior and modern computer decision-making programs, *The Biological Origin of Human Values*.)

The way in which such a process can stand in place of a true look-ahead planning system is apparent: if I had no such system, then I would have to employ a reasoning process to infer that if I walk ahead in the presence of a long drop I will probably arrive at a painful situation and therefore I should go some other way. Because this sort of thing is time consuming, and impossible for a simple brain, the reasoning process is rendered unnecessary by the attachment of a negative goal value to the visual stimulus of the drop and by then reacting to that stimulus in the present time. (It should also be noted that once such a system is developed, it can serve not only as a substitute for rational planning of action, but also to facilitate true look-ahead processes. That is, the planning process can quit upon reaching a prediction of attaining an emotionally "significant" state; it does not, therefore, have to carry the analysis to an arbitrary degree of refinement.)

What is needed for a truely rational process is the ability to attach motivational significance to stimuli or symbols in a *temporary* fashion. The mechanisms of emotional response may attach a weighting to otherwise neutral predictive stimuli through the process of conditioning, but this requires that the stimulus in question be frequently associated with the actual goal. If we want to plan rational acts, we will usually have to assign a goal status to some intermediary state that we wish to attain, which may have no necessary or lasting relationship to either a real or a secondary conditioned goal stimulus. For example, if you let me see you put a one hundred dollar bill into a red envelope and then mix it into a pile of other objects, I will probably devise all sorts of clever goal-oriented actions designed to obtain the red envelope. On the other hand, it would be a poor strategy for my brain to engage as a customary procedure in the pursuit of red envelopes. (Incidentally the one hundred dollar bill itself is a good example of a secondary reinforcer, or secondary value, which is strongly conditioned to elicit "joy" responses because of its repeated association with all sorts of primary reinforcers.)

A major requirement, then, for employing rational processes in behavior is the ability to attach goal significance to objects (or symbols) in an arbitrary and temporary fashion. This can be thought of as "symbolic desire." When in symbolic reasoning processes we say "Let X equal the desired quantity..." and then set off in the logical pursuit of X, we are doing essentially the same thing as when we perform a rational, although non-verbal, process in pursuit of the envelope. I shall return to this process after examining some of the other feedforward functions.

VII. OPTIMIZATION OF FUNCTION BY MOTIVATIONAL FEED-FORWARD

One area in which substantial savings in processing time can be effected on the basis of the predictive value of present information is the input analysis function. You will recall that I discussed an active form of object analysis in perception which involved eye movement and "attention shift" in the examination of the visual field. I also examined a static process, which did not involve such shifts of the region of examination, but which could be dynamically "tuned" to maximize the probability of identification for objects whose categories were suggested by the preliminary attempt at match to sample. The speed of these processes can be improved substantially if the system has some prior indication of its goal.

This concept forms the basis of our understanding of the effects of "set" on perception. Identification time will be vastly reduced if the object to be identified is one which is to be expected on the basis of our current situation. This is a "feedforward" procedure which can function in a number of ways. First, it can help in the dynamic tuning process by selecting the appropriate set of categories of objects for the initial try at matching. This process, called "framing," has been discussed (although not to my knowledge exploited) in relation to artificial perceptual systems. The savings in time introduced by such a procedure will be proportional to the power of the perceptual system. That is, the more objects it can potentially identify, the more time will be saved by scanning the most likely interpretations first.

The problem is how to specify the set of most likely objects. It is clear that the identification of a chair in your living room will probably be much faster than it would be if you and the chair were in the middle of the Gobi desert. But how does the visual system know that you don't expect chairs in the desert, but do expect them in your living room? This function would seem to require an input from some system which had knowledge of where you were and, at least, access to information about what had been encountered in similar situations in the past. The data available from the current perceptual world model, plus learned associations are sufficient. Note that, in a fashion similar to that discussed for the goal system, this would not constitute a true look-ahead procedure, but only one in which certain categories of objects had, through learned association, achieved the status of "best first guess." This is a kind of "secondary value" system for perceptual analysis. The effect of context on perception is in fact so powerful that under conditions of marginal input, you can (erroneously) see what you expect to see.

A similar class of perceptual feedforward function is relevant to the present context of goal-directed functions: a high command system, which has some goal in mind, can instruct the perceptual system to be attentive to features of the world which are important to the achievement of the organism's goals. This kind of influence can be exerted at all levels of the input system and can direct the active process by adjusting the center of gaze at selected features or objects. On a finer scale, it can adjust the center of "attention" within the area of the center of gaze (as by manipulating receptive field size and location in the manner discussed in the case of the cells of the infero-temporal cortex). In the case of the static process, it could actually manipulate the tuning characteristic of the input units. Notice that such a process is essentially a definition of what I have called "attention," at least in so far as it concerns the perceptual process. Thus "attention" is seen as essentially a feedforward mechanism which is active in preparing the system for maximum rapidity or selectivity of response to a set of features deemed to be important by some higher center that has access to goal system information. It is still not a true look-ahead system, but it does employ a present set of goal statements to define a set of "perceptual goals" that maximize the system's efficiency with regard to its current goal structure. In a simple reaction time test, a person will show a faster reaction time if he knows that only one stimulus may occur, but will be slower if he knows that there are other, incorrect stimuli that *might* occur. Only one stimulus and one response occurs in either case, so we assume that the difference must be due to the system being able to use the information about possible events to pre-adjust the response mechanism for maximum efficiency consistent with correct performance.

This feedforward control of processes in the input and other systems is exerted by portions of the prefrontal cortex and is manifested throughout the brain as an inhibitory effect which can suppress what is irrelevant. In its most general form, feedforward selectively controls local arousal which is exerted by the prefrontal cortex over all forebrain processes, including those of the input system. That is, given a state of arousal or alertness to begin with, it is possible to channel this activation into particular operations. The functioning of this mechanism is what we experience subjectively as "attention."

In figure 7.4, notice that the projections of the reticular formation are shown as having a facilitative effect on a relay neuron in the thalamic nucleus (lateral geniculate) which is part of the pathway from the retina to the visual cortex. It does this by inhibiting a set of neurons in the so-called "non-specific" nucleus of the thalamus which in turn normally inhibits the sensory relay neuron. Activity in the reticular formation thus promotes functioning in the visual perceptual system, as it does in other brain systems, and lack of reticular bias tends to shut it down. Now notice the input from the prefrontal cortex to the cells of the "non-specific" thalamic nuclei. These have the opposite effect; they counteract the reticular drive and shut down selected systems. In a similar fashion, the prefrontal cortex and the reticular formation jointly control the operation of most forebrain systems. The way the process is arranged, the prefrontal cortex does not directly activate any of the system, but it can selectively oppose reticular activation. This means that, given a general state of arousal activating the forebrain, the prefrontal cortex can manipulate the operation of various systems by cutting them out of the reticular activation or permitting them to run. Thus, the prefrontal cortex can "carve out" a pattern of specific

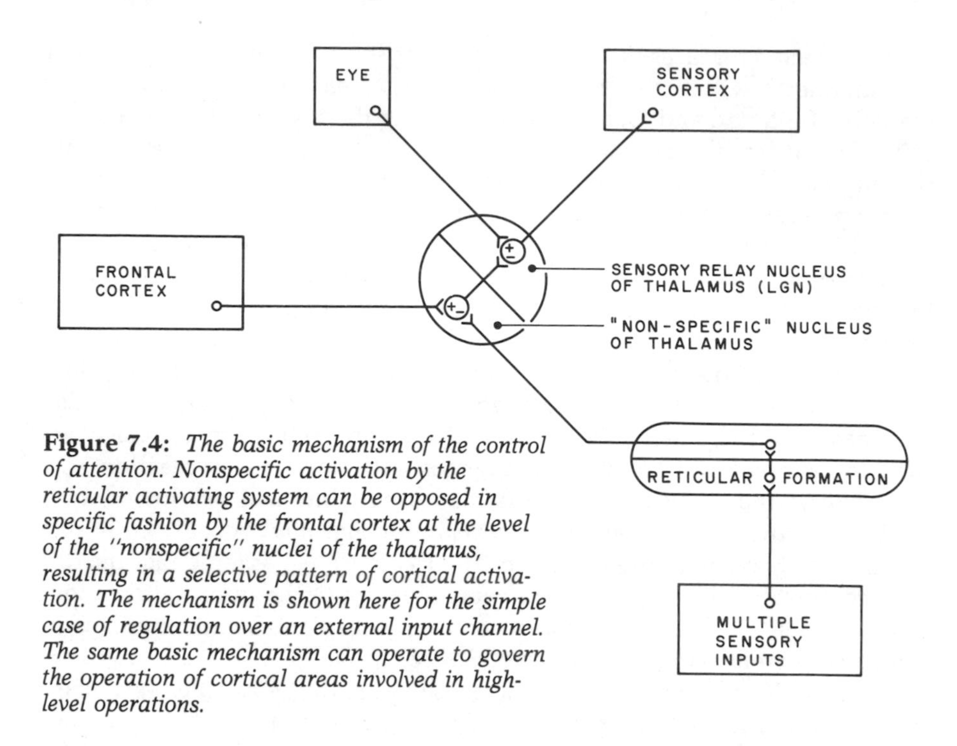

Figure 7.4: *The basic mechanism of the control of attention. Nonspecific activation by the reticular activating system can be opposed in specific fashion by the frontal cortex at the level of the ''nonspecific'' nuclei of the thalamus, resulting in a selective pattern of cortical activation. The mechanism is shown here for the simple case of regulation over an external input channel. The same basic mechanism can operate to govern the operation of cortical areas involved in high-level operations.*

processing actions by selectively inhibiting what is not wanted. It is thus capable of orchestrating the operation of the higher processing centers, given basic arousal of the brain by the reticular system.

This process of attention has its primitive origins in a hard-wired operation called the ''orienting reflex'' which shifts attention and has motor components which orient the receptors to any strong novel stimulus. You experience this primitive mode of attention focusing when you reflexively spin around to face the source of an unexpected sound. At a somewhat more advanced level, the prefrontal cortex can make use of the information from the motivational system to decide what stimuli or systems are ''relevant'' at the moment and focus processing activities accordingly. At the most advanced levels of operation, focusing of attention may be directed by the results of logical operations. Whatever the source of the decision, the process is similar. The mechanisms of figure 7.4 are somewhat simplified for clarify, but the principle is preserved.

The prefrontal cortex is one of the most advanced parts of the brain system, and one of the roles it plays is the organization of the activities of other systems to achieve high-level operations. Presumably, the reason it operates by opposing reticular drive, rather than activating the forebrain directly, is because it is itself activated by the reticular system under the control of the basic drive states or strong sensory inputs. There may be no reason other than evolutionary sequence for not combining the two functions, although the initial general activation as well as the response speed of the reticular formation are a safety feature.

There is a general inhibitory system which can oppose reticular activation at all levels, and the prefrontal cortex "attention control" is actually best viewed as the highest level of input to this general inhibitory system. Other lower processing units can contribute inhibitory control to this system at appropriate levels, and the actual pattern of activation of the brain at any instant is the result of an ascending activating influence from more primitive levels being modulated by a descending pattern of selective inhibition from more advanced levels. This is part of the general "fail safe" pattern of control that was described in the first chapter. Loss of high command does not incapacitate lower centers; it frees them to operate independently as best they can by removing this descending inhibitory control. This may also be an argument for the division of function between the reticular and prefrontal systems. If the higher centers are busy, the lower centers are simply temporarily deprived of higher control and are left to "mind the store" for a moment in a less optimized, more broadly tuned mode.

For refined types of control over input functions, the prefrontal cortex can apparently influence receptive field organization. The effects of stimulation of the prefrontal cortex on recordings of the receptive field structure of low-level units in the visual pathway show a pronounced influence of such stimulation on the receptive field which is opposite in nature from the effects of infero-temporal cortex stimulation. The prefrontal cortex is also associated with the "frontal motor eye fields," which are areas of motor control over the direction of gaze. These serve the function of "voluntary" eye movements, but that may be just another way of saying that they are employed to bring the system to examine our point of attention.

The ability of the prefrontal cortex to exert selective inhibitory control over output motor function is also of great interest. This control has enormous range and power; its outputs reach to the level of motor suppressor areas of the pons which in turn exert direct control at the level of the lower motor neurons! Although it can "carve out" patterns of permitted operation throughout the motor apparatus, it is not immediately clear why such a capability should be of value in a feedforward system. It can probably be best conceptualized as a kind of "motor attention" system which prepares the organism for rapid synthesis of specific goal-relevant actions. In a true look-ahead system, however, such a system is a prerequisite for the inhibition of responses to current world states while evaluating the future consequences of such responses.

VIII. LOOK-AHEAD OPTIMIZATION

What we see up to this point is the improvement of system performance in three major areas, motivation, input, and output, by the introduction of feedforward mechanisms which can prepare a "best guess" configuration of the system. This "best guess" is a preparatory response based on learned associations of present external stimuli or internal goal states with certain probable future actions or stimuli. The implementation of this feedforward preparatory process requires a supervisory system which can achieve a functional restructuring of all other systems. This "anticipatory system" is itself under the control of the motivational system, but can in fact be thought of as the highest level of that system because it restructures the other systems on the basis of goals defined by the motivational system. The prefrontal cortex, with its intimate connections with the limbic system and the motivational areas of the lower brainstem, and its profound control over I/O systems, as well as its extensive inputs from all higher level cortical processors, probably evolved as the high end of such an anticipatory processor. Subsequently, it has developed further along these lines in a way which permits it, in conjunction with other systems, to function as a true look-ahead rational processor, at least in its most developed form in man, and to operate probably to a more limited extent, in higher mammals. I will now review the processes by which this kind of operation may be carried out and examine some of the evidence.

In understanding the action of the prefrontal cortex with regard to true look-ahead operation, it is important to bear in mind, once again, the principle difference between the nature of the problems the brain faces in everyday life and the type of problem typically assigned to traditional computers. That is, computers are designed, literally, for problems like computation, performed via algorithms which arrive at THE solution. On the other hand, brain problems are typically those which require the attainment of a satisfactory solution which is "good enough." They must do it in a complex problem environment which requires a heuristic rather than algorithmic technique and which permits multiple possible paths to sufficient solutions.

Effective action in such an environment requires characteristics which, in human terms, would be called "judgement" and "insight." What judgement really means is the ability to choose the optimal approach from among the possible ones, and the ability to choose the best stimuli to attend to from among the many attractive ones. Insight, on the other hand, means the ability to "see" meaningful relationships between stimuli and events which are somehow related in a useful but not necessarily direct or even logical way. By the skillful exercise of these two abilities, a brain can function effectively in a heuristic problem-solving environment. If it can exercise judgement, it can choose the best of many possible paths and signposts available to it. If it can employ insight, it can identify relationships which have a high *probability* of indicating fruitful paths for heuristic approaches, even when there is insufficient data to be certain that the path is the correct one. Finally, it must be capable of predicting outcomes of actions without actually implementing them, so that it can engage in building models of future time for evaluation. While a traditionally designed computer might conceivably be programmed to function in this fashion, it is clear that no one has yet been able to design such a program capable of dealing with complex problems in reasonable time. In the following sections of this chapter I shall present a proposed functional, and where possible physical, model of the way in which the brain may accomplish these things. It should be understood that, on the available evidence, the model must be considered speculative.

IX. IDENTIFICATION OF TEMPORARY GOALS

The first major advance which the prefrontal cortex apparently makes over the optimizing feedforward mechanisms already described is the ability to command the temporary storage of certain aspects of its internal code. It apparently can command the temporary storage of information on the current state of the motivational and reward systems, and it can also command the temporary storage of information on the current states of its inputs from the perceptual and motor systems. The temporary retention of the motivation and reward states has been termed "allassostasis."

There are many forms of memory in the brain, including the so-called "short-term memory" which seems to encode current events for a brief period of time in a rather transient and easily disrupted fashion. It is not clear whether the temporary working storage of the prefrontal cortex is associated with this classical short-term memory or not, and it does not greatly matter here; all we need to consider is that the prefrontal cortex has access to a "scratch pad" memory system.

The event which causes the prefrontal cortex to trigger this information "hold" request is apparently the appearance in the perceptual input of events which are target objects of the attention-focusing processes initiated by the prefrontal cortex as part of its function in feedforward optimization of input processes for detection of drive-state relevant events. It is as if the lower motivational system tells the prefrontal cortex what is interesting, the prefrontal cortex adjusts the perceptual system to "attend" optimally to those stimuli, and then, when they are encountered, the prefrontal cortex "takes a snapshot" of the relevant accompanying details. This function, like so many others in the brain, initially evolved in the service of mapping the environment, in this case for motivationally relevant objects. If the brain can store the current world map at the time of sighting a goal object, it can then maneuver out of sight of the goal for a brief period in order to arrive at the goal later. Without such a mechanism, the object is "out of sight, out of mind"; more precisely, it is no longer included in the world model when it is not available to the senses.

The temporary retention of drive state information (allassostasis) has a similar original function. Grueninger and Grueninger (see bibliography) cite the following example: "A small mammal is about to emerge from the shadows of dense undergrowth into the sunlit glare of a clearing. Through the leaves it catches a momentary glimpse of a well-known and feared predator. Thanks to the "biological buffer register" the momentary flicker of sensory information is sufficient to produce and maintain the allassostatic response, which prepares it for maximal flight. It runs unhesitatingly to a nearby hiding place, ignoring the tempting berries it happens to pass en route. Safe and out of sight, its heart rate slows, its blood pressure returns to normal, and finally it relaxes its vigil and starts to groom. Picture the same animal without the "biological buffer register" to rely upon. The momentary sensory flicker would produce no lasting shift in the homeostatic systems. Habit might make it turn and start to run to the hiding place, but the sight of the berries would distract it and it would begin to eat, remaining in plain sight, awaiting the predator's pleasure. In this model the prefrontal cortex, as part of the biological buffer register, maintains the allassostatic readjustment, as well as the related behavioral plan until the appropriate physiological or behavioral goal has been achieved."

The immediate evolutionary advantage of such temporary buffer mechanisms for both spatial maps and motivational states is obvious. They represent a first step away from stimulus-bound operation by permitting temporary maintenance of goal-object location or motivational and reward system status while out of contact with the stimuli themselves. However, they also open the door to quite a different level of operation. Consider what has, for other purposes, now been acheived. We have acquired, associated together in temporary storage, information about which drive state specified the object of interest that triggered the attention-directed perceptual search, and information about the other objects, locations, and events physically associated with that object at the time of the observation. This is just the information required for the employment of *temporary* goals of the sort mentioned earlier. To take a concrete example, let us suppose that I am motivated to acquire money and am therefore attentive when I see you take out your wallet (a previously learned predictor, illustrating simple motivational feedforward). At the sight of you putting the hundred dollar bill into the red envelope, my prefrontal cortex takes its "snapshot" on the basis of the appearance of the bill (conditioned goal object), and this includes the red envelope (*particularly* the red envelope, because "contains" is a very powerful and highly predictive spatial relationship which I have learned to attend to early in life). The motivation (greed) once associated only with the bill is now *temporarily* associated in the "scratch pad" with the envelope. The system can thus proceed to employ the red envelope as a goal stimulus in further processes of attention, stimulus-bound approach, or more advanced learned strategies. Current drive state "greed" now points to "red envelope" in some temporary scratch pad. The process could be extended to higher orders of temporary value assignments as far as your neural mechanism would allow. In the case of humans, this is quite far. You put the red envelope in a brown one, and that into a box, which you put under the bed, and so on. Ability at this sort of problem increases in direct proportion of prefrontal lobe development as one ascends the phylogenetic scale. In solving the problem, note that I am using a primitive form of "judgement" with regard to what stimuli to attend to, and I am, in one simple way at least, operating in future time.

The major advance in this mechanism is the procedure for forming these associations on a temporary basis. Any lizard with no prefrontal cortex could *learn* to respond to envelopes in a stimulus-bound fashion if they were very frequently associated with food. Envelopes would simply become conditioned reinforcers for the beast, but the response would be "blind." If the lizard were so conditioned and then saw you put the food into a box instead, it would still go to the empty envelope because the food was out of sight and the envelope was a conditioned reinforcer, not a temporary one.

How can temporary goal stimuli be employed in a rational process? For the simple response just considered, all that is necessary is for the prefrontal cortex to supply the temporary goal to the pre-motor cortex/parietal cortex mechanisms as the "desired world state" (WS5 of my earlier example). In advanced brains, the prefrontal cortex seems to be the principle source of self-generated goal states for the pre-motor/parietal mechanism. (Although as we shall see, they also can be supplied externally.) Anatomically, this appears to be a simple extension of the principle that the more anterior portions of the pre-motor cortex deal with the goal in its most global form.

Quite probably this function was originally served by direct input from the limbic system's motivation and reward mechanisms to the pre-motor cortex, in the area where the prefrontal cortex has subsequently developed to mediate between the two. Thus, the goal stages originally supplied to the action-synthesizing processors were limited to primary or learned goal objects recognized by the limbic system. The emergence of prefrontal mechanisms permits the insertion of temporary goal states into the process; it defines them by their momentary relation to the original permanent goal states.

X. OUTPUT INHIBITION AND EVALUATION

The second major contribution of prefrontal function to rational behavior lies in its ability to selectively inhibit output operations. Given our hypothesis that the pre-motor-parietal mechanism generates its scheme of output as a parallel-coded sequence of operations and projected intermediate world states, the prefrontal cortex can have the whole sequence available for examination before permitting its translation into serial action sequences. This means that the intermediate states or actions can be evaluated for potential significance by the limbic "pleasure-pain" detection system. For example, the nonemotional, pre-motor action synthesizer might generate a "frightening" intermediary state, and the sequence could be rejected before execution. This is another example of simple "judgement." In addition, and more importantly, the *final* projected world state can be evaluated in terms of its similarity to the stated goal before execution is allowed to decide if the proposed solution path is "good enough." It results in a state sufficiently similar to the desired state to produce a strong pleasure response from the limbic system. The more strongly a projected world state resembles a goal state which can produce a pleasure or pain response from the limbic system, the more relevant features they will share, and the more strongly the "emotional feature extractors" will be activated by the projected state.

If the prefrontal cortex could compare these projected states with a *temporary* goal representation and, when a match or partial match was found, activate the reward system, then temporary goal states could be used in the same fashion to judge the sufficiency of solutions "proposed" by the pre-motor/parietal system. Connections from the prefrontal cortex to the final output elements of the reward system have recently been identified, and it appears that stimulation of these connections has the same results as stimulation of the older "reward centers" in the limbic system. Hence, the prefrontal cortex may thus have the capability to judge the adequacy of the projected state (resulting from a projected action) in terms of a temporarily defined goal state, and to initiate the projected action if a sufficiently good correspondence is found.

The operation to this point gives two principle ingredients of rational behavior: the ability to exercise "judgement" in selecting adequate approaches to temporary goal states, and the ability to investigate and reject alternative solutions on the basis of that adequacy. This last feature is a necessity for a heuristic mode of operation which must try out and evaluate alternative paths to solutions. What gives the system real power is that it can compare projected world states, or consequences of projected actions, with temporarily defined-goal states internally without committing itself to action and actually trying them out. The ability to do this rests on the adequacy of the projections, which in turn rests on the degree to which the system has adequately learned the sorts of transformations in world states which will follow different sorts of action outputs. This is why ability at rational thought grows with experience, both in our general operation in the world and in the development of expertise in any particular type of endeavour.

The discussion so far has been in terms of nonverbal representation of world states and actions, or, in other words, rational thought of a concrete form, suitable to real-world problems. However, a system with sufficient power to handle symbolic representation of world states and actions can deal with the symbols in essentially that same fashion as the states and actions themselves. Symbols can serve as temporary goal states and may be compared with the symbolically stated projected results of symbolically represented actions. The process is one of learning intermodal equivalents, as discussed earlier, and one of the chief benefits is the reduction in the amount of neural activity required to represent a state or action. This in turn permits the operations to be carried to a higher level. This mechanism is principally valuable in producing verbal or symbolic solutions to problems with multiple "good enough" solutions, just as in the non-symbolic case. When true algorithmic solutions are required, as in the solution of algebraic equations and the like, we must simply fall back on the serial execution of learned algorithmic steps and mimic in our own slow way the action of a traditional computer.

In point of fact, the mechanism so far described is only adequate to "solve" problems in which the definition of a temporary goal state gives the pre-motor or parietal cortical mechanisms a goal which can be reached by synthesis of some action plan. The success of the attempt will depend on whether or not the system has available a learned set of associations between actions and perceptual transformations, and between required transformations and actions, out of which it can construct a sequence resulting in a satisfactory approximation to the goal state. It is essentially interpolative. If the defined-goal state does not trigger an adequate solution sequence, or if the problem is so complex that the temporarily defined-goal state does not represent a full solution, the problem is not solved. In this case, one must resort to trial-and-error procedures to get to the temporary goal state and, if that does not represent a full solution, to define a subsequent goal state. Simple trial and error is not a very satisfactory state of affairs in complex problems; it would probably eventually produce a solution, but it might take intolerably long. This state of development essentially represents the limits of action of sub-human brains.

XI. INSIGHT

What is still missing here is the "insightful" aspect of brain operation. The system, as modeled so far, is indeed capable of judgement, simple rational behavior and solving problems in a heuristic manner. It is not very "bright" as there is no adequate mechanism for making good guesses about which of the possible paths of a heuristic solution procedure are most worthwhile for early examination. This is the problem that plagues activities such as chess playing on traditional computers. You either have to consider all the possible moves in turn, to whatever depth you want, and evaluate all the resulting positions, or you have to select a few that you want to try, according to some sort of algorithm based on the current situation. In situations less structured than chess games, the problem gets even worse because the rules get fuzzier; it becomes unclear what might indicate a "best guess" about promising paths to follow in the search for adequate solutions.

The problem is that there may be no logical way of deciding what constitutes the best guess at a trial solution. That is, there may be no algorithm for making such a decision, or, as is more usually the case, there may simply be inadequate information. In complex situations, the problem will more likely be one of discovering possible avenues of approach, let alone deciding amongst them. When the human brain tackles such a situation, it goes through some procedures which we experience subjectively as "considering" or "examining" the problem situation. We "turn it over in our minds,'' until quite suddenly we have a ''flash of insight,''; we ''see'' a possible path to solution and then proceed to work it out, verify it, and test its consequences. Much of this can be understood on the physical level in terms of mechanisms already discussed. The prefrontal definition of temporary goal states and focusing of attentional processes are the ''examination of the problem'' and correspond to the subjective phase of ''turning it over in our mind.'' Synthesis and evaluation of intermediate and final projected states in the path to solution correspond to the phase of working out the details of the proposed solution path once it has been identified. The missing link is the ''flash of insight'' which points to the action or goal state for further investigation.

In attempts at heuristic programming on traditional computers, various algorithmic feedforward procedures can be employed to maximize the probability that the next potential route to the solution that is evaluated is a good one. An example would be examining only moves that led to a capture or strong position in a chess program. In effect, what the brain is doing is quite similar, but it can employ the unique features of its basic architecture to do this in a way that is extremely powerful.

Basically the brain is screening the possible organizations of the current perceptual world state for patterns which are ''similar'' to the desired temporary goal state (or for temporary states which are ''similar'' to whatever is known about the desired state or solution, in the more complex case where the temporary goal state does not define the solution). This ''similarity'' is the feedforward algorithm. One tries for intermediate states that are most ''similar'' to the desired goal state. It sounds trivial, but it is not. Consider what ''similarity'' might mean in something as complex as the real world, or in a symbolic, logical environment. (Which is more ''similar,'' a red square and a green square, or a red square and a red circle?) How can you get a quantitative measure of similiarity between complex patterns, when the relevant similarities may be in quite abstract properties of those patterns? What we are asking the brain to do is to recognize among the many patterns of the current state of affairs those that bear resemblance to the temporary goal state, as defined along some important conceptual dimensions.

In the brain, a key to doing this may lie in the feature of its basic elements which I earlier nicknamed the ''ALMOST gate,'' and in the fact that the principle of selective convergence in the input system has already been made according to (or perhaps more properly, *defines*) our important conceptual dimensions. The ALMOST gate feature, you will recall, results from the fact that a neuron, unlike an AND gate, need not necessarily have all of its inputs active in order to fire. On the other hand, unlike an OR gate, it requires a substantial number of them to be active in most cases. The production of an output pulse from a neuron depends then on some *percentage* of its inputs being activated. The result is to allow a neuron to fire in response to a number of possible patterns of activation among its inputs. Given that the neural inputs to a feature extractor in the perceptual system will be related to one another in specific ways by the principle of selective convergence, all of the patterns of input which can fire it will have the relationship of "similarity" along some perceptual dimension. Let's say I have a hypercomplex feature extractor that will be maximally driven by a circular shape in its receptive field. A pattern of dots and broken lines in a circular arrangement will fire many of its inputs, but not all. Clearly, such a pattern is "similar" to a circle in a way which relates, perhaps usefully, to an essential conceptual feature of a true circle. If the hypothetical feature extractor can respond to less than the absolute optimal input pattern, it may fire in response to the pseudo-circular pattern.

This, of course, is just the process of "generalization" that we have already considered at the lower levels of the perceptual system. However, when the same process is applied, by the same type of circuitry, at levels of the system where highly abstract properties of the world model are encoded, it can become a very powerful tool for recognizing relationships between temporary goal states and particular aspects of the current perceptual environment. The more extensive the system's set of "extractors" for abstract relationships, the more powerful such a process becomes in suggesting profitable lines of approach.

There are two basic processes that can be involved here: the first identifies similarities to suggest profitable intermediate temporary goal states for trial in situations in which the temporary goal state, defined by association with a reinforcing stimulus, cannot be reached directly by synthesis of an action plan. That is, the desired state is defined by the "snapshot" of the world state associated with some goal recognized by the reward system, but the attempt to generate an action scheme leading to the temporary goal state fails. The "best guess" is to define, as an intermediate goal state, some state which *can* be reached and which has similarities to the desired state. Once there, it may or may not be found possible to reach the desired goal state. If not, further intermediates may be sought by similarity, or backtracking may occur. This sort of approach has its origins in simple problems of spatial and temporal relationships and maneuvering in the environment, but with the increasing complexity of the abstract relationships that can be extracted by the system, and particularly with the development of symbolic representation of relationships, the process can become much more general and very powerful. This is the fundamental mechanism underlying the experimentally well-documented "Forward Search" strategy that is at the heart of the human approach to problem solving.

The second problem that can be solved by the similarity "algorithm" is one in which the goal state is itself a recognizable rewarding state, and does not need to be defined by temporary association with a reinforcer, but the route to the goal cannot be synthesized in a direct manner. An elementary example of such a situation might be the case where the organism is hungry, but there is no food in sight. In an animal with an advanced prefrontal cortex, the hunger stimuli can act as cues to recall from memory lists of situations associated with food in the past, and these can serve to define temporary goal states which can be compared with the current situation in a search for similarities. Again, this mechanism has its roots in spatial search in the physical world. However, in a sufficiently advanced brain, the basic mechanism can be operated with more abstract goals and relationships, again even symbolic ones, and the goal state can be defined on other bases than past associations with basic drive states.

In this second process, an important feature is the production of a complex set of usable world states representing potential intermediate goals from some, possibly very simple, cue (such as hunger) on the basis of a memory process. Although we know very little about the brain's memory mechanisms, one well-known, and fundamental feature, is that the memory is ''associative.'' That is, a single item can serve to address (or ''recall'') not a single memory item, but a whole host of items which are associated with *or similar to* the cue in a variety of ways. If you know someone named John Smith, the sound of the name will bring to mind a great many different items: what he looks like, where he works, what his wife looks like, or the last time you went out with him.

It seems likely that the ALMOST gate feature of neural operation that underlies the similarity extraction process might also be the mechanism behind this associative recall process. That is, if an item in memory is The Color of John Smith's Eyes, you could recall that color by the exact cue ''the color of John Smith's eyes.'' It is clear that, however the recall is done on the basis of that cue, the partial cue ''John Smith'' can also bring the color ''blue'' to our awareness. It may be that the cue ''John Smith'' serves to activate some, but not all, of the inputs to recall the memory of his eye color, but this will be sufficient to activate the relevant circuits if the neuron's biasing level is set permissively. In any case, by whatever mechanism it is accomplished, it is clear that associative memory processes are a very important part of neural functioning, particularly in problem solving, and that the development of such memory systems in computer hardware will be a great step towards the development of brain-like operation. So-called ''associative'' memories are already available for computers, but at the present time these only function to address items in a location-free manner by specifying a ''tag'' code (perhaps the first few bytes of the item) which identifies the rest of the material to be returned. It would not be a great problem to implement an ALMOST-gate type of action into the circuitry recognizing the ''tag,'' so that items with *similar* tags would be returned as well. At any more brain-like level of development, the stringency of the similarity test could be a variable which is set appropriately for the task at hand. The difficult part in developing an artificial system would be to insure that the dimensions of generalization (similarity) were conceptually important ones, and that these were also the dimensions of the recall code. These problems may be handled in the brain by the way in which the perceptual code is developed on the one hand, and the use of the perceptual code as the storage code on the other.

The process I have now outlined consists of the ability to selectively attend to goal-relevant features of both input and internal states, to suppress output pending evaluation of projected outcomes of actions, to define intermediate goal states on the basis of similarity to goal states, and to define temporary goals on the basis of current or past associations with goal-relevant stimuli. This defines the basic equipment of the brain for rational thought. If it seems too simple to account for our reasoning processes, bear in mind that I am talking about the hardware. Registers and half-adders don't seem so impressive either, until you start running programs.

In fact, most of the real power of advanced rational thought is developed through the learning of strategies of approach to problems, and learning to recognize *similar types* of problems. Note that learning problem types is essentially a process of learning to recognize important similarities in the situation and to use this recognition to recall previously learned successful lines of approach. A great deal of the superiority in reasoning power that humans enjoy over animals is due to our enhanced cortical capability for symbolic representation, and this is not valuable only because complex items can be reduced to simple expressions for purposes of manipulation; discovered solutions may be recorded so that they can be taught to successive generations who will then not have to waste time rediscovering them, but can use them as a learned base on which to build. It is estimated that the basic human capability at rational thought and foresight, if recourse to symbolic representation is denied, is just about sufficient to play Tic Tac Toe. The limiting fact is the amount of information with which we can deal at one time in short-term (temporary) memory.

As I have said, the basic principles of operation are surprisingly (or perhaps not surprisingly) simple. However, in attempting to apply them to computer or robotics technology, there is a problem. The real power of the brain system lies in the similarity recognizing feature inherent in the firing of neurons by partial input patterns, and this in turn is only a valuable method where (1) the pattern of inputs to the element is already so defined that partial patterns are necessarily related along conceptually important dimensions, and (2) the capability for massive parallel processing exists. The essence of the problem is that you don't usually know what dimensions of similarity are going to appear, and therefore you must search them all. In the kinds of problems brains handle well, all of the data for the search are simultaneously available internally or externally to the system; with sufficient simultaneous parallel processing power, the search can be done very rapidly. In complex problem environments, it would be a practical impossibility to employ serial search techniques to such a quantity of data. Therefore it would appear that here, as in other features of brain function, adequate artificial analogs will have to be based on massive parallel processing approaches.

XII. EFFECTS OF PERFRONTAL DAMAGE

In animals with experimental damage to the prefrontal cortex, a variety of deficits in problem-solving ability appear. In particular, they are deficient in performance of delayed response and alternation tasks. In delayed response tasks, the animal is permitted to observe a reward such as food being placed in some position, say under one of several bowls, and is then required to wait for a period of time before being allowed to search for the food. In alternation tasks, the animals must make one choice on one trial and the opposite on the next trial in order to acheive the reward. All of these tasks require some ability at representing the relationship between the environment and anticipated rewards for a period of time.

It appears that memory *per se* is not the problem, and various workers have attributed the disabilities to difficulty in maintaining a conceptual problem in the face of new information input, and to difficulty in attending to relevant cues in the situation, increased distractibility, and the like. Any or all of these explanations might fit, and it is not clear that they are really all that different. Such animals also experience difficulty in situations where the solution requires them to inhibit an immediate response to a stimulus, and this too has sometimes been seen as a failure to ignore the irrelevant through failure of an attention mechanism and sometimes as loss of the ability to prevent responses.

Some of these explanations focus on problems in input control, some on problems in output control, and some on problems in control of goal-direction in processing. This basically seems to reflect the fact that different investigators have been impressed with different aspects of the prefrontal cortex's control. Undoubtedly, some types of problems will emphasize some aspects of this functioning, while others will emphasize different aspects. Therefore it would be surprising if any one aspect of prefrontal function provided a complete explanation of all the problems produced by damage to this structure. This is particularly true given the fact that the prefrontal cortex consists of several distinct regions with different connections and most probably different sub-functions, and that experimental results which are being compared do not always result from comparable damage. The deficits which have been suggested for animals with prefrontal damage include the loss of: the ability to inhibit responding to irrelevant stimuli; the ability to inhibit premature output; the ability to compare consequences with goals; the ability to formulate plans of action; the ability to attend to or recognize crucial details. Any or all of these can be understood in terms of damage to the model of pre-frontal function which we have considered here, and which one is prominent in any given case probably reflects a combination of the particular task and the exact mechanisms damaged.

In humans with prefrontal damage, similar sorts of disability at problem-solving tasks are encountered, as are drastic changes in personality and temperament. In the area of intellectual function, such patients are unable to form a strategy of solution to a problem. They skip the orienting phases of reflecting on the requirements of the task and proceed immediately to whatever feature first comes to hand. They do not seek out relevant cues and, very interestingly in terms of our hypothesis on the role of attention in perception, their eye movements do not reflect orientation to relevant cues. Luria has compared the eye movements of normal people with those of prefrontal damage cases when the subjects were looking at a picture, and being posed different questions about the picture. Normal subjects show an orientation of gaze in each case that depends on the kind of question asked about the picture and which reflects attention to the relevant set of details. Prefrontal damage cases have fixed patterns of inspection of the pictue which do not change with the kind of question asked.

Prefrontal patients are not deficient in strictly logical processes as such, but cannot operate in situations where the problem requires choices among several probable alternatives. They operate in randomly selected logical schemes, and are unable to make goal-directed choices from among the alternatives. They do not attempt to verify hypotheses or to compare the consequences of an action with the desired consequences. In other words, they seem unable to direct their logical processes in accord with the goals involved.

The personality changes which appear in persons with prefrontal damage reflect the same sorts of disabilities applied to the problems of everyday life. They become impulsive and irresponsible. They act immediately on whatever chance emotion happens to possess them. They become unable to forego momentary pleasures in favor of long-term goals. They become inattentive and easily distracted and unable to apply themselves to any task. They are continually losing the line of thought in their speech and wander from one irrelevant association to the next. In a sense, their behavior becomes child-like; they move quickly from violent rage to pleasure and seem to be more dependent on current input than on long-range plans in determining their behavior. This is probably a reflection of the fact that "child-like" behavior is seen in children due to a lack of learned strategies for implementing the rational capabilities of the prefrontal mechanisms, and the fact that the prefrontal cortex is among the last portion of the nervous system to fully develop, some pathways becoming functional only at about twenty years of age. (In general, the order of functional development of brain regions in the life of an individual follows the order of evoluntionary development of brain regions. The prefrontal cortex and the most advanced portions of the parietal cortex are very late evolutionarily and become fully functional latest in life.)

XIII. THE UNDERLYING ANATOMY

The detailed mechanics of the operation of prefrontal mechanisms are obscure. We know from a combination of effects on behavior and on the operations of other brain regions essentially *what* is being done, but just *how* is not clear. There are at least two major divisions of the prefrontal cortex which can be discriminated on the basis of their connections with the rest of the system and on the basis of their anatomy. These are depicted in figure 7.5. The portion labeled "DL" includes that part of the pre-frontal cortex which lies in the forward and outside portions of the prefrontal cortex. It has extensive connections with the perceptual systems of the cortex and is apparently the region involved in the control of attention. It is interconnected with the ''non-specific'' nuclei of the thalamus and acts as an ''input controller'' for the brain, providing feedforward to perceptual systems and inhibiting, perception of irrelevant stimuli. Given this relation to input processors, it is probably also the system responsible for initiating or controlling the temporary storage of goal-relevant input states, but because we know little about the storage mechanism, this is only speculative. I have indicated this function in figure 7.5 but marked it ''?'' to indicate the speculative nature of this connection. The other portion of the

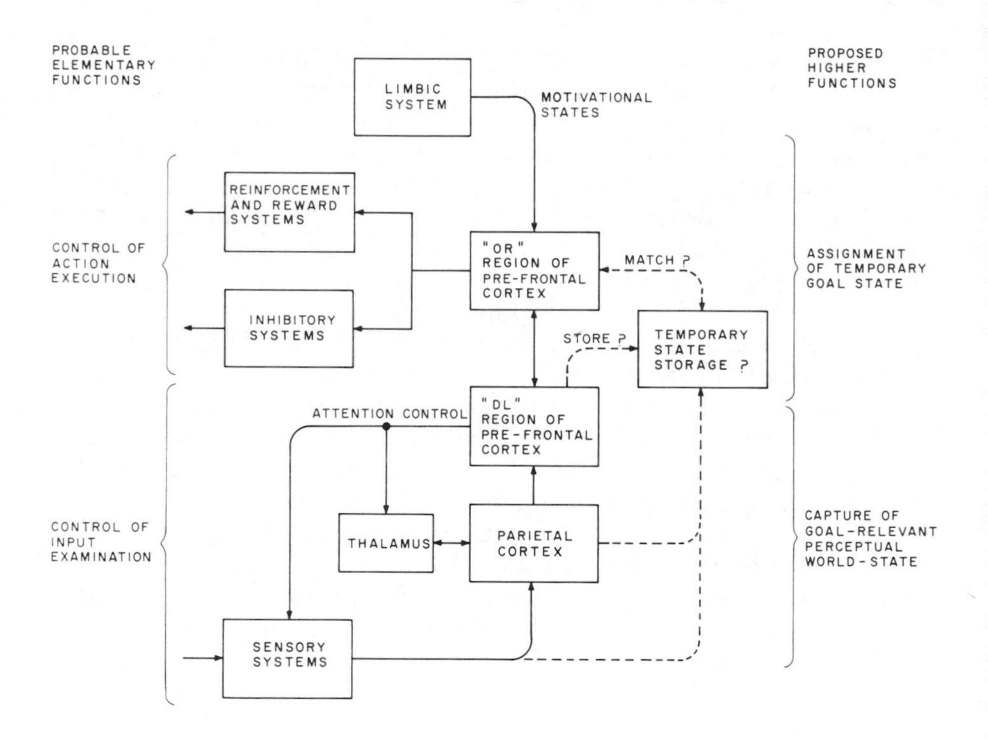

Figure 7.5: *A hypothetical model for the action of the prefrontal cortex in the control of goal-directed behavior. The two major regions (orbitofrontal* = OR and dorso-lateral = DL) are probably involved in the control of execution and the control of attention. Both of these functions are executed in the served of the motivational states expressed by the limbic system. In higher level action according to the proposed scheme, the mechanism not only locks onto immediate goal objects that can direct current action on the basis of innate or learned relations to drive states, but can also generate ''temporary goal states'' on the basis of momentary association between an innate or learned goal object and a neutral perceptual object (see text).

prefrontal cortex shown in figure 7.5 is marked "OR" and represent portions of the prefrontal cortex on the bottom surface of the brain and on the portion lying between the two cerebral hemispheres. This region has extensive connections from the limbic system and is the site of origin of two powerful output-control systems. It is the source of the extensive motor inhibitory connections which enable the inhibition of action of all levels of the system, and it is the source of the prefrontal cortical inputs to the output-enabling portion of the reward recognition system via connections to the mesencephalon. It thus has the appropriate circuitry to function as an output controller, either initiating or inhibiting actions. It also has inputs from the motivational system which can provide information about when such actions are appropriate.

The two systems of the prefrontal cortex probably represent initial lines of development of the feedforward function. They are heavily interconnected, and it is presumably in these interconnections that the real power of the system as a look-ahead processor develops. One supposes that in these interconnections we see the means by which the attention directing mechanisms are placed under control of the motivational system, and by which motivational cues or recognition of reward stimuli can trigger the storage of information for use in temporary goal definition. These interconnections also are well placed to permit the projected states originating in the pre-motor-parietal logic system to undergo comparison with goal states leading to activation of the reward system and ultimately the output controller. The cellular mechanisms of the prefrontal cortex tissue are the most likely site for our postulated "ALMOST" gate mechanism of comparison for similarity in abstract perceptual dimensions.

8 Hemispheric Specialization and the Higher Functions

I. THE NATURE OF THE PHENOMENON

In the preceding chapters, I have dealt with the higher cortical functions without mentioning the problem of hemispheric specialization. This was done in order to present some of the basic concepts of the higher functions before introducing this additional complication. I have dealt with the verbal logic and arithmetic functions, and with the spatial perceptual functions as though they were always performed in the same place and, indeed, I presented the former as dependent upon, or at least arising out of, the latter (following the argument of Luria). While I believe this to be fundamentally correct, it is nevertheless time to grapple with the fact that in the adult human brain there is strong evidence that the performance of these functions may be separated anatomically.

The evidence, which will be reviewed in more detail in a moment, basically suggests that in the adult human, language-based higher functions are carried out in one hemisphere (the left, in most instances), while spatial functions (picture recognition, for example) are carried out in the other. A great deal has been written about this phenomenon in the popular press in recent years, and you may already have encountered some of it. Most of it is the wildest sort of speculation, based on extravagant expansions upon data which are in fact slim and fraught with difficulties of interpretation. Nonetheless, there is substantial reason to suppose that the left and right cerebral cortices of man may be performing different functions.

In terms of the concerns of computer and robotics designers, there are several points of interest about this apparent division of labor. Some of the questions include:

- Are the observed functional differences the true differences in action between the hemispheres, or are they merely representative of some deeper functional distinction?

- Are these functional differences due to specializations of the "hardware" of the hemispheres for these kinds of activity, or do they represent only the assignment of incompatible functions to different regions of basically equivalent machinery?

- Why has the brain found it expeditious to separate these functions?

- What, if anything, can the nature of this division tell us about the underlying nature of cortical information processing?

It might be that the cortical structures on the two sides are fundamentally similar, but that there is some functional incompatibility between the two kinds of processing. A possibility is that one type of process may deal with information about the way in which the world model is changing with time, while another is dealing with the way in which static relationships are organized in space. It is possible that the two processes might interfere when they tried to utilize the same portions of the neuronal apparatus even though the brain is fundamentally capable of either type of operation. On the other hand, we could postulate that the linguistic functions were in some sense more highly evolved than the spatial-perceptual functions and that the hemispheric division of labor represented something analogous to the evolutionary process of "encephalization" in which newer, higher functions develop later.

There is some reason to accept this latter view. We have already postulated that the linguistic or symbolic operations are developed out of an existing older mechanism for spatial-perceptual analysis of relationships in the world model. If new hardware developments are found to enhance this later functional development, we might expect to find the brain evolving new, yet higher, levels of structure to accommodate the new processes. In the past, such development of more advanced structures has been in a "forward" direction and bilaterally represented. It is thought that an upper limit on brain size in relation to body size has been reached, and that this already requires such adaptations as massive folding of the cortex to achieve high surface area in a small volume. It is possible that lateral specialization of function might represent a new direction of evolutionary continuation of the progressive encephalization process. There would be disadvantages, such as loss of redundancy, but these might well be offset by the advantage of an entire "new" brain to work with.

In fact, it is not unlikely that we may be seeing an interactive combination of these effects working in concert. That is, it may be useful on purely functional grounds for the brain to divide certain jobs. Once this is done, it may be that actual hardware specializations in the two hemispheres can confer some advantage to one or the other of the processes, and this specialization may then be selected for evolution. There is some evidence that this may be occurring in the ongoing process of human cerebral evolution.

II. THE DATA

There are three primary types of evidence which pertain to the problem of lateral specialization in the human cerebral cortex. These include studies of the so-called "split-brain" operations, the effects of unilateral brain damage in the appropriate areas, and normal individuals using special techniques to assess the contributions of the two hemispheres. Each of these has its own particular difficulties of interpretation. For example, it is clear that the operation of one hemisphere by itself is not necessarily like its operation in conjunction with the other hemisphere. There have been only a handful of "split-brain" operations in history, and these have been performed as a last resort on patients whose brains were already in poor shape. Cases of damage to a single side suffer from all of the problems which were previously mentioned in conjunction with clinical studies of human material. With these cautions in mind, let us examine the major findings, which suggest lateral specialization in man.

The *corpus callosum* is the large bundle of fibers that runs between and interconnects the two cerebral hemispheres. The fibers which compose it are the axons of cortical neurons, and in general they run from cells in one cortical hemisphere, across the corpus callosum, to synapse on cells in the opposite hemisphere's homologous cortical region. To a first approximation, we may think of them as carrying information needed to coordinate the activities of the two sides. This fiber bundle is by far the largest of several such bundles called "commissures" which interconnect the two sides of the brain.

In the split-brain operation, this fiber bundle as well as three or four of the lesser commissures are severed. This operation, which in no sense actually "splits" the brain, is occasionally performed as a last resort in cases of life-threatening epilepsy. (Just why it helps is not crucial here.) The result is to sever the route of crosstalk between the two cortical hemispheres and between a number of subcortical structures of the telencephalon as well. On the other hand, most of the remainder of the brain is unaffected, and several indirect routes of communication remain between the cerebral cortices.

The superficial results of this procedure are negligible. It is generally not possible to tell by casual observation that a person has undergone such an operation. Despite the fact that the two noncommunicating parts of the brain control one half of the body each, there remains an absolutely astonishing cooperation of function in terms of general behavior. Indeed, one has to devise special situations to demonstrate that the two halves are capable of acting at cross-purposes. This superficial cohesion notwithstanding, it appears that the true result of such an operation is to produce two separate and individual conscious intellects dwelling within the same head!

The cohesiveness of the behavior can be attributed to the fact that only one of these individuals possesses much of the speech and language functions of the whole original, so that only one of the two conscious minds speaks for the pair. In addition, except under special experimental conditions, the two cortices have very similar sensory experiences from the receptors, because they share, for example, the common fixation point of the eyes. Further, their long-term memories, acquired prior to the operation, are the same (including learned inter-manual coordination). The still-connected lower brain centers, such as the cerebellum, act as a further unifying force by virtue of their roles in balance, coordination, and other housekeeping functions. There appears to be a unity of emotional experience, as we might expect from the shared operation of the lower limbic and diencephalic emotional and motivational systems (although one of the hemispheres may be in ignorance of the *cause* of the emotional response it is feeling under certain conditions).

On the other hand, when experimental conditions are properly attended to, it is possible to deal independently with the two hemispheres of such an individual. In a split-brain patient, it is possible to communicate visually with only one hemisphere, or in different ways with two hemispheres separately by projecting the visual communication onto the appropriate half of the visual field while preventing or controlling for eye movements. This is because the projections of the retina of the eye are arranged so that the image falling on the right half of the retina in both eyes is conveyed to the right hemishere's visual cortex, while that falling on the left half of both retinas is conveyed to the left hemishphere. (In the normal person the fibers of the corpus callosum would carry the information received by each hemisphere across to the other, so that each hemispherc would receive, directly or by callosal relay, information about the entire field.) In addition, at least with respect to voluntary or complex movements, the motor regions of each hemisphere control the limbs on the opposite side of the body so that independent responses by the two hands are possible in the split-brain patient.

By means of these and similar experimental manipulations, it is possible to show that there are indeed two conscious intelligent and separate minds operating within the same skull. It is most convincing to watch a film of such a patient carrying out two completely different tasks simultaneously! It can be demonstrated that one hemisphere can observe a stimulus and make a complex response to it without the other hemisphere having any notion of what is happening or what the stimulus was. It is possible to give opposite stimuli to the two sides so that they give opposite responses to the same problem.

In this fashion, it becomes possible to probe the intellectual capacities of the two hemispheres separately. When this is done in the split-brain patient, it is generally found that one hemisphere (typically the left side) is virtually unimpaired with regard to language functions such as producing and understanding speech, reading, and writing. In addition, this hemisphere has virtually unimpaired functions with regard to processes involving serial logical steps in reasoning, mathematics, and so on. It can easily master problems whose solution requires the manipulation of lists of items or naming of objects. On the other hand, the right hemisphere is severely deficient in these kinds of tasks. It shows only rudimentary verbal ability and has little skill at the kinds of problems at which the left hemisphere excels. However, it clearly possesses unimpaired ability to solve problems or perform tasks which require nonverbal reasoning, appreciation of spatial relationships, recognition of pictures, and patterns too complex or subtle for verbal description, and a host of similar, poorly categorized functions. These functions, in turn, are difficult or impossible for the "language" hemisphere to perform. Due to the difficulties of specifying the exact nature of the functions of the nonlanguage hemisphere, its true capacities are not well understood.

This picture of a "language/serial logic" hemisphere and a "spatial/parallel logic" hemisphere is borne out in various ways by studies in patients with damage to various cortical areas but who still have an intact corpus callosum. You will by now have recognized that many of the differences between the two hemispheres are in areas which are strongly reminiscent of the kinds of parietal lobe functions discussed earlier. Although I did not mention it then, it is indeed found that damage to the parietal lobe has an effect which depends on which side of the brain is involved. Two language "centers," one known as Wernicke's area (important in verbal processing) and laying at the temporal/parietal junction, and another, known as Broca's area, in the frontal lobe (important in the control of speech production) are found only on the left in most people. Damage to these areas on the left side of the brain produces profound deficits in verbal ability, while damage to the corresponding point on the right side leaves verbal ability largely unimpaired (although other functions, such as the ability to perceive musical form, may be lost). In general, the results of cortical damage tend to support the findings from the split-brain studies. Damage in left versus right parietal areas shows verbal reasoning deficits versus spatial reasoning deficits. Damage to frontal areas shows an interesting impairment in recent memory for order of occurrence of stimuli; verbal stimuli sequences are affected if the damage is in the left frontal lobe, whereas pictorial stimuli sequences are impaired if the damage is in the right frontal lobe.

An important point should be noted here. If the damage occurs in later life, these effects are striking. If the damage occurs early in life, little deficit is seen. Even in cases of complete loss of the left hemisphere in early life, the right hemisphere seems capable of taking over the functions of both. It is generally thought that the lateralization of function is incomplete until at least the age of five, and perhaps longer. A few cases of congenital failure of corpus callosum development are known, and these subjects appear to have a full set of intellectual capabilities on *both* sides!

There is also a procedure (the Wada technique) which is employed occasionally to ascertain the hemisphere of linguistic sensitivity in persons about to undergo brain surgery. An anesthetic substance is injected into the arteries feeding one side of the brain, which selectively depresses the operation of that side. Again, the results seem to support the idea that there is a division of function between the two hemispheres of the sort just described.

Finally, there are procedures for estimating the activities of the two hemispheres in the intact human. This is done by presenting in brief flashes different types of verbal and nonverbal stimuli to the appropriate parts of the visual field and measuring the reaction time (processing time) required to perform whatever task is employed in the experiment. The rational here is that the reaction time should be shorter when the material is presented to the hemisphere where the process occurs than when it must be first routed across to the opposite hemisphere. Similarly, one can present the same stimuli and look at the dependence of reaction time on the hand with which the response must be performed. When these kinds of tests are run, one indeed finds that reaction time is faster in the right visual field for nonverbal processing tasks.

Another type of experiment along the same lines is the "dichotic listening" technique, in which different stimuli are presented simultaneously to the two ears. Although both ears have projections to both hemispheres, the transmission time is longer for one. Additionally, phenomena such as "masking" in which the occurrence of one stimulus may prevent the processing of another, can be profitably brought into play. The results of experiments on intact individuals support the idea of functional differences between the hemispheres, although the results differ in important respects from those seen when the hemispheres are isolated.

III. THE NATURE OF THE DIFFERENCES

Clearly, the data raise many questions about the nature of higher functions in man. One of the first is the degree to which the real nature of the hemispheric differences in processing is linguistic versus nonlinguistic. This dichotomy has been at the root of a great many studies of this phenomenon, but it is not clear that the linguistic superiority of the left hemisphere results from a specialization for language, verbal logic, or sequential reasoning *per se*, rather than from some more fundamental difference which happens to favor linguistic processing. For example, it has been suggested that one hemisphere might be superior at dealing with temporally ordered stimuli, while the other might be superior at spatially ordered processes. Other positions have seen the language hemisphere as excelling at "analytic" processes while the nonlanguage hemisphere excels at "synthetic" processes.

In basic acoustic processing, it appears that there is no simple sort of superiority of one hemisphere over the other; in fact, one of the areas in which the nonlanguage hemisphere excels is music. This notion is worthy of deeper examination for it is possible that the processing of complex verbal material might well require some sort of special auditory adaptation, which might even be incompatible with the operations of the sort carried out in the nonlanguage hemisphere.

A more careful examination of the auditory responses of the two hemispheres reveals some interesting peculiarities. If we do not look at whole verbal elements, but break the stimuli down more finely, we find that the left hemisphere is superior only in the processing of consonants; vowels are handled equally well by the right hemisphere. It is easily shown that the majority of the information content of speech is in the consonants. Try dropping the vowels out of a sentence, nd y wll fnd tht t stll mks sns. Better yet, substitute a single vowel as a separater ond yoo woll fond thot ot moks ovon mor sons. Doing the same thing with the consonants does not work.

Most of the information in speech is carried in an acoustic entity called "formant transitions" which are formed principally during the pronunciation of vowels. If this information is presented to the left hemisphere, a consonant is heard. If it is presented to the right, a chirping tone is heard (which is what would be predicted strictly on the basis of the frequency contents). Moreover, if the frequency spectrum is continuously varied, the right hemisphere hears a changing complex tone, whereas the left hears a constant consonant up to a point at which it abruptly shifts to another consonant. Without going far into the complex area of verbal acoustic spectra, it seems clear that the left hemisphere may be treating the auditory stimulus in a manner designed to provide special processing for the information-carrying aspects of speech.

One small difference in the basic auditory capabilities of the two hemispheres has been noted: the language hemisphere is found to be slightly superior in fine temporal resolutions of click stimuli. At first glance, this would hardly seem sufficient to account for the differences between the two hemispheres, but it is possible that this small specialization might give an advantage in just the kind of auditory discriminations that are important in the information carrying aspects of verbal material. If the processing of verbal material excludes, for whatever reason, the efficient operation of other processes, such a small advantage might be sufficient to determine the development of the hemispheres. Given that many kinds of logical processes, especially those involving verbal tokens, rely on the basic linguistic function, and that written language seems to be processed through the same mechanisms as spoken language after an initial equation with verbal constructs, the subsequent specialized development of the hemispheres might be the result of such initial small differences.

Another very interesting hypothesis on the nature of the processing differences between the two hemispheres has been offered by Liberman (see bibliography). He proposes that the language hemisphere is specialized (not yet specifying whether functionally or physiologically) for grammatical conversions. Here the word "grammar" is being used in a rather special sense not restricted to the ordinary grammar of daily use. His basic notion is that the language hemisphere provides a set of grammatical conversions, in several steps, which serve to interface between the incompatible formats of the acoustic stimulus and the deep structure of sublinguistic cerebral processes.

He argues that the information in verbal material must exist at various times in at least three forms: an *acoustic* form for transmission through the air, a *phonetic* form appropriate to immediate neural perceptual processing and temporary storage, and a *semantic* form related to manipulation by the long-term storage and logical processes of the brain. These three forms are quite different, and the process of transcribing one into the other is by no means a one-to-one recoding. At first glance, it might seem that the transformation from the acoustic to the phonetic form was fairly straightforward and might involve only forming a representation or encoding of the acoustic material into particular patterns of neural activity; although it is not known in detail how the phonetic form is represented in neural terms, it is clear that this recoding is quite complex.

What is meant by the phonetic form is essentially an exact replica of the message intended by the speaker; ie: a word-for-word translation from your brain to mine, mediated by a temporary encoding in acoustic waves. Clearly, we can carry verbal information in this fashion for short periods. If you speak a sentence, I can repeat it back to you exactly, at least for a while. What I have is some pattern of neural activity which encodes all the phonemes (consonants and vowels essentially) of your message. However, this set of phonemes is a smaller set of items than the set of acoustic forms out of which they are produced. Leaving aside the question of different pitch ranges, accents, and speech styles of different speakers, there is the fact that the actual acoustic forms of the phonemes will depend not only on their identity, but also on their permutations with onc another. Phonemes have different acoustic forms when preceded and followed by different sets of other phonemes. Worse, acoustic forms do not appear in discrete time intervals because at any instant, the acoustic waveform contains information from several phonemes. Clearly, the set of acoustic events which encodes the phonemes as speech is much larger than the set of phonemes which encodes the message as we would write it or think it. The encoding process, or "grammar" of translation between an acoustic event's effects on the receptors and the neural representation of the phonetic material, is not necessarily a simple one-to-one correspondence. This is why it is so very difficult to build a good general-purpose speech-recognition apparatus.

In the process of reading, essentially the same sort of transformation takes place between the visual forms of the words and the phonetic level of representation. From the phonetic level on, the two processes appear to employ the same operations. (The translation of visually encoded speech is not trivial either; consider the variations of handwriting that we can recognize.)

If the complex translation from the acoustic to the phonetic form is necessitated by mismatch between the format of immediate storage of verbal material in the brain and the physical constraints on the acoustic transmission formats, a similar kind of mismatch subsequently occurs between the format of phonetic representation and the format of the "deep structure" of the brain's sublinguistic levels of action. Let's examine two arguments advanced by Liberman: first, it is clear that the format of long-term storage of ideas retains the logical relationships involved, while it dispenses with the exact phonetic format. You can reproduce for a while the exact phonetic reconstruction of something I've said, from short-term memory, but, after several days, it is lost. However, if it was an important idea which you retained in long-term memory, you could easily produce some paraphrase which would convey the logical implications of the idea quite accurately. The implication is that the long-term storage format stores logical relations and entities in a form which must be translated from, and retranslated into, the phonetic verbal form to or from which the acoustic translation is made. The fact that the translation process produces paraphrases, rather than reproductions, demonstrates that this encoding is not a one-for-one translation process.

Secondly, Liberman argues that it is clear that long-term memory and logical processes antedate language. There is no reason to suppose that the formats appropriate to these processes should resemble those developed to interface to acoustic language transmission. Indeed, it may be that the sublinguistic deep structure of thought is a property arising out of the brain's hardware, or at least out of the interaction of the hardware with experiences common to all brains interacting with the physical world. Also, the phonetic structure is different from one language to another. When humans provide good idiomatic translation, the process probably involves a ''down'' translation from the phonetic encoding of one language to the deep structure of thought, followed by an ''up'' translation to the phonetic encoding of the other language. It would be, in principle, exactly like the process of paraphrase within a language. (And you wondered why computerized translation is hard!)

This ''deep structure'' or sublinguistic level of thought is what Liberman referred to as the *semantic* level of representation. Neither he nor anyone else has any clear notion of just what the semantic level may be like in detail, although some theorists such as Chomsky have discussed its nature if not its physical mechanisms. This point is not essential to Liberman's hypothesis, and I shall return to the subject of deep structure later. What concerns us here is Liberman's idea that the functional specialization of the language hemisphere is that of making the acoustic to phonetic level of translation. In this view, the relevant specialization of the language hemisphere is not one of acoustic processing, but one of grammatical (in the broad sense) processing of acoustic material. Each hemisphere is seen as being essentially equivalent in terms of basic acoustical perceptual processes, but the left hemisphere performs an additional (perceptual?) step in translating the sounds into phonetic levels of encoding.

The basic logical process is common to the two hemispheres, but the translation from acoustic to phonetic form is for some reason restricted to a single hemisphere. As mentioned above, written material is also encoded into the same phonetic code after perceptual equation of the raw visual forms with auditory equivalents, so visual verbal abilities are likewise found in the language hemisphere.

Liberman notes that language is somewhat different than other sorts of perception in that it tends to be categorical. That is, one tends to perceive a given word as such, regardless of distortions, up until the point where it is either abruptly perceived as a different word or else as wholly unrecognizable sound. Nonverbal perceptual processes, such as musical sounds, tend to be non-categorical; a continuously varying tone is perceived as such, not as abruptly shifting tone. (I do not find this argument too compelling, because it seems to me that a number of other processes in perception, such as visual object identification, tend to be categorical. It may be argued that this really represents application of a verbal token to a perception, but I am convinced that I can see the cow in a categorical fashion before I verbally tag it.)

This hypothesis attempts to explain what is different about the kind of processing going on in the language hemisphere. It enables us to understand why and how logical processes can go on in either hemisphere, yet only the language hemisphere is capable of dealing with the material at the verbal level. Moreover, to the extent that some logical or memory processes can only occur, or only occur efficiently at the phonetic level of coding (ie: by employing verbal tokens to represent items), it can explain why this hemisphere appears to be required for certain kinds of logical operations. Quite possibly a fuller form of the theory would explain that the nonverbal hemisphere is also specialized for some sorts of transformations between deep semantic-level constructs and some sort of nonverbal surface structure. However, at the present time, we know too little about the nature of the nonverbal hemisphere's special abilities to conjecture on that point.

What this hypothesis does not do is suggest how the specialized grammatical conversions of the verbal hemisphere are carried out, or why they need to be restricted to a single hemisphere.

A further hypothesis of Liberman's suggests an approach to these questions: if the translation from the phonetic level into the acoustic product is considered, what is seen is the necessity for precisely controlling the movements of a number of organs (tongue, lips, diaphragm, vocal cords, etc) which participate in the formation of the sounds. This acoustic encoding process is far simpler than the decoding process, because most of the interactions of phonemes which result in permutation-dependent variations in the acoustic entity produced arise as a result of physical (or low-level neural) interactions between the final phases of articulation of one phoneme and the initial phases of articulation of the next. Thus, the complications are dealt with more or less automatically in the production apparatus, and encoding is largely a process of submitting the stream of phonetic code to an appropriate I/O (input/output) driver interface.

What Liberman proposes is that the decoding process relies on a model of the encoding process, which he terms the "production model" of linguistic decoding. He cites the fact that the perception of consonants follows the variations in articulation of the production apparatus much more closely than it follows the variations in actual acoustic form (which may be affected by a variety of other factors). Further, patients with damage restricted to the portion of the language hemisphere concerned with the motor control of speech production show problems of consonant recognition, even though other verbal processes and other auditory abilities are unaffected. He does not specify exactly how this presumptive model of the production process is to be employed in the decoding process. Its existence makes a potentially useful tool available to the decoding machinery. While it is unlikely that the process is anything as simple as a trial-and-error or match-to-sample procedure, it is clear that information present in the encoding scheme might be usefully employed.

If this were so, it would give us a rationale for the lateralization of the language functions. I have not discussed the point until now in order to avoid confusion, but the functional lateralization I have been considering is usually not absolute. Some degree of all functions is frequently found in either hemisphere, and the degree of lateralization is variable among individuals. (The data on this are confounded by time and degree of brain damage in many of the studies, and, as we shall see, there are discrepancies between normal and abnormal sources of evidence.) However, of all the functions which show a greater or lesser degree of lateralization, the motor control of the language apparatus is by far the most strongly lateralized. It has been hypothesized that this may simply be due to the undesirability of dual control over structures involved in precise rapid movements. These are midline structures, unlike the limbs for example, and both hemispheres would share control of the structure if lateralization of control were not imposed.

If, as Liberman supposes, a production model of acoustic language were employed in the decoding process, we might expect to find this process lateralized on the basis of the underlying lateralization of the motor generators responsible for that model. Other functions which depend on the linguistic decoding function would then follow.

IV. INTEGRATION OF FUNCTION IN THE NORMAL BRAIN

In considering the role of lateralization of function in the normal brain, one of the problems which arises concerns the degree to which the apparent lateralization may be an artifact of the situations in which it becomes manifest. In other words, to what extent is the apparent division of labor, seen in abnormal situations, a reflection of the true state of division of labor in the normal brain? Although visual hemifield projections and dichotic listening tests can demonstrate certain speed advantages for one side or the other according to the type of processing required in the normal brain, it is also clear that either side can get the job done. There is no failure to perform as in the case of split-brain subjects or cerebral damage cases. A number of workers (see particularly the Broadbent reference) have studied this problem. The concensus seems to be that in the split-brain cases, we are looking at fragments of the total intellectual process, whereas in the normal brain a continuous integrated operation is being performed. The two sides are each seen then as contributing in an interactive fashion to a more global task which may involve operations for which each is particularly suited.

Broadbent argues that if the two hemispheres were truly independent processors, then they should be able to perform independent tasks simultaneously. In fact, he contends, such things seem possible only in split-brain cases. When it is attempted in the normal human, there is always some degree of interference between the two tasks no matter how different or how tailored to the two hemisphere's specialties they may be. This suggests that in fact the cortical operation is a unitary phenomenon, at least at some point, so that the machine must deal with simultaneous tasks on a ''time-sharing'' basis which sacrifices some speed or efficiency.

He argues further that this is necessary given that brain operation is not a simple stimulus-to-response operation. That is, each operation is computed not on the basis of a simple single stimulus, but on the basis of the total world model. Any stimulus, even one which is unrelated to the task at hand, alters the world model. If one tests for reaction time in a simple fashion, one finds that the hand on the side of the stimulus shows the quicker response. A simplistic argument could be made that this is due to the stimulus being processed immediately in the hemisphere that controls the hand. However, if the starting position is set up with the hands crossed, the reaction time of the hand nearer the stimulus is *still* better! Clearly some sort of feedforward control, based on the global world model, is being employed to alter the nature of hemispheric control of processing.

By the use of dichotic listening techniques in conjunction with masking effect, it is possible to investigate just which parts of a processing task can be performed in one hemisphere without interfering with operations in the other hemisphere, and which ones cannot. Broadbent interprets these data in the following way. There are three major stages in the processing of a response to a stimulus:

- storage of the recent past (ie: the neural ''trace'' of the stimulus which has just been received for processing);

- a decision process which allocates the correct response to the stimulus on the basis of the stimulus and any relevant past context;

- a production process which carries out the response.

It appears that independent tasks may be carried out without interference by the two hemispheres during the first (stimulus storage) phase and during the final (response production) phase, but that even when the two tasks are very different and appropriate to the specializations of the two hemispheres, they must interfere during the decision process.

This would suggest that the decision machinery is common to the functioning of the entire brain in the normal individual, even when the relevant decisions in the two tasks are unrelated. Given that these tasks can be carried out (without interference) when the two hemispheres are separated, as in the split-brain studies, it would seem that either hemisphere is potentially possessed of the decision-making machinery, but that somehow in normal operation they are not permitted to employ it independently. This is consistent with the notion expressed earlier that the logical and long-term storage processes are sub-verbal and therefore potentially available in either hemisphere.

The question which remains is why, and by what mechanisms, are they normally forbidden from independent operation save in more superficial processes? The "why" may simply be that there is a danger that they would come to different decisions and produce conflicting behaviors. Even when the independent hemispheres of a split-brain case decide on the same eventual outcome, they frequently employ different strategies. Instances have even been reported of the hand belonging to one hemisphere slapping back the hand belonging to the other when it attempted to operate in a problem area in which the first hemisphere was more proficient!

The mechanism by which the hemispheres are kept in harmony is more problematical. The evidence suggests that most processes are represented in both hemispheres in split-brain cases. Even the linguistic disabilities of the right hemisphere have recently been demonstrated to result more from an inability to produce speech than from an inability to comprehend speech. Sperry has remarked that it seems unlikely that we can produce a mind with a knife cut; the implication being that the apparatus of a full consciousness is present in both sides and we have simply removed the coordinating influence that normally integrates them into a unified whole. But if this is true, the unification process must essentially be one of suppressing one or the other at appropriate times. It is believed that the principle action of the fibers that run in the corpus callosum is inhibitory, and it may be that when a region of one cortex is in operation, it can suppress the activity of its opposite number. This explains, perhaps, the immediate mechanism, but it does not explain how the decision is made concerning which side shall dominate in any given task or sub-task.

This job may represent one of the higher levels of operation of our ubiquitous "attention" function. It is certainly just the kind of operation that we saw earlier under that heading: a feedforward operation which does not process the data, but which sets up the data-processing machinery in a way most appropriate to the expected job requirements. Further, this kind of action, by suppression of unwanted data, is just what we have seen in other instances of attention processes. Thus, when the attention processes "know" that the hands are crossed in the current world model, the new situation with regard to appropriate hemispheric competencies could be set up in advance of the decision process. In the intact individual, we would expect to find that the interplay of the two hemispheres was then regulated according to their special competencies for the requirements of any given phase of the problem by a feedforward process which could operate on the two hemispheres as a unit. Just such a unitary process may be the vulnerable point in attempts to generate two simultaneous operations without interference.

In the split-brain cases, the suppression would be removed, and each hemisphere would be free to display its maximal capacities. Even in this case, it appears that the subsystems within a separated hemisphere which are responsible for its operations involving same or opposite sides of the body are subject to alternate activation by an attention-like process as the appropriate sphere of action shifts.

The picture which emerges from these considerations is that of two very similar hemispheres with some functional specialization, integrated in their action by a suppressive process which enhances their functional distinctions. This action provides for their noncompetition when necessary, while permitting them to operate in parallel on different portions of their specialized tasks which do not interfere. From this model, it is possible to understand the fact that in situations (such as early childhood brain damage) where the functions are all forced into one hemisphere, there is a loss of power and efficiency, but no complete incompatibility. The parallel operation feature is lost, but the basic mechanism can still support all of the necessary processing.

V. THE PHYSICAL BASIS OF THE SPECIALIZATION

This raises the question of how and why one hemisphere becomes better at some tasks than others. If, as seems clear from the cases of childhood damage, either hemisphere can support either or both kinds of functions, why doesn't this occur in the adult following damage, and why isn't the distribution of linguistic versus nonlinguistic capability random between right and left in the population, even granting that the brain might want to specialize the hemispheres for efficiency alone? The facts of childhood damage suggest lack of physiological specialization, while the facts of adult damage suggest that it exists. The preference for language on the left, and the fact that even when it occurs on the right, it seems to be a genetically transmitted trait, suggest also a physiological difference between the hemispheres.

These are not necessarily incompatible results. In the first place, the recent findings of substantial linguistic comprehension in isolated right hemispheres, emphasizing their major deficit as one of speech production, suggests that, even in the adult, the capabilities of the two hemispheres are not so very different once they are released from the inhibitory control that governs their interactions (note especially the Searlman reference). In fact, in the child, it seems that this inhibitory control of interaction is much less extensive, and that the operation of the child's hemispheres is not nearly so well divided as in the adult. We are led to suspect then that the developing specialization occurs during maturation, and that by adulthood it may have proceeded to a point where the hemispheres have achieved an irreversible degree of specialization. This might represent simply a learning process in which the hemispheres simply do not "know" how to do each other's chores, but this seems unlikely in view of the lack of recovery in cases of adult damage.

It seems more likely that an actual physiological differentiation takes place between the hemispheres. We know that throughout childhood and adolescence, the higher cortical systems are developing and being made functional through myelination of fiber systems and quite probably through continuing formation of functional synaptic connections. If structures on one side are removed experimentally in animals, we see a compensatory invasion, during development, of structures on the opposite side by neural connections that would normally have gone to the missing structure. It takes no great stretch of imagination to envision such a process yielding a progressive physiological distinction between the hemispheres during development, or to imagine it being redirected out of its normal course if the necessary damage occurs while substantial plasticity still exists.

This would not conflict with the evidence which suggests that, all other things being equal, there may be an inherent genetically controlled physiological basis for the choice of hemisphere in the undamaged case. Careful studies have shown that there are indeed detectable differences between the two hemispheres in terms of relative size of certain areas. In the case of the language hemisphere, this is presented as an increase in the size of the auditory and surrounding areas exactly where we would have predicted it. In the case of the nonlanguage hemisphere, there are similar regions showing an assymetrical increase in size, but at present we do not know enough to say if their occurrence is related to any unique functions of hemispheric specialization. Neither do we understand what is signified by increased size, but it would not be unreasonable to think it associated with particular richness of neural machinery and connectivity. Such distinctions between the hemispheres could certainly account for the propensity for functional differentiation to occur along genetically established lines of preference. This could be true without detracting from the fundamental similarity of the processing power of the undeveloped hemispheres.

It is possible, given that linguistic ability and hemispheric specialization in general are very recent developments, that we are witnessing here the first steps in the direction of further evolution of the human brain. It is a truism that ontogeny recapitulates phylogeny, and the functional development of specialization occurs late. It may be that given the benefits of hemispheric specialization, there is a continuing selective pressure for evolutionary physiological specialization which will eventually cause the hemispheres to become increasingly different from the beginning and hasten the process of functional differentiation. Ultimately, man may be born with very different hemispheres and, who knows, perhaps language? One day our brains may be regarded as very primitive in their almost total symmetry.

VI. SURFACE PROCESS AND DEEP PROCESS

I have postulated, following Liberman's arguments, that there exists several levels of neural representation of cerebral material or the "content of conscious thought." At the bottom, the deep structure, there exists the basic logical-mnemonic substance of world model relations and similar constructs, encoded in some fashion in neural activity patterns which differ in some way from the encoding of the phonetic representation of them in the language hemisphere (and presumably also from the corresponding I/O related surface code patterns in the nonlanguage hemisphere). In reality, there are probably several "levels" of representation intervening between the deep and surface representations. Given the existence of fundamental deep structure entities, whose encoding scheme is probably a function of the apparatus of their generation, I may say a few words about the process by which they are converted to surface structures appropriately coded for I/O operations.

Most of the speculation on this subject has been the work of linguists such as Chomsky, who have arrived at their conclusions more from the study of language than of neurophysiology. It is of some interest to have an outline of this approach. In simplified form, a Chomsky-like hypothesis of the conversion strategy from deep structure to surface structure might be as follows: let us say that at the deep level, we have the codes for a number of concepts such as "the dog is black," "the dog is hungry," "the dog is old," "the dog bites the man," "the man is fat." We can easily imagine these kinds of facts being encoded as aspects of the perceptual representation of the world model discussed in earlier chapters. Chomsky sees such primitive ideas as being the meaningful content which must be converted first from this "semantic" level to a "syntactic" level incorporating the rules of grammar in our language, and then to the "phonetic" level discussed earlier which incorporates it further into the representation of speech. This process would first convert the factual content statements of relationships and description to a grammatically correct form encoded at some intermediate "level" not in the form of the deep structure, or yet in the phonetic code. This might produce "the hungry old black dog bites the fat man." This re-encoding produces a syntactically correct representation of the information contained in the separate but related elements of the deep structure, which are in turn facets of the world model representation. Such a re-encoding would be necessary to impose the arbitrary rules of syntax in any language upon the deeper level code which was constrained by the apparatus of its formation. The nature of the neural mechanism underlying such a re-encoding is, of course, completely obscure, but it is interesting to note that similarity; hence, the ALMOST gate could have a role here. A second conversion from the level of a properly formed sentence to the level of a phonetic code (in a form suitable for presentation to the I/O drivers) would have to be made according to the rules of pronunciation as discussed by Liberman.

We are left with a number of tantalizing glimpses of possible relations between world model, deep structure of thought, surface structure, and hemipsheric specialization for I/O operations. At no point can the connections that we sense be established, from Hubel to Luria to Chomsky to Liberman, and yet there is a provocative feeling of underlying continuity that demands further research.

9 Storage and Retrieval

I. THE KINDS OF MEMORY

Throughout the preceding chapters, I have alluded to memory, learning, and storage processes in the brain, with a promise to deal with the issue later. Alas, later has arrived. I come to this point with some reluctance, not because memory is an uninteresting subject, but because of the abysmal state of our understanding of the physiological basis of this complex function. Fortunately, a good deal is understood concerning the *functional* organization and operation of the brain's storage systems, and it is this material, rather than the actual physical storage mechanisms of the brain, which should be most inspiring to the computer or robotics designer. In fact, our existing computer storage media, from the various types of random access memory (RAM) through magnetic disk and tape materials, already possess many of the desirable properties of the brain's storage mechanisms. We cannot fault them for information density, because the size of a modern memory "bit" is comparable to that of a neuron. Their access time, at least in one sense, is similar to that of neural systems. There are, however, certain features of the brain's data storage systems which we have not yet incorporated into our artificial systems, and there are many feats of human memory which modern computers, for all their speed and capacity, cannot match.

To begin a study of the brain's data storage, it must be recognized that storage exists in a variety of different forms which differ functionally along several dimensions; each type probably represents a differing physical system in terms of its ultimate physiological substrate, location, and operating principles. Storage systems may be distinguished by the kind of material they contain and by the time frame in which they operate. First, let us consider some of the types of material.

In earlier chapters, I discussed two fundamental learning situations:

classical or "Pavlovian" conditioning, and "operant" conditioning. At one point, I suggested a mechanism by which the outcome (or reward) dependent operant conditioning process might be understandable in terms of a classical conditioning process of a reward system, which in turn acted to "gate" behaviors. Whether or not this view validly pertains to data storage mechanisms, these two kinds of learning processes have certain fundamental similarities. In particular, they are usually considered to represent a sort of connection (not in a literal sense) between a stimulus and a response. The response may be broadly conceived and may include internal perceptual processes, but what is stored is still an association between some stimulus that the system can be aware of and some definite internal or external system response. Such learned stimulus-response behaviors are clearly a form of stored data. Even if the ability to recognize the stimulus and execute the response predate the learning operation, the connection between them is still something which the organism derives from experience and must store in some form for future use.

The laws governing the operation of such learning processes have been studied in great detail, usually in circumstances such as maze learning in the rat, eye-blink conditioning in man, and similar situations in which some overt behavior, usually nonverbal, is learned as a response to some previously neutral stimulus. Because these studies have successfully generated laws of broad applicability, a number of theorists have made attempts to bring all other types of data acquisition in the organism under the purvue of such laws, by reducing seemingly more complex kinds of data acquisition to stimulus-response learning processes. While the issue is still debated, many of these attempts are more noble than compelling; it seems plausible that other sorts of storage mechanisms are required.

In particular, we tend to think of stimulus-response learning situations as being somehow different from "memory" of the sort implied by "What did you do last Tuesday?" Certainly, they are different from the temporary storage we use to momentarily hold in mind a telephone number that we never intend to call again. Possibly there are many kinds of learning; I have mentioned perceptual learning, and there is the learning of motor skills. All of these processes require data storage of some sort.

It is not as easy as it might seem to ask what is similar or different about the storage mechanisms employed in these assorted processes. Seeing only the behavioral end-product, we may confuse different methods of encoding or retreiving data with differences in form of storage. As a beginning, let us examine several rather different memory retrieval tasks.

Visual Imagery

We know that humans, and presumably animals, call up visual images of past situations. Visual imagery is a "picture" or scene that one can create and examine out of one's memory of a past situation. *Something* must be stored to enable us to do this, but this something may not be the actual image experienced in the act of recall. When I said "create and examine," I meant exactly that. It appears most likely that some sort of nonvisual representation is stored which describes a previously experienced scene, and in the recall process, something like the original visual experience is reconstructed from this information. There also exists an immediate memory trace of sensory input which persists for a moment after the actual stimulation has ended, but this is a different phenomenon, to which I shall return in the next section.

Given all of the work that the system has to go through to reduce the original sensory input to object-level identification and to then categorize the objects, it would seem terribly wasteful to turn around and store the original sensory data only to have to repeat the process in order to examine the remembered scene. More likely, the system stores a compact description of the scene in terms of the analyses it has had to perform and then reproduces an image composed of these results. This notion is supported by a number of observations. First, it is notoriously easy to "remember" details of a scene incorrectly. These errors take the form of images that are reasonable expectations in terms of the rest of the scene, or which are in accord with the person's beliefs about the contents of the scene. Very frequently, people will insist that they have a perfectly clear remembrance of the nature of a visual scene, when in fact it can be demonstrated that they are in error. Further, we are aware that such remembered scenes are less clear and detailed than the originial, but we are frequently not aware of just how restricted their content really is. If you try to remember a scene, you will find that you seem to have a fairly clear and detailed picture of it. This is only because you are capable of generating those details to which you are attending at the moment. If you now pose some question about a detail of some portion of the scene, you will find that you cannot "see" the rest of it clearly while you are examining the detailed portion of the scene. There is a fixed limit to the quantity of detail that your remembered image can contain. This limit on the contents of attention will reappear in other memory contexts, but its importance here is to emphasize that the original total scene only had the appearance of rich detail and you were fooled into thinking that it had anything like the richness of a real visual image. A little experimentation should convince you that you cannot "see" particular details until you actually actively search for them, and that others disappear to make way for them. This fact, together with the ease with which the system can fabricate portions of the image to fit with expectations or beliefs about a past scene, suggest that this process of "attending" to a detail of a remembered visual image is closely related to the actual recall process, and that it is an act of construction of a perceptual experience out of some metaperceptual level of data representation.

Similar kinds of image memories occur in modalities other than vision: auditory image memories are not uncommon, and tactile memories may occur, too. Descriptions of true memory images from other senses (as opposed to recognition of previously encountered stimuli) are rare.

Verbal Memories

The great majority of studies of human memory processes is concerned with verbal memory. I will consider the implications of some of this material shortly. Here, I will just try to establish what is meant by the term. If I present you with a list of unrelated words to learn, you will be able to reproduce it, after some rehearsal, with more or less accuracy. Clearly, something is remembered. Unless you are a very unusual individual, your retrieval of the information as you reproduce it will not be anything like the original sensory experience that constituted the list. You will find yourself neither looking at a visual memory image of the list and reading off the words, nor listening to an auditory "playback" of a spoken list. The memory of the items in the list will simply arise from somewhere in your memory store as information capable of driving your verbal processes without further recognition being required. There are people who can reproduce a visual image of a page in sufficient detail to actually read backwards, but they are extremely rare and it is not considered to be a normal process. The ease of remembering such verbal material is heavily influenced by the degree to which the material conveys some sort of meaning. The hardest material to remember is a set of nonsense syllables; the easiest is sentences forming complete stories. (For this reason, pure nonsense syllables are used in research, because "meaning" is hard to quantify except in the case of no meaning at all.)

Motor Learning

We learn many different skills during the course of our lives. We improve our abilities at various sports, play musical instruments, operate machines, and engage in a host of other activities. Almost every new behavior that is learned is improved by practice. The acquisition of these skills is a change in our information content; the proper way to perform is stored as information somewhere in the system and in this sense, it is a memory.

Such data storage nonetheless seems to be very unlike the experience which we associate with memory of past events. When we execute a well-learned task, there is no subjective sense of remembering how to do it: we simply note that we now perform skillfully what was once clumsy. The point of difference between "motor memory" and "remembering" seems to be something like the presence or absence of a perceptual aspect to the data retrieval process. If the result of the information retrieval is a sensory-like experience, we are more likely to think of it as memory, while if the result is an output operation we tend to think of it as the expression of a skill and deny that we "remember" anything. Actually, the difference could be simply in the way we experience the effects of stored information on the operation of the resultant perceptual or motor process, rather than in any real difference in the form of the data or the mechanism of its storage.

There is evidence to indicate that there may indeed be different kinds of data storage in the brain, but the distinction between data base types may not be related to differences in the nature of the stored material. The true differences may have more to do with such matters as storage of "averaged" representations of prototypes, as in motor skill learning and word list learning, as opposed to storage of "unique" information, as in the recall of individual events. I shall examine this distinction in section III, but first let us look at another possible way of categorizing types of memory processes.

II. THE KINDS OF STORAGE

One profitable line of approach to the understanding of memory processes is to examine the differences which exist among the memory mechanisms which deal with different time frames of the storage process. As information enters the system, it passes through a succession of storage mechanisms with differing properties. It is not clear that all of the kinds of material to be stored must necessarily pass through all of these mechanisms, but much of it clearly does.

The first stage of storage in the memory system is frequently referred to as *sensory memory*.This is a temporary store into which the incoming information in its entirety is placed for a brief period. The apparent function of this stage of memory is to retain a record of the sensory event long enough for perceptual mechanisms to operate on it. In this regard, it might be thought of as analogous in function to an input latch which captures a byte from a keyboard and holds it while the processor examines its flags and places the data on the bus. It does not really hold data until cleared, nor so far as we know set flags, but the function of capturing transient data is similar.

You can observe the operation of this sensory memory by rapidly moving something back and forth in front of your eyes. As a finger or other object moves, you will observe that it leaves a shadowy "trail" behind. If it is moved back and forth rapidly enough, this trail becomes a steady blur lying in the path of the moving object. (A good way to observe this is to light a moving finger from the side and observe it against a dark background.) The explanation of this phenomenon is that the image of the finger is retained for a moment in sensory memory after it has actually moved away, so that a shadowy remnant seems to remain. One of the distinguishing properties of such storage is that the information spontaneously decays with time; older images appear fainter, which explains the disappearing trail following the image.

In addition to spontaneous decay with time, there are several other features which distinguish sensory memory from other types of neural storage. One is that all of the information present in the input is stored. As noted above, there is reason to believe that typically this wealth of detail is not stored in later stages of memory. However, if the system is to select features for permanent storage or perhaps generate higher level encodings of objects, it must first capture all of the data and hold it for the time required to perform these analytic operations.

A second important feature of sensory memory is that it is "erasable." Obviously, the requirement that information be retained long enough for processing is balanced against the requirement that the duration of storage must not interfere with the intake of new data. The nervous system seems to approach this dilemma by allowing sensory memory to decay with a fixed time constant, as long as no new data appears, in case new data is present, the sensory memory is updated immediately.

A few classic experiments illustrate this interaction well. One method of attempting to measure the time constant of sensory memory is to very briefly flash a set of letters or numbers onto a screen and after a variable interval, to flash a marker onto the same visual space to indicate which item the subject is to report. Obviously, if the sensory memory of the screen full of characters has already decayed by the time the marker indicates the desired item, the subject will be unable to report it accurately. Using this technique, the duration of sensory memory has been found to be about 250 milliseconds. Other methods of estimating this duration give about the same value and show that the decay of the memory trace begins as soon as the stimulus disappears and continues in a more or less exponential manner for about a quarter of a second.

However, it was found that if the marker occupied exactly the same visual space as one of the target characters, it simply erased the memory of the character rather than being superimposed on it. This occurred even if the marker was a circle designed to circumscribe the character. We can understand this last result in terms of the earlier discussion of extrapolation between boundaries in the visual system's feature extractor network. The visual system sees the center of the circle as a solid patch of background between two boundaries and interprets its contents as background level illumination. The original sensory memory image (which is not at some "higher" level of the system) cannot "download" a character into the center of the circle presently undergoing feature extraction. Accordingly a "solid background" is placed over the character as the new sensory memory contents are added to the old. The fact that the characters presented to one eye can be erased with an encircling form presented to the other eye suggests that the sensory memory is indeed a central process, rather than simple persistence of excitation in the retinal receptors.

It is not the case that all of the information in the sensory memory is lost after a half second or so. Obviously, if we are shown a pageful of characters in a brief flash, we will remember some of them a few seconds later. Moreover, the fact that a later probe marker can be used to indicate which ones we are to report suggests that there is some limited capacity mechanism for longer term storage, into which we can direct a selected portion of the sensory memory information for further examination. This more persistent form of storage can contain less information than sensory memory, but through an attention-like process, any item of information in the sensory memory may be transferred into it. This second stage of the memory process is commonly called short-term memory for reasons that will become apparent shortly.

Several facts about this short-term memory are easily established. One is that its capacity is about seven items. Another is that the addition of new items erases old ones in a temporal sequence, so that adding new items causes it to drop the oldest ones "off the stack." As our attention shifts to new items, the content of short-term memory is continually being updated from sensory memory. Items can be maintained in short-term memory indefinitely by "rehearsal" or repetition. Anyone who has maintained a telephone number or address in short-term memory while looking for a pencil knows that the way to do it is to mentally repeat the information. This keeps the contents of short-term memory refreshed. The attention process is now focused on short-term memory itself rather than on the contents of sensory memory, and the short-term memory is not updated from new material. It is difficult to establish whether the contents of short-term memory disappears with time, as is the case with sensory memory, or whether all of the decay is due to overwriting with new material. Certainly, the contents of short-term memory usually vanish within a fraction of a minute if they are not subject to rehearsal. However, it is very difficult to design an experiment that will prevent both rehearsal (which would preserve short-term memory contents) and entry of new material (which would destroy them). The best current guess is that short-term memory may be like sensory memory in that it has both a time-decay erasure process and an active erasure process that operates when new material is entered.

Besides the time course and the storage capacity of the two memories, there are major differences between sensory memory and short-term memory with regard to the kind of material stored. Typically, short-term memory is verbal memory. That is, regardless of whether the content of sensory memory is the sound of words or the sight of printed letters, the material in short-term memory seems to be in the form of words. When a person is asked to inspect a scene and then remember what was in it, he will ususally proceed by naming the items he sees and holding the names, as verbal material, in short-term memory. We rehearse short-term memory contents by repeating words to ourselves. Of course, one can try to recall the visual scene and describe it later, but for most people this is very difficult. It seems impossible to "rehearse" such a visual memory and so keep it in mind, as we can do so easily with a verbal description of the same material. It is possible that there is a short-term storage for images analogous to the short-term storage for verbal material, or it is possible that such images may go directly into permanent storage from sensory memory without passing through an intermediate attention-selected stage of short-term memory.

There is not a great deal of information on this point, but it appears that if there is some form of short-term storage for nonverbal material, its properties are somewhat different than those of the better understood verbal type. It is clear that to the extent that nonverbal materials can be temporarily retained, they are held in either a separate storage system or represented in a form so different from that of verbal materials that they do not interact in temporary storage.

It is easy to show that using material from short-term verbal memory is almost impossible when some other verbal task must be conducted at the same time. As we have already seen, new verbal material entering short-term memory erases the old. However, if some other concurrent task requires nonverbal operations there is no interference with recall from short-term verbal memory. The converse relation exists in attempts to recall material from nonverbal memory. If I ask you to inspect a figure and then to perform some task which requires you to use information from your visual memory of the figure, there will be no difficulty if your concurrent task is a verbal one, but recall will be greatly impaired if your task is to make visually directed pointing responses. These findings seem to imply that either the verbal and nonverbal materials are handled by different mechanisms, so that the displacement of old items by new affects a different pool of storage, or that the materials are coded in such different fashions that they cannot replace one another in temporary storage. It may also be that separate verbal and nonverbal attention processes can coexist in time. (See Chapter 8.)

The next stage in the sequence is usually referred to as long-term storage. Specialists in the area have not agreed whether there is a clear distinction between short- and long-term memory systems on the basis of retention time, or whether they simply shade into one another. Others postulate a whole sequence of longer term storage mechanisms with slightly different properties. For our purposes, the common usage of a single, long-term memory system will suffice.

It is clear that there are substantial differences between short-term and long-term memories in terms of their organization of data, retention times, and type of material stored. Long-term memory refers to the more or less permanent form of information storage in the brain; it differs most obviously from short-term memory in that material, once entered, is apparently retained indefinitely. The phenomenon of forgetting is really a problem in access, which I shall consider later. Material seems to enter long-term memory as a result of repeated rehearsal of the contents of short-term memory. It may take many repetitions or activation of data in short-term memory to establish it in long-term memory; this is the process of learning or memorizing as those terms are applied in everyday usage. Similarly, when material is brought out of long-term storage, it is placed in short-term memory storage where it is retained during the subsequent examination or other use for which it was retrieved. Thus, the amount of material which may be "displayed" from long-term memory is equal to the amount of input which may be examined at any one time.

It is as though the short-term memory storage system formed something analogous to a computer register or buffer through which information had to pass on its way to or from storage and input/output systems. The analogy may not be pushed too far, but to the extent that it is valid, it helps us understand why, at least in regard to the processing of verbal material, the brain seems to exhibit a property like computer word length. That is, although the perceptual and motor systems may deal with massive amounts of material in parallel, and the content of long-term memory storage is vast, it may be that verbal or symbolic material is only available to our awareness in a limited capacity; short-term memory storage system, like a computer's working registers, has a relatively small and fixed capacity.

In the average case, this "word length" is about seven verbal or symbolic items. At the beginning of this volume, I remarked that we were not particularly well constructed for verbal or symbolic logic and arithmetic operations. The limited word length of verbal working storage, suggested by the role of the short-term memory system, appears to be one of the bottlenecks. No such restriction appears to be involved in the operation of nonverbal information processing of many sorts, which helps to explain our facility with such data. Indeed, the number of digits you can hold in short-term memory and repeat back after hearing them once (short-term digit span) is a standard item in intelligence tests. On the basis of the above argument, we can see that it may well provide one index of the verbal/symbolic logical processing power of an individual brain. The average range of digit span is from five to nine items, with a mean of seven. Even in the best brains, digit span rarely exceeds twelve items unless special mnemonic "tricks" are employed.

The capacity of long-term memory is very large. There is some reason to believe that even a great many seemingly trivial items, which you would not believe are remembered, are in fact retained. This is not to say that we can always get them out again. In most cases, forgetting is actually a problem with locating items in long-term memory and, in particular, involves locating the desired item among items of a similar nature. This interference in retrieval between similar items immediately suggests that information is stored in long-term memory in some manner related to the nature of the item rather than to its order of arrival, as in short-term memory, or its position in the sensory field, as in sensory memory. This is apparently the case, but it only represents a part of the remarkable way in which long-term memory encodes and stores information.

Sensory input is progressively encoded, or analyzed during the perceptual processes; the raw data are first organized into features, then into objects, and finally into parts of a world model. You will recall that I also emphasized the fact that as each stage of analysis is accomplished, the additional identification is concatenated to the analyzed data. The process of input in the memory system follows a similar course, in that different levels of abstraction are found in the successive stages of storage. Indeed, it may be that the perceptual process, or at least portions of it, employs one or more of the mnemonic storage pools as holding registers during its analytic operations. In any event, it is clear that sensory memory contains something akin to the extracted features of the sensory field, while short-term memory contains something more like the object identification level of analysis (ie: phonemes in the case of verbal material or spatial objects in the case of nonverbal material).

The principal content of long-term memory seems to be material at the semantic level of analysis. In Chapter 8 it was noted that short-term memory allows us to recall what was said verbatim, while recall from long-term memory generally involves paraphrase. The suggestion is that the material in long-term memory is not the verbal or symbolic object, but rather a semantic or "meaning level" representation of the ideas and relations conveyed by the words. We might suppose then that the transition from short-term to long-term storage also involves a further step in perceptual analysis from the verbal object to the representation of meaning and relationship. Among these items might be the perceptual codes for spatial or logical relationships which I discussed in relation to the world-model building processes of the central parietal cortex. We saw there that such relationships were apparently treated as high-level perceptual features.

Cognitive psychologists have proposed a variety of constructs for representing information at the semantic level in long-term memory. In addition to the notions of relationship and membership in object class, these data bases include components of "meaning," such as the assignment of agent and recipient roles in temporal action sequence, as well as a primary concept of an object itself in the true internal perceptual code. The relationship "name of" equates this perceptual construct with a verbal object (ie: a name) at the level of phonetic representation. In this type of analysis, the set of constructs at the semantic level, including the representation of their various relationships to one another, is envisaged as forming a complex construction known as "semantic network." (For an excellent introduction to the details of semantic networks in human data processing, the reader should examine Chapter 10 of the Lindsay and Norman reference in the bibliography.)

Such semantic networks are very similar to the kinds of complex data structures actually employed in the storage of large computer data bases. Indeed, it is probably the case that the semantic networks of cognitive psychology were first suggested by computer analogies rather than the other way around. Anyone who has had some experience with data structures, in which the nodes or atoms are linked by multiple types of pointers that may be pursued to locate atoms or further pointers bearing particular types of relationships to the parent node or atom, will feel at home with the semantic networks of cognitive psychology.

However, it should be borne in mind that although these representations of the data in long-term memory may adequately serve to represent or describe the data and its internal relations, they probably do not bear close relation to the actual physiology of the brain. This type of relational data structure is well adapted to the operation of typical serial processing computers. In this operation, some initial information generates a set of pointers which may be followed down branches of a tree or graph structure to find other information and associated pointers, until the desired information is assembled. This is fine as long as all the machine can do anyway is to pursue one pointer at a time from one address to another. In the brain, other possibilities are available, and in fact the magnitude and rich connectivity of the data base of the human mind requires that more efficient operations be employed. In particular, it is likely that the brain does something more like following all the possible branches of the tree simultaneously. It is unlikely that there is really any "tree" or set of pointers to be followed, or even that "following" as such occurs, but the general scheme of what happens is more nearly captured by that comparison than by the sequential search implicit in semantic networks in which the relational pointers are pursued by a serial machine.

In addition to this parallel approach to organization, the data structure of long-term memory is also characterized by an associative property which allows the (simultaneous) retrieval of many pieces of information related, by similarity along various dimensions, to the precise search key. I will consider both the associative and parallel features of data retrieval from long-term memory in later sections of this chapter. The important point to be made here is that the basic organization of data in long-term memory is made in terms of the meaning, content, and category of the data, rather than in terms of temporal relations, as is the case with short-term memory, and that the nature of the item represented undergoes a transformation from object or verbal level to semantic level of representation. Thus, long-term memory probably encodes data in the fundamental internal code of the central machine discussed in earlier chapters.

An interesting question immediately arises. I have discussed semantic networks as ways of representing the organization of knowledge in memory, but I have done so in terms of verbal memory. If the semantic level of representation in long-term memory is truly in a unified internal code to which equivalents from all modalities may be reduced, what is the fate of nonverbal information? This question is not nearly as adequately studied as the problem of verbal memory, but it is clear that we can remember over the long-term ideas which were not presented to the senses in verbal form. We can remember relationships among external world events that we witness, and we can employ in our reasoning concepts of spatial relation which we have learned from experience. The question is, do we handle the long-term storage of this information by first creating a mediating verbal description of it, or can nonverbal information be stored directly at the semantic level? Clearly we do sometimes describe to ourselves in words what we observe, and we also may state verbally what we remember about a nonverbal experience. Yet, it is also certainly possible to recall from nonverbal memory a purely visual experience, or to remember information which we certainly have not verbalized. Nonverbal animals can obviously employ long-term storage of meaningful information. This leads us to ask whether there is a separate long-term storage mechanism for verbal or symbolic information at the semantic level, or whether the semantic level in long-term storage is the same regardless of the route used to generate meaning.

This is not an easy question to approach. The paraphrase argument provides an intuitively convincing basis for the existence of a semantic level of represention of symbolic material, but it is not easy to conceive of paraphrase in the nonverbal realm (although you might think of it as reproduction of a nondetailed or incorrectly detailed sketch of a previously observed scene). A few experiments seem to indicate that nonverbal material is stored at the semantic level and, in addition, is placed in the same semantic data base as material that entered through the verbal-symbolic route. Such experiments present ideas both verbally and non-verbally; the subject is subsequently asked to identify verbally and non-verbally presented materials in terms of whether or not they were presented in the earlier set. Suppose both verbal and nonverbal representations of an idea are reduced to the same item in semantic level storage. We should then expect a greater tendency to identify a verbally presented idea from the latter set as having been presented in the first set, when in fact its nonverbal equivalent was presented. Conversely, we might expect to identify a nonverbally presented idea in the second set as having been presented in the earlier set, when in fact its verbally presented equivalent was presented. People do make such confusions more frequently than would be predicted by chance. One might regard this as an instance of paraphrase between the verbal and nonverbal modes of object level representation which is based on a common semantic substrate. This technique can also be used to show that there is mixing between verbal materials presented in different languages, which would be mapped into the same semantic space, and thus be subject to confusion when the ideas are reconstructed into language from semantic storage. (An interesting set of experiments in this vein is found in the Rosenberg and Simon article referenced in the bibliography; this article also presents a computer model of the operation of semantic memory which can simulate the observed results. This is, of course, a functional, not a physiological or hardware model, but it is instructive.)

It would seem that even a common pool of semantic storage for symbolic and non-symbolic modes of data entry is still not a sufficient construct to explain all of the phenomena of the brain's data storage. For example, it seems difficult to place the learning of motor skills or stimulus-response relations, such as Pavlovian or instrumental conditioning, into this framework, because they do not seem to involve a semantic intermediary. One does not "remember" the appropirate response to the bell and then decide to salivate! It seems likely that some additional long-term storage mechanisms may be required to account for this kind of information storage. Perhaps these should be thought of as being more nearly equivalent to incorporating the results of experience into the machine by wiring changes, as opposed to making new entries in a data structure.

The examination of memory pathology reveals certain aspects of how we remember. In the most common type of amnesia, there is a loss of the ability to enter data into long-term storage. The patient can retain items in short-term storage through rehearsal, but when the data is lost from short-term storage, it has left the system. Although this is the generally accepted view, based primarily on studies of the patient's verbal responses, there is evidence that this amnesia is principally verbal in nature. The patient is certainly unable to make any intellectual use of information presented earlier, yet there are certain things that can be learned. Motor skills can show improvement with practice; the patient will show increasing skill at traversing a "finger maze," which is essentially a kinesthetic stimulus-response learning situation. He will show such improvement despite his verbal protestations that he has never even encountered the maze before. We are thus led to the suspicion that semantic long-term memory may be only one of the forms of available long-term storage.

In many cases, different storage formats may be required by the means of access which are appropriate for different kinds of information. Semantic level representation is deeply involved with the notion of category and relationship, and these, together with similarity along various semantic dimensions, are appropriate keys for retrieval and therefore constitute appropriate principles for the organization of storage. Such a scheme may be entirely inappropriate in other cases where categorical relations are irrelevant and speed of response is paramount.

III. THE ROLE OF CONTEXT

In the discussion above, you may have noticed that I sometimes referred to memories as requiring practice, rehearsal, or repetition in order for them to be acquired. This is certainly true whenever we want to learn a motor memory or skill, or when we want to memorize a set of facts for an examination or remember our new telephone number. On the other hand, we can clearly recall events that may have happened to us incidentally and which we have not attempted to learn in the sense that we did when memorizing a list. In fact, this is usually what we mean when we speak of our "memories" of our lives. How can we account for the fact that it is so seemingly easy to retain such material and yet so hard to retain a set of facts or vocabulary words?

It has been recognized for some time that these two aspects of memory represent distinct phenomena, and there is accumulating evidence that there are underlying differences in the physiology of the storage or retrieval mechanisms. Some authorities believe that they represent two different aspects of the same fundamental storage process. In physiological studies of the mechanisms of learning, these aspects of memory have been referred to as context-free and context-dependent storage processes, while in the study of human verbal memory, they have frequently been referred to as semantic and episodic memories. Whatever differences may exist between these naming conventions need not concern us here.

The basic difference between the two types of material is in the extent to which the information about the item to be remembered may be averaged by the system. Consider a rat learning to run a T-maze: he runs down the stem of the maze and must then remember whether the food is on the right or the left, in the black arm or the white arm, or some other such simple, learned response. In each trial, many things are not quite the same as in the previous trial; the exact angle from which he sees the visual world, the exact state of his internal stimuli, the exact nature of the sounds in the room, and a host of other incidentals may differ. In this situation, none of these differences carries any information relevant to the decision he must make. In this trial and in all the preceding trials the result of attending to a particular stimulus, such as the color of the alley, or noticing one particular aspect of the stimulus is independent of minor fluctuations, is constant. Perhaps the problem is harder: maybe the rat must attend to an aspect of the stimulus, such as size or brightness. Still, we have seen that these are simply higher order features of the perceptual world model, and they do not differ in their relation to the proper behavior from one trial to the next. In this situation, a good way for the rat's brain to proceed is to simply average all occurrences of the situation. Those that are randomly related to the outcome (ie: noise) will average out. Those that have a constant relation to the situation will be strengthened and become (appropriately, because the rat has no notion of cause and effect here) a part of the memory trace which guides his choice of response. Such a memory problem is called context-free because what has to be recalled is always true regardless of any other factors in the current situation. From the rat's point of view, the correctness of the black maze arm is a universal truth of nature. It is not a thing which depends on the present context; it is a property of black maze arms to contain food.

This sort of relationship has a parallel in human verbal memory. The semantic network in human long-term storage is essentially the representation of knowledge at the semantic level. This data base contains knowledge of objects or concepts, their properties and relationships and class memberships, and their correspondences with certain verbal objects (called their names) on the phonemic level of perception. These things are fixed qualities of the semantic constructs of knowledge. They are universally true and either do not change, or change only rarely, in which case we have to relearn some fact about them. This is the kind of information which is referred to in human memory as semantic memory. Examples are your name, the diameter of the earth, the mnemonic for "exclusive-OR" in your favorite assembly language, and an uncountable number of similar statements about the world.

In distinction to this, consider the rat in the maze, with one slight change in his situation. We shall make the correct arm on the present trial the opposite of what it was on the previous trial; if black contained the food last time, the rat must go to white this time. This relationship can be expressed as a semantic construct in context-free memory. A little more complex perhaps, but it certainly is a universal truth when applied as a description of the experimental world in which the rat finds himself. It can also be constructed by averaging as before. However, in order to apply this information once it is learned, the rat must perform a very different kind of recall. He must remember where the food was on the last trial. Only then can he apply the context-free statement of the relation between food and arm color. The location of the food on the last trial is something that cannot be averaged. If he were to do so, success would average to zero if the food were presented equally often in the two possible conditions, or he would develop a preference for the more probable connection if the food distribution was biased to one arm of the maze. Thus, the rat must remember something correctly, which is brought into the context in which the present trial occurs. The conditions prevailing on the last trial are a part of the current context of this problem and do not relate to the nature of the context-free statement of the rule stored in semantic memory. Yet, they are part of the information necessary for solving the problem and must be remembered.

These contextual conditions are unique to any individual instance of the rat's performance in the maze, and so the outcome of every trial must have a separate location in memory and remain there at least until the succeeding trial has made use of the information. We might suppose that the animal may hold such information in short-term storage, and this is probably true in the simple situation just described, but situations in which this would not suffice can be devised and we still find that the problem is solvable. In animal experiments this is frequently called context-dependent memory, because the nature of the response depends on the context in which it is occurring. A similar process in human verbal memory has been called episodic memory. It is, basically, our ability to remember individual events or episodes which are unique in our experience and not part of any averagable representation of knowledge that is independent of individual instances.

Examples of such episodic memories would be "what Fred said at dinner last night," or "what I did last Thursday," or "who I saw at the theatre yesterday." Such things are facts or events that are associated with a particular time and place and might not be true at any other time or place, or might be different in the same place at another time, and so on. They do not reflect semantic knowledge. "The color of Fred's eyes is brown" is a statement of semantic knowledge; you would expect it to be true under any circumstances (ie: in any context) in which you encountered Fred. (Indeed, our world model is such that we suppose it to be true even when we aren't encountering Fred.) On the other hand, "What Fred said at dinner last night," while still information about Fred, is not something that would be a part of the semantic network that would represent Fred. If it was typical of him, or told us something about his character or his customary behavior, this episode might enter into the formation of some abstraction about Fred that would become part of our semantic knowledge about him, but our knowledge of Fred as a semantic-level concept would not include the content of his dinner conversations. These episodes must therefore be stored as individual memories of events rather than as part of general knowledge about something, just as the rat's memory of what happened on the last trial must be represented as individual and discriminable, independent of his knowledge about the consequences of the fact in general.

There are some interesting differences in the operation of the two classes of memory. Semantic or context-free material is usually remembered easily and quickly. Fred's eye color, your name, the first few digits of π may all be recalled quickly. In general, the ease of recall of such material is a function of how frequently it is used. For things that are used often, recall is almost instantaneous. Episodic or context-dependent memory, on the other hand, is only easily recalled immediately after the episode (a probable short-term memory process), and it becomes harder as time passes, until distant events are scarcely ever retrievable without much effort.

Are these two classes of recalled events stored the same way in the brain? Some theorists believe that they represent two distinct forms of long-term storage with different properties, while others believe that there may be explanations of the differences other than the basic storage mechanism. The latter position holds that the underlying data base is the same for the two types, and that the apparent differences stem from the way the data is organized and retrieved. Long-term memory is not simply a temporal, serial store like short-term memory; data appears to be organized in long-term memory according to its class membership and relation to other items. This is the essence of the organization conveyed by the semantic network concept. The retrieval mechanism from long-term storage seems to depend on finding the right pathway through this information in a fashion analogous to some linked list or tree structures in computer data bases. The argument is then made that semantic elements in memory are constructed out of a large number of episodic elements which may link to the semantic concept in a variety of ways. As the semantic element is ever more richly bound into the structure by widening circles of association with other semantic constructs through individual episodic instances, it becomes easier and easier to retrieve because many stimuli functioning as retrieval keys will be associated with it, and many branches will lead to it.

This certainly fits with a number of facts about long-term storage and recall. Certainly we do build semantic data by an averaging process that functions not so much to drop out the noise, as to form a montage of pathways leading to a heavily burned-in, common intersection of the semantic representations of the episodes. It is a well-known fact that retrieval of long-term stored data is aided by bringing to mind as many things as possible that are associated with it. The richer the associations, the easier the retrieval. An isolated episode with few associations would thus be more difficult to find. In fact, forgetting is thought to be a retrieval problem rather than a loss of stored information. It is thought that data are forgotten because insufficient retrieval keys are present and because of interference from other similar items which share similar associations and which are compatible with the available associations. This latter effect would be more detrimental to the recall of episodic or context-dependent memory than to the recall of context-free or semantic data.

In the context-free case, there is no problem when "all roads lead to Rome." That is, if the data desired are context-free, failure to discriminate their recall by association with one episodic "overlay," from recall generated by a another route, is of no consequence. The item to be recalled is the same in either case. If, on the other hand, we are trying to find a particular episode, it may be very difficult to discriminate it from a host of similar eposides, because some very specific retrieval keys are required to distinguish it from all the others. If you want to recall some fact about cocktail parties in general, that fact is the only key you need in order to retrieve the information. You immediately get a recall of the essential features of cocktail parties you have attended or read or heard about. Contrastingly, if you want to recall what the hostess was drinking at a party you attended on the 15th of February, 1975, you will find the task very difficult. In fact, it would be hard to recall any specific aspect of a specific party. It turns out that the way to approach such a recall problem is to try to bring to mind all of the data you possibly can that were associated with that particular party. In other words, reconstruct the stimulus context with which the data was associated in memory. The more such contextual data you can assemble, the greater the likelihood of finding the desired episodic information associated with it. It does not have to have any meaningful association, as long as it was presented to the senses contemporaneously with the desired information. (For this reason, the best way to improve on recall on an exam is to study in the same room in which you will be examined. The incidental stimuli of the room discriminate those data from all others and therefore help to retrieve it without interference from other data accumulated in normal study surroundings.)

One might view semantic or context-free storage as a construction which emerges out of the accumulation of episodic or context-dependent data storage and which gains in recallability at the expense of particularity. At one extreme is a piece of information which was only true of a specific occurrence in a particular setting, and at the other are general truths, such as the definition of a word or the sum of two numbers, which will be true in all the instances encountered. Some kinds of information develop naturally into semantic or context-free forms with increasing experience. In this case, the difference in the two types of data would indeed be more in the circumstances of recall and access than in the basic nature of the storage.

As pleasingly parsimonious as this view is, there are other data which suggest that the two forms of information may be represented by different mechanisms of storage. One such line of evidence comes from the cases of amnesia. There are a number of kinds of amnesia, and it is not certain what their exact mechanisms are, but it is clear that semantic recall is often unaffected by brain damage that eliminates episodic information. Thus, the typical amnesic patient will not lose semantic information: he does not forget the words or grammar of his language; he knows what objects are and what they are good for. His deficit is in the particular events of his experience; he can tell you what a car is and how to use one, even drive one, but he cannot tell you about any particular time in his life when he has personally taken a ride in one. He could not describe any episode involving a car.

The corresponding animal studies are beginning to demonstrate a similar sort of effect in the sphere of context-dependent, information-based learning situations. A structure of the limbic system known as the hippocampus and its related connecting structures have frequently been implicated in instances of human memory disorder. However, in animal studies, hippocampal damage is usually reported to have little effect on learning. Recently, a number of studies have clarified this discrepancy by recognizing the distinction between context-dependent and context-free recall in animal learning situations and demonstrating effects of hippocampal damage on context-dependent learning situations. While the final word is not yet in on this approach, it seems that this distinction will help to bring the animal experiments into line with the human clinical studies.

The importance of this in the present context is that it suggests the dissociability of the two kinds of retrieval process on the basis of the underlying physical operations, and this in turn suggests that they are different entities rather than one data base manifested through different approaches. It is difficult to see how the arguments presented earlier could allow a separation between the two extremes of retrieval requirements on the basis of differing structures involved in the physical retrieval. (A threshold argument might be advanced, however.)

One consequence of the distinction between context-dependent and context-free storage may help us to understand some of the seemingly different kinds of memory tasks discussed at the outset of this chapter. Some tasks require that a particular series of operations be run off in a particular sequence every time, independent of extraneous information. As an example, the performance of a skilled motor task, such as playing a piano, is a sequential process. Verbal tasks may also be organized in this fashion; saying the alphabet is easy when we start with "A" and proceed in sequence. Try starting with "R" and proceeding backwards, however. Try starting with "B" and reciting every other letter. Most people will have to mentally recite things in the ordinary sequence and select what they want for verbalization. One of the standard tasks used by the psychologist when he wants to interfere with mental processing of something else is to have the subject count backwards by threes. The difficulty we have with tasks done in an unusual fashion seems to stem from the fact that they are organized in our memory as a linked, sequential list in a computer data base. Each item contains a pointer to the next item. If you fall out of sequence you can't find any of the items in the list without going back and starting over, and you can't skip over the retrieval of any item.

Other kinds of tasks involving context-dependent information resemble multilinked data structures with random access entry points and many kinds of linkage pointers. The type of pointer used at any branch may be determined by contextual cues peculiar to the current situation, so that the same data may be accessed in different ways at different times; getting lost does not require a rerun of the entire sequence. In fact, there is no necessary sequence and data may be accessed without reference to any other item. We may have to run through the alphabet in order, but we don't have to look at all the preceding words in a mental dictionary in order to find the meaning of a word. Both kinds of data arrangement have appropriate uses just as they do in computer data bases.

One of the consequences of this distinction is that context-dependent and context-independent tasks should generate different patterns of responding when an error occurs (as a result of having to restart the sequence in one case). The hippocampal damage mentioned above tends to affect the error distribution in such a way as to indicate a deficit in context-dependent processing.

IV. CONTENT AND ASSOCIATIVE ADDRESSING

In a standard computer memory, addresses are used to access the storage location of a particular byte of information; ie: the address must be known in order to obtain the byte. In more recent developments, content addressable or, occasionally, associative memory systems have appeared. In either of these cases, the underlying principle is the provision for data retrieval by a keyword which is associated with the desired information, either by being an actual portion of it, or simply by being associated with it in storage. In such a system, a typical operation might be the return of a multi-byte record when the first several bytes were provided as the address. This permits the storage of information anywhere within the physical memory; a program only needs to know what data or keywords are required and does not care where they are located in the physical memory. There are many applications for such a system, and it is no surprise that the brain employs a similar method. Indeed, the brain's action in this regard was known long before the advent of these new memory systems and may even have inspired their development.

I have already implicitly considered this associative property of the brain's memory system; I have observed that the way to obtain data from long-term memory is to assemble as many stimuli as possible which were associated with the desired data at the time of its entry into the system. The more explicit the set of retrieval keys, the better the chances of accessing only the desired information. There is a further aspect to the addressing of brain memory, related to these considerations, which takes us beyond the realm of current computer techniques. This is the property of brain memory which is implied when the term "associative" is used in its psychological sense.

In earlier chapters, this associative property of neuronal memory was invoked to explain certain hypotheses regarding the ability of the brain to generate intuitive solutions, as well as to generate paths to solutions on the basis of reasoning by analogy with similar known situations. As we used the term there, associative referred to the property of biological memories which allowed them to return items *similar* to the one specificially addressed. Now this aspect may be considered in more detail. In the brain, long-term memory appears to store information according to its semantic category relationships, rather than in terms of sequential order, as in short-term memory. Hence items which share some semantic-level properties will have similar pointers or retrieval keys for their access. The utility of this fact lies in what we mean by similarity and in the underlying nature of the semantic conceptual organization.

Similarity is a concept which is intuitively meaningful, but difficult to define. It is probably the case that similarity is a statement about our perceptual experience which reflects something about the organization of our brains, rather than about the nature of the objects in the external world which we are calling similar. The dimensions of perceptual experience on which we choose to rate objects as similar or dissimilar are purely arbitrary and may even vary according to the purpose which we have in mind at that particular time. These dimensions of perceptual experience may be at low levels of abstraction, as when we see a dinner plate and a wheel as similar because of geometric shape, or at high levels of abstraction, as when we see a wheel and a tank tread as similar in function. In general, the more dimensions in which perceptual entities are of the same class, the more similar we perceive them as being. A wagon wheel and an automobile wheel are seen as very similar because they are members of several of the same perceptual categories at several levels of abstraction in several perceptual dimensions. Some of these perceptual categories seem to be wired in: color, size, shape and so forth, for example, seem to be primitive features to which all nervous systems attend. Other categories are learned, but this does not preclude their being perceptual dimensions of the world which we experience.

I would like to suggest that similarity is dependent upon the way in which the brain defines perceptual dimensions, including the way in which it selects important ones in relation to the goals of the moment. As discussed in earlier chapters, it is probable that the internal code, in which the central machine represents the world model and operates upon it logically, is the same as the high-level, perceptual representation code. This reflects the fact that perception evolved within the world which it models and has incorporated the logic of its relationships into the brain's own processing. An immediate extention of this line of reasoning suggests that the code of the semantic level of representation is fundamentally the same. Certainly, the non-real-world items which we might wish to represent at the semantic level are a very late evolutionary option, and they form only a tiny fraction of the material which even human brains are called upon to represent. If this picture is substantially correct, then the code in which semantic items are represented in long-term memory would probably either be identical with or at least be organized along the same dimensions and categories as the perceptual representation of the world model and its temporal and spatial (hence logical) relationships.

Of course, in a trivial sense, we know that we do operate this way. When I recall items from memory, they are organized and classified along lines of similarity that are essentially perceptual if referring to features of objects, or which reflect the logical relationships of real-world spatial relation or causality. I am suggesting, however, that we do operate in this fashion, because those are the dimensions of the code which represent and retrieve the items. If this is true, then there is a very simple explanation of how human associative memory might function.

According to this view, the retrieval key would be represented at the semantic level in the perceptual code just as any other semantic construct would be. Thus, the representation of the key in the system, whether generated internally or presented from the external world through the perceptual process, will automatically activate the appropriate permutations of fibers for addressing the information associated with that code. Perhaps some sort of memory "read enable" signal is required, but that is not important here. This mechanism gives us an operation that is reminiscent of content addressable memory in computer systems. In order to obtain the kind of associative action that is seen in brain memory, we need to invoke another concept which has already been introduced—the variable threshold aspect of neural operation that allows the "ALMOST gate" type of action when applied to decoding networks. If this feature of those neurons performing the perceptual decoding function is applied in the retrieval situation, it will generate a "fuzzy address." Any given line or set of lines encoding a feature of the key will be selected not only by that key, but also by others that are less than the optimal stimuli for those neurons. How much less would depend on the level to which the cell's internal bias was raised. Conversely, any given key would decode not one specific line or permutation of lines, but a whole set of such codes including both the one for which it was optimal and ones for which it was only almost optimal.

If the storage code for semantic items is the perceptual code as suggested above, then this kind of action will cause the recall of the specific item required, as well as a number of other items related to it *by similarity along perceptual dimensions*. This is the desired operation. It accounts for the fact that associative action in memory recall gives us items which are similar to the ones we have requested. Thus, by combining a very common feature of neural operation with the hypothesis that the semantic memory code is the same as other encodings of information in the brain, we can generate a hypothetical system which will have properties like those actually observed. In addition, it is easy to understand how the recall process could be made more or less selective. If there was a widely distributed bus exerting a biasing action on all of the neurons performing the decoding of the recall key into memory, then we could require that only very similar items were to be accessed, or we could conduct a broad search.

It is necessary to discuss the reason for organizing the associative recall procedure around similarity in perceptual dimensions, or, indeed, for having associated items returned in the first place. Here, as in all other areas, it is important to bear in mind that the brain is a machine for solving real-world problems. For the most part, human beings reason by analogy. We do this because it works. It works because we are reasoning about the real-world, and in the real world causality functions such that similar things very frequently operate in similar ways. If you are doing formal logic or algebraic operations, analogy is not a very productive way to proceed. Minor differences between two formulae or between two logical statements may make a crucial difference. Of course, similar things don't always work the same way in real-world problems either, but reasoning by analogy is generally a very useful first attempt. To the extent that our perceptual systems have evolved around the definition of perceptual dimensions, which are related to the causal connections and physically important properties of the world, reasoning by similarity is frequently very powerful. To do this we must be able not only to recognize similarity, but also to recall similar situations from our past experience. Points of similarity between current events and past events whose outcomes are part of our store of information must be noted. One way is to allow the perception of the current stimuli to function as recall keys in an associative or fuzzy memory system organized around perceptual encoding schemes.

Another function of this type of memory system is in the recognition of final or intermediate goal states which are usefully similar to those involved in a current problem and with which we have had experience in the past. I may never have had to find a path to the present goal state before, but its representation in a memory recall operation may produce a similar past situation, which in turn has a remembered path for its production. If it is similar enough to the desired state, it may constitute a sufficient solution; if not, it may prove to be a valuable intermediate. Associative memories, in our sense, are thus fundamental to the operations that adapt brains well to real-world problems.

V. THE PHYSIOLOGICAL BASIS

With so much known about the details of organization and types of operation in the brain's memory systems, it is remarkable that so little is known about the underlying physiological mechanisms. This difficulty is underscored by some of the early experiments in the physiology of memory, when it was discovered that the degree of impairment in learning seemed to be related to the amount of cortex removed, without regard to what areas were involved. This led to the formulation of the so-called law of mass action, which held that memory was somehow homogenously distributed throughout the brain. Although this view has been severely restricted in recent years, there is still enough truth in this interpretation of the results to warrant support for some kind of distributed storage process. We shall discuss one sort process in the next section.

A fantastic variety of hypotheses have been offered in explanation of the fundamental physical nature of storage; ranging from encoding in the molecular sequences of large molecules through encoding in changes in the cell's electrical activity to changes in the connections among neurons. Some evidence has been produced for most of these ideas. It is generally assumed that the basis of long-term storage is somehow involved with an actual structural or chemical change in the neurons, whereas short-term storage is probably retained in patterns of ongoing neural activity. The reason for this is that short-term storage is generally lost whenever the brain is subjected to any process which would tend to disrupt or confuse patterns of neural activity. Thus, a severe blow to the head, an epileptic seizure, the passage of strong electric currents through the brain, or the administration of convulsive drugs will all produce a type of amnesia called retrograde amnesia, a loss of memory for those events immediately preceding the disrupting event. On the other hand the contents of long-term storage are not affected by such procedures.

This idea fits with the conception of the operation of the short- and long-term storages; short-term storage is seen as temporarily holding the information during the time required for its translation into a more permanent physical record. It is easy to understand how short-term storage might be accomplished with maintained patterns of neural activity; the material to be stored is already presented in that format. The only problem would be a way of maintaining the activity, which hardly seems difficult. The real problem is how to transfer that activity into some sort of structural change in the system, and how to get it back out into neural activity again.

In work done on very simple learning, such as habituation of a reflexive response to a stimulus in simple invertebrate organisms, it has been possible to show that the change in response occurs by virtue of a change in the ability of one neuron to excite another across a synapse. This is equivalent to a change in the synaptic weighting function in our electronic model of the cell. This type of action might be produced by any number of mechanisms, chemical changes, changes in contact area, and the like, but it is interesting just to know the site of the action. Unfortunately, it is a long jump to extrapolate from such cases to memory in the human brain. We might well suppose that similar operations, such as conditioned reflexes, could store the necessary changes in function at the interactions between the neurons involved. Indeed, we have already speculated that many kinds of sequential stimulus-response sorts of learning may involve changes in the hard-wiring of the circuits that actually perform the response. This seems less likely to be the case in more elaborate types of data storage in the brain, particularly in context-dependent operations, although it is possible. The complexity of keeping separate all of the stored data-based operations that use the same apparatus is a formidable problem. Consider how the storage of verbal memory might be encoded in changes of the neural weightings involved in the apparatus that produced the speech, or the apparatus that generated the content of the speech. A great deal would seem to have to funnel through these synapses to account for all of human verbal output, and it seems intuitively unlikely that it could all be encoded by the weightings of the synapses involved in the basic processing.

It seems more reasonable that, in this sort of storage, the brain has resorted to the computer-like option of providing functional systems specifically concerned with the storage and retrieval of information. I have already mentioned that some areas such as the hippocampus of the limbic system may be concerned with such operations, and evidence from the stimulation of human cortical areas also suggests that some regions may be specifically related to episodic memory retrieval. In particular, the cortex of the temporal lobe has been reported to produce the experience of fragments of long forgotten memories when stimulated electrically during brain surgery in conscious patients. It seems that this is due to the area's connections with some distant region, rather than to a storage function in temporal cortex, for the same effect is obtained when the cortex has been removed and the underlying connecting fibers are stimulated. Thus, temporal cortex may be concerned with organizing keys for recall rather than with actual storage.

In animal studies, damage to a great many areas impair learning or retention, as assessed by performance on various learned tasks, although it is extremely difficult to be certain that memory functions themselves are involved. Such impaired performance can easily result from deficits in sensory or perceptual abilities, in motivation, in attention, in ability to generate the proper response, and in almost any function involved in the performance of a complex task. Ingenious controls have been devised to sort out these possibilities, but we have yet to demonstrate conclusively the necessity of *any* structure for a definitive storage function. One complicating possibility is that the brain, when deprived of one kind of storage, may make do with other kinds to solve the problem. It is an amazingly flexible machine.

On the chemical side, we seem to have made more progress. There appears to be substantial evidence to implicate various chemical neurotransmitters in the process of consolidation (the name for the process of transferring information from short-term to long-term storage) and in recall. Unfortunately, we do not know the exact nature of the mechanisms that employ these transmitters, and therefore, their existence is of little value in the present context.

VI. THE FREQUENCY CODING MODEL

While dealing with the higher perceptual processes, I devoted some space to the discussion of spatial frequency-based models of image processing. At that time, it was noted that spatial frequency features were no different in principle than other, more geometrically intuitive features as ways of describing or encoding the patterns of activity on the receptors. It was also pointed out that such an encoding scheme possessed certain advantages peculiar to Fourier-like processes. Among these was the ease of separately representing form and location, thereby allowing location-free recognition. I noted that the spatial frequency-coded representation of the perceptual world might be an appropriate form in which to access memory for the recognition tests necessary to the active re-tuning process.

Considerations of the properties of physical data storage systems have also struck many neuroscientists as highly reminiscent of the properties of memory systems in the brain, and a good deal of work has been devoted to the development of possible neural analogs to these processes. For example, a holographic negative is a memory of sorts, and it is based on the Fourier Transform process which imparts to it the property of distributed information storage. That is, destroying parts of the photographic memory only serves to degrade the overall quality of the subsequently reconstructed image, rather to eliminate specific portions of it. Additionally, holographic information storage systems can be easily employed to emulate other facets of brain memory, such as the associative recall and cross-correlation (hence recognition) functions.

The hologram is only one member of a broad class of transforms involving Fourier-like procedures; any number of such schemes could be devised to produce desirable information storage and recall systems. The extreme flexibility of neural processing systems also lends them well to hypothetical systems for employing these approaches. The holographic model, however, has received the greatest attention, and there are real physical analogs between typical neural systems and the optical systems used to generate holograms. (For a brief introduction to this topic the reader should note the Pribram, Nuwer, and Baron reference in the bibliography.)

In essence, it is a property of lenses that a Fourier Transform relation exists between the geometric object at the front of the lens and a spatial frequency transform of that image at a plane behind the lens. This spatial frequency transform cannot be captured directly on film because the phase relations of the light are averaged over the exposure time. However, with a slightly more complex procedure, employing a coherent reference beam of light, an image of a transform can be recorded which will be re-transformable into a geometric image when illuminated by the reference beam at a later time. This image recorded on film is the now-famous holographic image, and it has a number of remarkable properties. Aside from the distributed nature of the information mentioned above, it is the case that when a properly prepared type of hologram is illuminated with another image, rather than with the reference beam, the output is the cross-correlation function of the two images. If they were identical, this would simply be a bright point. If not, the intensity of the output at that point is a measure of the similarity (at least in geometric terms) between the two images. The applications to quick recognition are immediately obvious and have been exploited in holographic information storage systems. Additionally, such optical holograms display a property similar to associative memory. If a hologram is made with a reference beam and a particular image and is then illuminated with a portion of that image, the remainder of the image tends to be reconstructed. This is clearly a form of content addressable memory. However, it is also associative in the strong sense mentioned earlier, in that some response will be generated not only to incomplete images (which are, of course, similar in one sense to the whole image), but also to other images. These images are similar to the image which produced the hologram but in ways other than simple incompleteness. The intensity of the activated image will be proportional to the geometric similarity between it and the illuminating image.

These considerations, not the more familiar property of creating a three-dimensional image in space, make the holographic model so attractive to neuroscientists as an analog for brain storage processes. Indeed, there is no reason why the image ought to be reconstructed in the recall process (who would look at it?). It is only necessary that some set of neural activities which encode it in unique form be activated. This consideration considerably frees up the constraints on the model, for there are many transform processes which have desirable properties, but which do not have the property of reversibility.

All of this is interesting from the point of view of brain modeling only because there is a real possibility that neural systems may be wired up to operate in this fashion. The essential feature for the functions just described is inherent in the transformed representation of the image in the frequency domain. The physical system accomplishing it is not crucial. As it turns out, the actual connections in neural networks are such as to permit this kind of tranform. Basically the process is one in which interaction by lateral inhibition between adjacent pathways conducting geometric information performs the role played by phase interference in the optical system. Moreover, models have been devised in which the only requirements made on the neurons as processing elements invoke well-known properties of real neurons. Finally, in many of these models, the actual physical change in the neural system which accomplishes the storage can occur as a change in weighting of synaptic transmission efficiency. We have already seen that this appears to be the actual physical substrate of some kinds of simple information storage in real neural systems. (Not that those systems are thought to be holographic, but there is an interesting parsimony in employing the same basic substrate of plasticity in more complex storage arrangements.)

Neural systems even have certain advantages over optical systems as ways of storing holographic information. It is quite possible to encode the information corresponding to the mathematical imaginary part of complex numbers in separate neural lines from the mathematically real parts. The result of this is that neural systems could dispense with the reference beams and coherent light required of optical systems. The photographic plate requires these because it cannot make this separation.

There is little hard evidence that the nervous system does use frequency-coded or Fourier-transformed approaches to data storage and retrieval. There is circumstantial evidence in the observation that many of the real transfer functions which have been observed in neural networks, such as those of the receptive fields of the visual system, are susceptible to explanation as Fourier transformations of the data. Unfortunately, some very difficult observations on minor aspects of the receptive fields are required to differentiate between the predictions of these and other explanations.

Whatever the true nature of the brain's storage and recall system, it is clear that the spatial frequency transform approach has much to offer. In particular, its operation is very well suited to parallel-processing systems of the sort that we are proposing here. The transforms could be computed on a serial machine, but it would be prohibitively time consuming. The present approach is inherently parallel. All parts of the holographic image are created simultaneously, and in most of the neural models the transformation may be accomplished in one pass through a single layer of laterally interacting neurons acting as simple parallel processors. This is not the place for a detailed consideration of the mathematical underpinning of this process, but the reader interested in applying these procedures to brain-like parallel machines is urged to peruse some of the references in the bibliography.

VII. ARTIFICIAL SYSTEMS

No one familiar with computer hardware can fail to have been struck by some of the similarities between the various functional memory processes of the brain and well-known forms of data storage in present day computers. I have remarked on the resemblance of sensory memory to an input latch and on the similarities between short-term verbal storage and a working register (or perhaps a stack area in core). Long-term memory, with its permanent nature, longer look-up time, categorical organization, and necessity for being accessed via return to short-term memory is very like a file-oriented mass storage device such as a disk.

Some of the properties of brain memory are not so obviously analogous to common existing counterparts in the computer world; some of them should be examined as potentially interesting processes for electronic emulation. The first of these is the role of the attention process in selecting that information from sensory memory which will be transferred into the short-term storage mode. In a traditionally programmed computer, this function would have to be built into the program. The program would have to know what it was looking for and when and what to examine. I have remarked earlier on the similarity between the brain's goal-directing system and the inherent purpose of a traditional program. We have also examined the close relationship between the attention function and the goal-directing system as it applies to the direction of the higher processes and the operation of the perceptual systems. It would seem that in this instance as well, the attention function acts as a mediator between the systems that specify general goals for the organism and the systems that execute functions in the pursuit of those goals. Attention insures that relevant material is retained by the system for further examination. This might be as simple a matter as a default decision to retain and examine novel stimuli in the case of no other overriding need, or as complex as capturing the information needed for a particular logical response. The role of attention here seems very like, and heavily intertwined with, its role in perception; a full explanation of what we usually call attention would probably have to include our experience of temporarily storing relevant things that "caught our attention."

This function does not seem necessary in the typical computer applications seen today. We don't want to give our big business machines the liberty to decide what data to examine when they are doing the payroll. In fact, in most current uses, the computer has such a restricted means of getting data that such an attention function would be useless. Everything that comes in through the card reader is relevant. In these cases, it is quite appropriate to let the fixed program decide what shall be stored and where.

In machines designed for the future, we will want to be able to specify only general goals, to give them the sensory capacity to receive large amounts of unstructured data from the environment, and to decide for themselves what is worth processing further. If they are brain-like in their ability to perceive truly vast quantities of environmental data quickly, they will face the same problem that the brain does in deciding what to store and operate on, and what to let pass without further action than recognition. Assuming that these machines will not have active storage to waste, they too will want to limit retention to things of interest. In the case of a machine designed to operate in a generalized environment, it will be impossible for these decisions to be made in advance by the programmer, and we will have to rely on an attention-driven process for filling active storage from raw input as in the brain model. In a brain-like machine which has developed an attention system as the operating arm of its goal-direction circuitry, this is not a difficult extension.

Another potentially interesting consideration is the application of parallel architecture to memory searches. Although the brain's long-term storage mechanism acts as if it were organized in a semantic network like a more traditional computer data-base system, the brain's parallel architecture enables it to search such networks differently. Even leaving aside the possibility of some very esoteric holographic recall system, the search procedure could be conducted in a brain-like machine by moving simultaneously through large numbers of branches of a tree or network. In this regard the search process would be quite analogous to the feature extraction process, in which a very large encoding job is handled in a few steps. If the material were in fact stored on the basis of some sort of Fourier Transform-like operation as the coding scheme, this statement would still apply; we would simply have changed the rules for the code that placed the data into the tree or network. The only difference would be that the item would not reside in one node, but would be in a code which was distributed over many nodes.

The Fourier approach certainly seems feasible when the task is to store sensory input. However, when the task is to store more abstract material, it is not readily apparent how we might proceed. To be sure, the mathematics of Fourier Transformation is completely oblivious to what it is representing, yet we must somehow put such material into the proper relationships for the transform to act on it in a way that is meaningful as well as mathematically correct. In the case of visual input, an appeal to the relation between the geometry of the real world and its spatial frequency description suffices, but the procedures for dealing with purely cognitive categories will clearly require more extensive processing of the data. It might well be that the optimum trade-off would involve the spatial, frequency-coded, distributed storage route for some kinds of data and a more intuitively coded semantic network for other kinds of data. Either coding scheme would benefit from employment of a machine with parallel-search mechanisms.

Associative recall by similarity has also cropped up repeatedly in our discussions of various types of processing in the brain, and it certainly seems to be a property of brain memory systems. In this chapter, I discussed the possible mechanism by which neurons might employ the "ALMOST gate" mode of function to achieve such a fuzzy address mechanism, provided that the storage code were built up in relation to meaningful dimensions of relationship between the materials. It is not difficult to imagine an artifical system which could work in this fashion. It might accomplish the necessary search by actually computing numbers to represent the desired intermediate values of coupling in the system, or by employing, as in the neural case, a continuous rather than binary logic in its basic gates. Multi-valued logics are not new in electronic logic, and an analog machine is essentially a continuously valued logic machine. A mathematically similar machine has been proposed by Dr James Albus (see BYTE The Small Systems Journal, May thru August, 1979). This machine is based on his conception of cerebellar function and has a number of interesting properties. Among these is the ability to deal with fuzzy addressing in multidimensional vector spaces.

To employ any such techniques for fuzzy addressing, except in the smallest cases, would seem impractical in a serial machine. Even in a parallel machine, doing simultaneous searches of the points in the multidimensional vector space defining the address, there is the problem of how to simultaneously return all of the items found within the fuzzy address region at the same time. The busing problems are enormous if we assume that the treatment of a single complex memory item probably requires a large share of the system's bus space in its representation. This may, in fact, be a limiting factor in the number of items from memory with which the brain itself can deal at one time. In a machine which made use of the brain-like spatial-temporal byte scheme, it might be possible to multiplex items on the bus in different frequencies contained in the temporal byte. Not using a clocked system and substituting multiple parallel components for speed may make this practical. There is at least some evidence that the brain might be using such a frequency-multiplexing scheme in memory recall.

One problem occurs in attempting any such brain-like scheme for associative addressing. That is the problem of treating recall in categories whose perceptual dimensions are not continuous. If this occurs, that is, if storage categories do not transform smoothly into one another along the storage dimensions, then the fuzzy address scheme runs into trouble. It is the same sort of problem noted in the Fourier scheme when it was asked to deal with abstract categories. Here we have an easier time, at least conceptually, in dealing with abstract dimensions, but these are not always continuous categories. It may be that the best that can be said is that (1) most similarities in the real world are not discontinuous, and we are dealing with a machine to solve real-world problems, and (2) we have already accepted the idea that reasoning schemes relying on association by similarity are going to make errors anyway as the price for the power to reason by analogy.

10 The Minds of Men and Machines

I. REVIEW OF THE BRAIN MODEL

In the preceding chapters, you have seen something of the way the brain is put together and how it functions. At least you have seen one possible approach to understanding the brain. There are many ways in which the functions of the brain might be conceptualized, even given that you and I were in basic agreement about what it does. The brain is so complex that any model used to deal with it in our thinking could only be an approximation, and it might emphasize different aspects or represent them differently than the model of another.

The paradigm developed also suffers from being incomplete; I have had to pick and choose from among a wealth of functions and topics of interest those facets that might have some bearing on artificial systems. Further, I may have missed the overwhelming importance of some facet of the brain for illuminating an important approach or ignored new technology which may tomorrow make relevant what today seemed unrelated. Finally, there are many missing links in our understanding of the brain. In many cases, I have had to present one theoretical approach for clarity where many exist and contend over minutiae of evidence; in some instances the model is necessarily more speculation than fact. In other cases, the constraints of an introductory treatment have forced simplification of complex issues, perhaps overly so at times.

Yet I come to this end of the book with some sort of model of brain function. If you look back again at figure 1.5 of Chapter 1, you should now see it in a new light. Where before the figure was an empty shell, your mind's eye should now be able to flesh it out with a wealth of detail. Into its abstract structure, you should be able to place the real anatomical pieces (although many of them reside in more than one place) and be able to see in many of its relations the actual neural connections at work.

In preparation for some of the notions in this chapter, I would like to briefly review, in global terms, this model of the brain. Some remarks about figure 1.5 from the vantage point of the intervening chapters will suffice to make the necessary points.

First, note that vertical orientation in the figure represents level of processing within the major functions. To a first approximation, it also represents an anatomical and an evolutionary direction. To the extent that the levels of processing can be characterized as reflexive, emotional, and learned, there is also some relation in this dimension to the ideas developed by Paul MacLean and popularized by Carl Sagan in *The Dragons of Eden.*

Ascending from the bottom to the top of the figure, notice that the columnar cross sections become narrower. This reflects the nature of the data base on which different levels of the major functions are operating. At the bottom, the input structures are dealing simultaneously and in parallel with an entire retina, the whole body surface, or the basilar membrane. Each point in one of these planes is a piece of data ascending through the system in parallel with its neighbors on all sides, generating interactions with them as it goes. In the output column, the structures at the bottom are producing simulataneous commands to uncountable individual muscle fibers. As the output information descends from above, each part of it is elaborated into yet finer grained commands to yet more particular elements. As this elaboration proceeds, it retains the basic plan from which it sprang, albeit modified at each level by corrections for local conditions and by interactions among adjacent processes.

Most data are handled at the lowest levels. The majority of jobs can be performed by functions that require the data to rise only one or two levels before being sent to the output side. At best, it requires modification by interaction with steady state setpoint commands received from a higher level. Some data rise much higher in the system. As it does so, it is organized to become at once more abstract and global in its representation of the world and, at the same time, more restricted in quantity. At the top, the paths are reduced from square areas representing vast arrays of parallel data points to lines representing (in the case of short-term memory) seven items of high-level symbolic thought. From the top, a small number of powerful global statements of an action plan can set in motion the whole cascade of output processing that drives the motor act.

All these operations could compose a fascinating machine, different in architecture from those of our experience and possessing awesome power for dealing with the kinds of problems for which it was evolved. Yet this device is insufficient for dealing with the real world: it would never be possible to build into its structure the required relations between input and output to handle all of its possible interactions with the unstructured environment.

To adapt our machine to the generalized environment problem, we must add a goal-directing function, represented here in the center column. At the bottom, this system takes as input a mass of information about internal states of the organism, and at higher levels, information from the input system about states of the environment. All of these it processes in its own way, quite differently from the way the input system does, thus producing a motivational world model. This is a representation of the system inputs interpreted in terms of their relevance to the fundamental goals built into its architecture.

These fundamental goals may not be changed. They are the final authority against which all other things are judged. The machine would have no basis against which to judge the desirability of changes in the fundamental goals. Located here are such goals as defense of the organism, obtainment of energy, and other aspects of survival. If these were all that the system could specify as its goals, it would not be capable of very advanced behaviors. The real complexity of the central column system is not simply for checking inputs and outcomes against these simple goals, but for elaborating more complex secondary goals, validated against primary goals through past experience of utility or by the logical inference of the outcome. There is no limit to the intricacy of this secondary goal structure. It may be elaborated or changed and refined by the machine at will, because the primary goals always exist as standards against which to judge the outcome of specifying newly contrived elaborations. It is this operation which allows the machine to respond with superbly flexible behavior in the face of a complex and changing world, while still being fundamentally directed by a handful of simple primitive goals.

At every level, this goal-directing system uses these constructs to specify, direct, and alter the operation of the input and output systems. Hence it is shown standing between the two and interacting with them at each level. Its operation may be as simple and general as bringing the system to a state of active processing when conditions require attention. It may be as complex as judging the relevance of the predicted outcome of an action to an abstract derived goal state. At intermediate levels, it may define categories of stimuli for special processing or modulate in real time the intensity of behavioral outputs on the basis of a goal-relevance analysis of their results. In one sense it acts as a living program for the rest of the device by constantly modifying its own operations and those of other systems; it is a dynamic program growing out of a very small base of fixed and inflexible statements.

Although we can now understand actual mechanisms to some extent, there is another aspect to the operation of brains which has, as far as we know, no counterpart in our current computing machines. This is the phenomenon of mind. The operation of the machine diagrammed here results in our experience of things called consciousness, subjective experience, emotion, and the like, in addition to the behavioral outputs with which I have been primarily concerned. This final chapter gives some consideration to the operation of mind in biological brains and explores the implication for mind in advanced artificial systems.

II. THE MIND-BRAIN PROBLEM

On one level the question of the relation between brain and mind is insolvable. Mind is a private experience, and there is no way of knowing the contents of any minds save our own, nor even of knowing of the existence of any minds other than our own. That you tell me about your mind is irrelevant. It only shows that you have certain kinds of verbal output behavior in response to certain stimuli. The fact that we both use the same words to describe our subjective mental responses to the same sorts of stimuli only shows that we have learned to associate certain words with what we feel, whatever that may be, when the particular situation is presented. It in no way proves that what we experienced was even similar. Given this, there is no way that we can test assertions about minds in general.

Worse, if we try to talk about the relation between mind, which is not a part of the physical universe, and brain, which is, we are immediately confronted by the problem of how to set the two in any sort of correspondence. If the mind is not physical, then how could we establish laws which would allow it to act upon or be acted upon by the physical brain? It is, and can only be, an act of faith on our part to assume that the operations of our minds, which we experience directly by immediate introspective experience, have any relation to the operations of the physical brain or the physical behaviors which we observe as aspects of the physical world. The two modes of observation are different, and we cannot place events in the one realm into correspondence with events in the other.

Having accepted this apparent limitation, there are two possible courses of action. One is to give up the whole enterprise. In this view, it is the proper place of neuroscientists to explain physical behavioral phenomena on the basis of the physical operation of the brain and its physical inputs; somebody else can try to explain the relationships that exist between sequential states of their mental experiences. The other possibility is to accept some things on faith. We all accept the idea that other people do have minds, at least in practice, if not in some strict logical sense.

In the same way, most people are ready to accept the proposition that mind and brain have something to do with one another. They may want to quibble about which one is the dominant member of the pair, but, by and large, everybody feels that their mental states have something to do with their physical behavior, and most would not object to acknowledging that the brain was involved in the process. If we will at least go so far as to accept this much, then there are many interesting issues that can be raised concerning the relation between brains and minds and the potential relations between minds and machines.

One question that can be posed is whether causality runs from mind to brain or from brain to mind. Does our brain work as it does because the mind directs it, or do we experience the phenomena of mind as we do because the operation of the brain generates them as a byproduct? (One cannot appeal to the subjective sense of free will here, because if brain determines mind, it is determining that impression of free will as well.) Another question is whether the mental or the physical realm is the "real" one. We only know of the external world through our senses, and we only know of what comes in through our senses from our subjective mental experiences. It might even be possible that we made the physical world up entirely as a figment of our imaginations. (Don't think so? Remember that moving piece of paper!) On the other hand, it might be that there is only a physical world, and our minds are really somehow a part of it.

Any library contains hundreds of pounds on speculation on these subjects written over the last four millenia. I do not propose to settle the issue here. It is not strictly necessary to even take a position on the issue of which of the two worlds is "real," but I will offer my own view: there is no distinction to be made between brain and mind, and the appearance of two worlds of experience, the mental and the physical, reflects only two different methods of observing the same phenomenon. According to this view, that event which we observe through the senses as neural activity in a part of our brain and that which we observe by introspection at the same time as part of our direct mental experience are the same event. This view does not specify whether both are really physical or both are really mental, or whether mental and physical are only names we have given to events as observed in these two different ways. It might well be that the underlying event is something else, out of which we construct both the apparent physical world and the apparent mental world, according to the way we observe it.

This view is convenient in several respects. First, it does away with the problem of how to correlate events between the mental and the physical worlds. There is no need to; they are only different reflections of one thing—they are connected by identity. Secondly, it frees us from the necessity of deciding whether the brain runs the mind or the mind runs the brain. This becomes a pseudo-question; there is only one operation, and we are free to seek its set of governing laws without worrying how their outcome is made to prevail in the "other" world. There is no necessity to make causal connections between the mental and the physical realms.

I find this a compelling view which arises naturally out of the experience of neuroscience. It seems clear from a study of brain function, and particularly from the study of human reports of subjective experiences attendant on brain manipulation, that a one-to-one correspondence exists between brain function and mental experience. Activity in the visual cortex is always correlated with subjective visual experience. Activity in the parts of the limbic system dealing with emotional processes in behavior is accompanied by reports of subjective feelings. Coming from the other direction, reports of subjective decisions to act accompany the activity in motor systems which produce the actions. True, this one-to-one correspondence might be due to two worlds of experience existing separately and running in parallel, but it seems a gratuitous complication to assume so and thus take on the burden of explaining the connections that produce the parallelism. A more comfortable assumption, once one has considered the effects of brain stimulation, is that the activation of neurons in the visual cortex *is* the subjective experience of sight, that activity in the limbic system *is* the experience of emotional feeling, that activation of pattern matching gates in the prefrontal cortex *is* the feeling of "Aha, I've got it!"

The principal result of such a position for the present discussion is that the apparent parallelism between mind and brain represents a real relationship. Therefore, we may profitably ask what aspects of brain operation are identified with a particular mental experience (or *vice versa*), and what properties are required of a brain-like organization of matter in order for it to demonstrate mind-like properties. The first part of this undertaking, given the basic assumptions, is not so difficult. It is easy to point to good correspondence between subjective experience and physical brain operation. Indeed, it has often seemed to me that the brain is much more likely to be understood in terms of the *constructs of subjective experience* which we use to describe the mind than in terms of the constructs of behavior theory which we use to describe physical actions of the organism. The anatomical divisions of the brain seem to have a closer and more nearly one-for-one correspondence with the former than with the latter. This, of course, is an expected result if the brain and the mind are really one.

The remaining two sections explore these two issues: the relation between functional brain systems and mental experience, and the necessary physical substrates for mind-like events.

III. THE CONTENT OF CONSCIOUSNESS

The subject of consciousness may be divided into two questions. The first concerns the mechanism of consciousness *per se*, and the second concerns the mechanisms that determine the content of consciousness at any one time. It is commonplace in our experience that consciousness in ourselves, and seemingly in others, has at least two distinct dimensions. It comes and goes, we can have it or not, and we can have it in varying degrees from intense, alert attention and concentration to drowsy relaxation verging on sleep. Additionally and independently of the level or degree of consciousness, our conscious activity has content. Our minds are filled with mental events of all sorts. Some of these may serve as stimuli to change our level of consciousness, as when I suddenly remember a forgotten appointment, but consciousness and the mental phenomena which occupy it seem to vary independently.

First, I shall address the existence of the state of awareness or consciousness itself. No one can say that "I am not," or deny the experience of being conscious. Consciousness has as a requirement the functioning of the brain, and probably the forebrain. Activity which seriously interferes with the brain function, such as a blow on the head or reduction of oxygen or blood supplies, eliminates conscious awareness. Damage to the brainstem reticular formation also eliminates consciousness in the absence of any damage to the forebrain. I have discussed the mechanisms by which the reticular activating system provides a sort of enabling gate function to the forebrain as a whole, thus controlling its operation. It would thus appear that the operation of the forebrain is the necessary substrate for consciousness and that its occurrence is regulated by lower brainstem centers.

I suggested at the time we discussed the reticular formation that its two forms of activation had implications for two types of conscious process. We know that the activity in the forebrain produced by reticular activity can be of a "tonic" type, which is a continuous, maintained state, or it can be of a "phasic" type which is a stimulus-bound state of activity. By stimulus-bound, I mean that the activation is dependent upon and persists only for the duration of some external or internal signal which drives the activating system. Such phasic action is the normal mode in lower organisms and is also seen in some cases of brain damage in humans and mammals, the subject may be momentarily aroused by an environmental stimulus, but in the absence of strong stimulation, sinks back into a state of unconsciousness. This stimulus-bound mode is characteristic of reptiles and lower vertebrates. In the mammals we see maintained states of activity (probably made possible by a warm-blooded energy economy). This maintained or tonic state of arousal seems to result from the ability of the cortex to maintain activity in the reticular activating system, hence forming a closed feedback loop.

Given what is known about the relationship between reported states of conscious awarness and arousal and the physical measures of forebrain activity such as the EEG (electroencephalogram), it seems reasonable to identify the existence of a state of consciousness as a primitive, mental experience with the operation of this reticulo-cortical circuitry, and a state of maintained conscious activity with the physical activity around the feedback loop. On the other hand, the various activities going on in the forebrain under this regulatory control seem to determine the *content* of the conscious experience. This is not to say that activity around a feedback loop is sufficient for a state of consciousness; obviously something about the activity in the forebrain at one end of the loop is required. The feedback loop itself accounts for the maintenance of consciousness as a self-regulating state of activity in the system. The forebrain's activity, which seems to correspond to a minimal state of subjective consciousness, is a state of preparedness to respond to incoming data.

This state of readiness to respond, without any differentiation of the activity of the higher systems, apparently corresponds to a conscious but "empty" mind. This is a state which we almost never achieve. It is virtually impossible to be conscious without attending to or thinking about *something*. The nature of the something seems to depend on what particular patterns of processing activity are set up within the forebrain. If I produce activity in your sensory systems, then perceptual experience enters your consciousness. If I produce activity in your limbic system (eg: the amygdala), then you experience an emotional feeling subjectively as a part of your conscious state. Similarly, activity in the logical association areas of the parietal lobe seems to fill consciousness with an awareness of certain relationships, and lesion of this area prevents consciousness from containing such entities. Lesion of the speech areas denies consciousness access to verbal representation. On the output side, note the description of the subjective experience of the Parkinsonian patient: when his basal ganglia are inactivated, he experiences a "paralysis of the will," which is his way of expressing the loss of the ability to perform voluntary motor activities. Thus, even the experience of the will is apparently only one of the potential items of content of the conscious state which can arise from differential activities within the underlying state of preparedness of the forebrain.

The particular content that consciousness will have is determined in several ways. Clearly, stimuli from the external world can force themselves upon us. On the physical level, this is simply the ability of strong or novel stimuli to produce activity in the senses, even when these systems are "tuned" for other things or set to a high-level of threshold gating. On the other hand, much of the content of consciousness in human brains is internally directed and not stimulus dependent. I have already examined the role of the brain's goal-directing system in overseeing the activity of all parts of the apparatus in the service of the primary goal structure, and I particularly noted the role of the prefrontal lobes in directing those activities of the higher processing levels whose operations constitute much of the normal content of human consciousness. I pointed out then that this directing function in operation is apparently what we experience subjectively as attention.

As the functioning of the forebrain diminishes, we seemingly impoverish the contents of consciousness. As more and more functional physical structures are disabled, the contents of consciousness become increasingly restricted. It is probably in this way that we can best try to understand what the consciousness of other animals might be like. Mammals (eg: a dog) probably share with us most of the essential perceptual apparatus, some of the logical apparatus, and virtually none of the verbal symbolic apparatus. A dog's conscious experience is probably similar to what we experience when we are attending to things in the external world without thinking about them verbally, anticipating consequences for the future, or categorizing them in any complex way. Its emotional experiences are probably similar to ours, although much more tied to immediate stimuli. It is probably a world of the present. A dog may have conditioned emotional or behavioral responses which anticipate future consequences on the basis of some present stimulus, but it probably does not have the apparatus to conceptualize a future and manipulate it symbolically. A dog probably does have a limited self-concept, if that can be identified with a representation of one's physical being in the perceptual world model, but it too is probably limited to immediate aspects.

How far down the scale might we go before we would not venture to grant any sort of conscious experience to an organism? One approach is to look at the relation between anatomical level and contribution to consciousness in our own brains. If figure 1.5 is examined again, it will be apparent that we could draw a vertical line somewhere around the middle of the figure and propose that activities occurring above the line enter into consciousness, while activities occurring below the line do not. Thus, clearly the seven items of short-term verbal memory at the top levels are usually part of the content of our consciousness, and clearly the individual modifications of muscular output involved in correcting for a shift in center of gravity are not. Where does the line get drawn?

On the input side, higher levels of categorization and abstraction which enter into our world model are usually found in consciousness, while the detailed analysis of the features and even objects which give rise to them are not. For example, think of the demonstration with the folded paper. All of the computing activity that was required to produce that phenomenon went on in some sort of preprocessing stage which did not contribute to the content of consciousness. You were not aware of making the necessary calculations to produce that movement. You were only aware of the output of that process, which entered into the contents of your consciousness as a finished perception of a moving object.

It appears that we can, via the attention function, move this cutoff level around somewhat. There is a fuzzy area around the cutoff line where we can either attend to or ignore detailed operations. On the average, I would venture that the line should be drawn across the input system's column somewhere around the primary sensory cortex at its lower end of range and at the tertiary levels of cortical processing at its upper end of range.

On the output side, we experience the planning of actions, but are not aware of the details of their execution. Here, too, the line has a range which is alterable by attention. We can take notice of and assume control of fairly detailed sequences of operations, although we cannot consciously experience the processing of centers that perform routine reflex operations. At the other extreme, we can concern ourselves only with general goals of activity and leave their execution to well-learned subroutines in processors which we cut out of the realm of conscious experience. We can easily drive while unconscious of anything other than our general plan of travel, or we can carefully attend to squeezing into a narrow space without scratching the car, while planning every move. On this output side, I would guess that the range of the line falls across the frontal lobe, with the primary motor cortex at its bottom end and the forward portion of the frontal lobe, located just behind the prefrontal cortex, at its high end.

Thus, on the input and output sides, I would venture that our consciousness seems to rely on cortical mechanisms. There may be some subcortical participation of rudimentary experiences, but they seem experimentally to be minor. In the central column, representing the goal-directing function, I would guess that the line must be drawn lower down. Here I would guess that the subcortical systems of the telencephalon participate in the content of conscious experience. Stimulation of the limbic system will result in displays of emotional behavior and reports of subjective emotional experience. Stimulation of other centers lower down will result in displays of emotional behavior, but will not produce reports of subjective feeling. (However, there is some reason to believe that conscious experience of the emotions generated by limbic stimulation may in fact require prefrontal mechanisms.) On this basis, we might guess that nervous systems possessing no telencephalon would be simple machines, entirely analogous to the machines of today in the way that we think about their potential for conscious experience. Brains above this level are no less machine-like in terms of their detailed circuitry, but they seem to possess some degree of complexity of organization, or other property, which is accompanied by conscious experience. The content of that experience seems to reflect the particulars of what is going on in that level's current operations.

In summary, man postulates the existence of a state of operation of the forebrain, which constitutes consciousness, and a varied set of possible activities of the structures of the forebrain which constitute the content of conscious experience. Together they constitute respectively the substance and the form of mind.

IV. THE ROBOT MIND

A machine built along the lines of our present computing machinery would be unlikely to have conscious experience on the basis of any similarity to the brain. Considered from the standpoint of hardware, it would rank very far below our hypothetical cutoff line for the location of conscious activity in the brain. Could a program be written of sufficient complexity to create mind in such a machine? Obviously we could write one which would cause it to behave intelligently and talk sensibly. If we didn't care how long it took to run, we might even do it on the same sorts of serial processing computers which are used today, although the difficulties might prove insurmountable in practice.

However, this begs the question; simply behaving in an intelligent fashion is no guarantee of mind. Again, we will never know for certain, but there are interesting aspects to consider. Accepting our article of faith about the unity of brain and mind in biological organisms permits us to ask what aspects of biology we might need to copy to produce a copy of mind. It seems unlikely that the constituents at the elemental level would be important. It is hard to conceive that the nature of mind is such that carbon, hydrogen, and oxygen as opposed to silicon, copper, and gold determine its occurrence or nonoccurrence. It seems more likely that some aspect of the brain's larger scale construction is essential, but which one? Complexity is a possibility. It may be that mind is a property of self-sustaining, self-organizing data-processing systems sufficiently complex to support it. This possibility is probably the one that most people who have considered the issue regard as the principal candidate; one reads of "emergent properties" of sufficiently complex organizations of matter, and the like.

Therefore let us consider this issue of complexity for a moment. There are two possible ways in which a system could be complex. It can be complex in space, as the brain certainly is, or it can be complex in time, as both the brain and the computer are. Complex in space means simply that it has a large number of parts with high connectivity. Complex in time means that its successive states shall be highly varied and dependent upon earlier states in intricate and lawful ways. At least compared to a brain, a computer of today is not complex in space. It can, however, be infinitely complex in time, given sufficient time. If we want to allow complexity as the source of mind, then complexity in time is the crux of the question of whether or not we could program a Turing machine to have a mind.

If we attend for a moment to the subjective experience of consciousness, we notice that almost all of the things which we want to call the content of consciousness are processes in time within the brain. It is difficult to conceive of what a static consciousness would be like; the nature of our mental experience is deeply involved with the occurrence of an ongoing process. Our thoughts, our sensations, our plans, our actions, are all processes in time. It is probably safe to conclude that the content of consciousness is not determined by a single static state of the forebrain, but by our subjective appreciation in time of the temporal pattern of reorganization of the states of the forebrain components. A thought would then be a temporal sequence of spatially ordered events in the brain.

By this reasoning, one might speculate that at least the content of consciousness should be possible to a system sufficiently complex in time. This reflects simply the fact that we can ultimately make a machine run a sufficiently complex program to imitate anything that a brain can do, given enough time for it to operate. However, content of consciousness is not the entire mind. I have spent some time here building the notion of the content of consciousness as something separate from the background state of awareness. It seems at least plausible to me that this background state, the substance out of which mind is formed, may require complexity in space. If we could eliminate the content of consciousness, if we could achieve a completely empty mind, it would seem that it would be very much like the hypothetical *static* state of mind that we just considered. It would be a state of temporally unchanging awareness, perhaps similar to what some religions or systems of meditation attempt to achieve.

Such a basic, undifferentiated state of awareness would not seem to rely (ie: in any complex way) on temporally organized processes. It might well be identified with the static state of "preparedness to process data" that was seen in the forebrain contingent on reticular activation. If such a background state of awareness is the substance of mind, which becomes aware of the content of mind by being modulated by temporally changing states of the brain's components, then it is difficult to see how this background state could be realized in a merely time-complex system. In our experience of consciousness, we are simultaneously aware of the moment-to-moment state of many components of our total conscious experience. The direction afforded by the appreciation of their temporal sequence is an important component of mind, but so is the ability to appreciate the simultaneous relationships of the parts. Perhaps we could program a machine to encode the temporal relations contained in this simultaneity experience, and perhaps it could have a sequential consciousness in which the background state of awareness was organized by the simultaneity code. If so, I think it would be a very different sort of consciousness than the one we know.

I suspect then that spatial complexity may be necessary for the substance of mind, while temporal complexity may suffice for the content or form of mind. If we remove the various parts of the forebrain, we do not so much eliminiate consciousness; we impoverish its content progressively. There does not seem to be one area that "watches" the rest. Within the remaining parts, however, the content runs on as usual in time, save where interactions with the missing parts are required. Thus, we might expect that an extremely impoverished spatial organization, deprived of much simultaneous complexity, would be able to be conscious of only a very restricted operation. In a computer of the present sort, the registers of the central processor can be thought of as being analogous to the extent of simultaneous self-organizing processes in a very restricted forebrain. Obviously the instantaneous content of consciousness in such a machine would be so restricted as to be unworthy of the name, even though it might sequentially carry out all of the processes involved in a brain's operations by virtue of the reprogrammability of this limited structure.

I propose then that a Turing machine type of computer will be able to contain the temporal *content* of mind and therefore imitate mind in its behavior, but it will not in any meaningful degree possess the substance of mind necessary for the appreciation of that content in the sense that we do.

All of this depends on certain assumptions: the identity of mind and brain, the separability of substance and form in mind, and the necessity of complexity for mind at our level. Unfortunately, these assumptions seem in principle untestable. It may be that consciousness resides in a soul which permeates my body, and that when I have thought I was unconscious it was only because I failed to remember the conscious experiences that I had at that time. It may also be that the entire universe was created, together with all of my memories, only five minutes ago. It may also be that I am followed everywhere by a invisible and intangible purple unicorn. My inability to disprove such things is not sufficient cause to believe in them; neither is my inability to prove something necessarily sufficient cause to discard it as an operating hypothesis. If we are to deal with machines that appear to be intelligent and conscious, we need operating hypotheses.

We behave towards one another and towards animals on the basis of similar operating hypotheses. I cannot prove that you have a mind, but I believe it and act on it because 1) you behave as I would in similar situations, at least to a first approximation, and 2) you are physically similar to me, at least with respect to your possession of a brain. We have just seen that brains reason by analogy; ergo, you have a mind. We treat animals in the fashion that we do on the belief that at least the higher animals have consciousness, and, in particular, emotions, but that the content of their minds is more limited than ours. Again, these operating hypotheses are based on assumptions made by analogy on the basis of similarity between their form and behavior and ours.

Turing long ago proposed his famous test for consciousness in a computer, by playing a game. As almost everyone who deals with computers knows, the object of the game was for the computer to deceive a human into believing that it was the human, in a forced choice between it and a real human. Most people concern themselves immediately with game strategies for the computer and the like, without pausing to reflect on the extremely insightful nature of Turing's solution to the problem. He recognizes immediately that we *do* assign consciousness to others on the basis of their behaving as we do. He recognizes that we also use analogy by form, and so he hid both the computer and the human player in other rooms with communication by teletype. This may be why most people instinctively reject his test as an interesting piece of pedantry; they are intuitively rejecting the computer on the form test of analogy. I don't mean that a computer should look like a human, but that I feel it must bear some sort of structural resemblance to one in functionally important ways. That essentially means possessing a brain. We would allow consciousness to an artificially maintained brain in a glass jar if it could be interfaced to a speaker and make sense. Might we not be much more inclined to intuitively accept the results of Turing's test if we knew that the computer employed a brain-like architecture and modeled its operations on ours? It would not be logically necessary to do so, but neither is it logically necessary to allow for mind in other humans.

This will ultimately decide the issue. If our computers or our robots behave in ways that we would expect ourselves to behave, and if we have reason to suspect that their operating principles mimic ours in a major fashion, then we will believe that they have minds and treat them so. It is simply in our nature to do so. (I once caught myself apologizing to a computer in all seriousness because it had behaved in a human-like way in a circumstance where a human would have felt annoyed. It was not even a verbal behavior; think what I would have done if it had discreetly harrumphed!)

It should be obvious from the preceding chapters that when building machines to deal with real-world problems in the general environment, we are going to build them to behave rather like we do, and that we will probably find that the most expeditious way to build them is to incorporate some of the basic design features of our own brains. Under the circumstances it is inevitable that we are going to accept them ultimately into the family of sentient beings. That does not disturb me. Other people, however, may react differently. Not because they really mind the idea of conscious machines *per se*, but because they fear a *different* kind of consciousness, particularly if it is powerful (see the Weizenbaum reference).

It is not hard, for example, to find fears of emotionless, coldly logical devices dealing with humans in an inhuman fashion. I would like to point out two things in this regard. The first is that advanced devices like ourselves have emotional systems for very good reasons. We need them in order to be very powerful systems, and so will our robots. The second is that in a properly designed system, the primary goals must be unalterable, and the emotions are either primary goal systems in operation or a function of secondary goals derived from and checked against the primary goals. (See the George Pugh reference for a discussion of the unalterability of the primary goals in a goal-directed processing system.) Given this, we need only design our machines to have primary goals that are in accord with our own. What if they then build themselves and decide to build themselves differently? This would simply be a violation of the primary goals. What if one of them malfunctions and the primary goal structures lose control of its behavior and it develops a positive emotional response to killing people? Well, when that happens to one of us, we call him crazy and lock him up. You can't make everything foolproof, but that is not now and never has been a valid reason to avoid the unknown. To fear it in a healthy fashion, yes; to proceed with caution, certainly; but not to turn away.

There is more reason to expect that such machines would be our powerful allies rather than our dangerous enemies. If, when we build them, they convince us of their sentience, I can think of few more exciting things that we might ever accomplish. Why are people so fascinated by stories, movies and other fictional forms of sentient machinery? You will have to look into your own reactions for the answer, but I think that behind many of the other reasons, we as a species are beginning to look at our isolation as the only truly intelligent minds on Earth, and we are beginning to feel lonely.

Let me end with these few summary remarks. There are some things that seem definite; among them is the great likelihood that we can find a variety of applications for machines that have brain-like capabilities. There are other things that seem very likely; among these is the probability that we can usefully employ the actual structure and function of the brain as a model for such machines in many respects. There are many things that are unclear; among these is the possibility that such machines might have a subjective conscious experience such as ours. I think I would want to talk to one to decide, but I might want to treat it kindly anyway, just in case. Finally, there is the question of how soon. Can we build one in a few years? A few decades? At all? I don't know. In large part, that is up to you. I hope I have been able to give you some useful models to consider.

BIBLIOGRAPHY

The following section contains an annotated bibliography of suggested readings for those wishing to pursue any of the topics discussed in greater detail. Only a few of the works cited are direct references from the text (although many of them have been of great value to me in locating or synthesizing this material). Rather, they represent a selection of textbooks or review articles which will be of value for building a vocabulary and a familiarity with the literature in various areas. In many cases these books and articles contain extensive references to the primary research literature which may be attacked after the reader has gained sufficient fluency in the field.

I have arranged these sources by major topic (some will appear in more than one place), and I have attempted to classify them according to level of technicality. In general, a good plan would be to begin with a good general textbook in physiological psychology, such as Carlson, and then to proceed to the review articles in the areas of interest. The reference texts in neuroanatomy and neurophysiology have been selected to provide excellent brief treatments suitable for looking up terms and concepts. More extensive texts are available in these areas.

In writing this book, I have had to pick and choose among competing theoretical interpretations of the data. I have attempted to correct any resulting imbalance here by including references to points of view which I may have neglected in the text.

I. General Reference Works

These volumes include several light, introductory works that provide overviews of the brain from the perspectives of various authors.

Blakemore, C. *Mechanics of the Mind.* Cambridge University Press, New York, 1972.

Hubbard, J.I. *The Biological Basis of Mental Activity.* Addison-Wesley Publishing Co., Reading MA, 1975.

Wittrock, M.C. *The Human Brain.* Prentice-Hall, Inc., Englewood Cliffs NJ, 1977.

Wooldridge, D.E. *The Machinery of the Brain.* McGraw-Hill Book Co., New York, 1963.

This book, while an introductory level text, covers only certain selected topics. It is interesting as a control theory and frequency-domain oriented work at the introductory level.

Pribram, K.H. *Languages of the Brain: Experimental Paradoxes and Principles in Neuropsychology.* Prentice-Hall, Inc., Englewood Cliffs NJ, 1971.

Intermediate and Advanced Works

As an intermediate and advanced work this text is a good standard introduction to the brain, emphasizing the behavioral level of organization.

Carlson, N.R. *Physiology of Behavior.* Allyn and Bacon, Inc., Boston MA, 1977.

The following texts will provide good sources of reference for the background material of behavioral level brain operation (ie: for the neuroanatomical connections of the brain, the physiology of its neurons, and the pharmacology of their action). These are good sources for acquiring the vocabulary needed to read the more advanced works on behavioral level brain functions. The Herrick and Netsky references are included because the study of the evolutionary course of brain development is often of great value in grasping the organization of more advanced brains. The reader desiring a single introductory text in these several areas will find the Gardner reference suitable. The Shepherd work is excellent and a must for anyone seriously studying the detailed operation of the brain at the level of local circuitry, but it is rather advanced and should be attempted only after some preliminary reading in anatomy and physiology of the brain. The Lindsay and Norman text is an introductory reference in basic psychology with an emphasis on cognitive processes. Individual chapters from it are referenced in other sections, but the work as a whole is worth reading for those not familiar with the scientific approach to human thought processes.

Chow, K.L. and A.L. Leiman. "The Structural and Functional Organization of the Neocortex." *Neurosciences Research Program Bulletin.* Volume 8, Number 2, 1970.

Cooper, J.R., F.E. Bloom and R.H. Roth. *The Biochemical Basis of Neuropharmacology.* Oxford University Press, New York, London, Toronto, 1970.

Crosby, E.C., T. Humphrey and A. Lauer. *Correlative Anatomy of the Nervous System.* The Macmillan Co., New York, 1962.

Gardner, E. *Fundamentals of Neurology: A Psychophysiological Approach.* Sixth Edition. W.B. Saunders Co., Philadelphia PA, 1975.

Herrick, C.J. *The Brain of the Tiger Salamander: Ambystoma Tigrinum.* The University of Chicago Press, Chicago, and London, 1948.

Lindsay, P.H. and D.A. Norman. *Human Information Processing: An Introduction to Psychology,* Second Edition. Academic Press, New York, San Francisco, London, 1977.

Sarnat, H.B. and M.G. Netsky. *Evolution of the Nervous System.* Oxford University Press, New York, London, Toronto, 1974.

Schmidt, R.F. *Fundamentals of Neurophysiology.* Springer-Verlag, New York, Heidelberg, Berlin, 1978.

Shepherd, G.M. *The Synaptic Organization of the Brain.* Second Edition. Oxford University Press, New York, 1979.

Stevens, C.F. *Neurophysiology: A Primer.* John Wiley and Sons, Inc., New York, 1966.

The three volumes of the Neurosciences Study Program (a fourth is due this year) provide one of the best sources of up-to-date and authoritative reviews of current thinking in most areas of brain science. In the remainder of the bibliography, many individual references are made to various works in these volumes. The reader attempting these volumes should be familiar with basic neuroanatomy, neurophysiology, and physiological psychology as offered in some of the textbooks above, and should definitely read the relevant sections of these volumes before attacking the research journals themselves.

Quarton, G.C., T. Melnechuk, and F.O. Schmitt. *The Neurosciences: A Study Program.* The Rockefeller University Press, New York, 1967.

Schmitt, F.O. *The Neurosciences: Second Study Program.* The Rockefeller University Press, New York, 1970.

Schmitt, F.O. and F.G. Worden. *The Neurosciences: Third Study Program.* The MIT Press, Cambridge MA, and London, 1974.

The following articles appear in the Neurosciences Study Program volumes and are of general interest in regard to the structure and operating principles of the nervous system.

Bullock, T.H. "Operations Analysis of Nervous Functions." F.O. Schmitt (ed), *The Neurosciences: Second Study Program.* The Rockefeller University Press, New York, 1970.

Gerstein, G.L. "Functional Association of Neurons: Detection and Interpretation." F.O. Schmitt (ed), *The Neurosciences: Second Study Program.* The Rockefeller University Press, New York, 1970.

MacLean, P.D. "The Triune Brain, Emotion, and Scientific Bias." F.O. Schmitt (ed), *The Neurosciences: Second Study Program.* The Rockefeller University Press, New York, 1970.

Nauta, W.J.H. and H.J. Karten. "A General Profile of the Vertebrate Brain, with Sidelights on the Ancestry of Cerebral Cortex." F.O. Schmitt (ed), *The Neurosciences: Second Study Program.* The Rockefeller University Press, New York, 1970.

Perkel, D.H. "Spike Trains as Carriers of Information." F.O. Schmitt (ed), *The Neurosciences: Second Study Program.* The Rockefeller University Press, New York, 1970.

Scheibel, M.E. and A.B. Scheibel. "Elementary Processes in Selected Thalamic and Cortical Subsystems—The Structural Substrates." F.O. Schmitt (ed), *The Neurosciences: Second Study Program.* The Rockefeller University Press, New York, 1970.

Segundo, J.P. "Communication and Coding by Nerve Cells." F.O. Schmitt (ed), *The Neurosciences: Second Study Program.* The Rockefeller University Press, New York, 1970.

Wilson, D.M. "Neural Operations in Arthropod Ganglia." F.O. Schmitt (ed), *The Neurosciences: Second Study Program.* The Rockefeller University Press, New York, 1970.

II. Sensory and Perceptual Processes

As introductory materials, the two readings below are highly recommended, the first for a very readable introduction to the area and the second for some elementary yet profound insights into the nature of the data processing problems of the brain.

Gregory, R.L. *Eye and Brain: The Psychology of Seeing.* McGraw-Hill Book Co., New York, 1973.

Ditchburn, R.W. "Sight and Survival." Address reprinted in: *The Advancement of Science,* 26, 1969-70.

This volume is a collection of articles on sensory and perceptual topics reprinted from the Scientific American. *The Hubel article in particular should be perused by the reader without the background to handle more technical references to feature detector neurons. Many other excellent articles are here as well.*

Perception: Mechanisms and Models. W.H. Freeman & Co., San Francisco CA, 1972.

As intermediate and advanced material, a number of the textbooks and monographs listed below are suitable for a reader who has mastered the basic vocabulary of brain science and wants to pursue particular topics in the area.

Dodwell, P.C. *Visual Pattern Recognition.* Holt, Rinehart and Winston, Inc., 1970.

Leibovic, K.N. *Information Processing in the Nervous System.* Springer-Verlag, New York, Heidelberg, Berlin, 1969.

Poppel, E., R. Held, and J.E. Dowling. *Neuronal Mechanisms in Visual Perception.* The MIT Press, Cambridge MA, 1977.

Schmidt, R.F. *Fundamentals of Sensory Physiology.* Springer-Verlag, New York, Heidelberg, Berlin, 1978.

Uttal, W.R. *Sensory Coding: Selected Readings.* Little, Brown and Co., Boston, 1972.

Uttal, W.R. *The Psychobiology of Sensory Coding.* Harper & Row, New York, Evanston, San Francisco, London, 1973.

Wathen-Dunn, W. *Models for the Perception of Speech and Visual Form.* The MIT Press, Cambridge MA, and London, 1964.

(The above work is a collection of numerous hypothetical models for both visual and auditory perception. Many of these are delightfully intriguing as food

for thought on artificial systems, although most of them are improbable as real brain mechanisms. Among those which I enjoyed are the following.)

Blum, H. "A Transformation for Extracting New Descriptors of Shape." W. Wathen-Dunn (ed), *Models for the Perception of Speech and Visual Form.* The MIT Press, Cambridge MA, and London, 1964.

Clowes, M.B. "An Hierarchical Model of Form Perception." W. Wathen-Dunn (ed), *Models for the Perception of Speech and Visual Form.* The MIT Press, Cambridge MA, and London, 1964.

Deutsch, J.A. and L. Traister. "Lateral Inhibition as a Mechanism of Shape Recognition." W. Wathen-Dunn (ed), *Models for the Perception of Speech and Visual Form.* The MIT Press, Cambridge MA, and London, 1964.

Freeman, H. "On the Classification of Line-Drawing Data." W. Wathen-Dunn (ed), *Models for the Perception of Speech and Visual Form.* The MIT Press, Cambridge MA, and London, 1964.

Leibovic, K.N. "Geometrical Probability in Visual Perception." W. Wathen-Dunn (ed), *Models for the Perception of Speech and Visual Form.* The MIT Press, Cambridge MA, and London, 1964.

Marill, T. "A Model for Visual Scene Analysis." W. Wathen-Dunn (ed), *Models for the Perception of Speech and Visual Form.* The MIT Press, Cambridge MA, and London, 1964.

Tenery, G. "Information Flow in a Bionics Image Recognition System." W. Wathen-Dunn (ed), *Models for the Perception of Speech and Visual Form.* The MIT Press, Cambridge MA, and London, 1964.

The following articles are theoretical review works appearing in the Neurosciences Study Program series. Many of these are excellent sources of ideas on artificial systems approaches, as well as good discussions of current ideas about natural information processing, but they should be approached after preliminary study of brain mechanisms.

Bishop, P.O. "Beginning of Form Vision and Binocular Depth Discrimination in Cortex." F.O. Schmitt (ed), *The Neurosciences: Second Study Program.* The Rockefeller University Press, New York, 1970.

Blakemore, C. "Developmental Factors in the Formation of Feature Extracting Neurons." F.O. Schmitt and F.G. Worden (eds), *The Neurosciences: Third Study Program.* The MIT Press, Cambridge MA, and London, 1974.

Bullock, T.H. "Introduction: Signals and Neuronal Coding." G.C. Quarton, T. Melnechuk, and F.O. Schmitt (eds), *The Neurosciences: A Study Program.* The Rockefeller University Press, New York, 1967.

Campbell, F.W. "The Transmission of Spatial Information through the Visual System." F.O. Schmitt and F.G. Worden (eds), *The Neurosciences: Third Study Program.* The MIT Press, Cambridge MA, and London, 1974.

Creutzfeldt, O.D. "Some Principles of Synaptic Organization in the Visual System." F.O. Schmitt (ed), *The Neurosciences: Second Study Program.* The Rockefeller University Press, New York, 1970.

Erickson, R.P. "Parallel Population Neural Coding in Feature Extraction." F.O. Schmitt and F.G.

Worden (eds), *The Neurosciences: Third Study Program.* The MIT Press, Cambridge MA, and London, 1974.

Gross, C.G., D.B. Bender, and C.E. Rocha-Miranda. "Infero-temporal Cortex: A Single-Unit Analysis." F.O. Schmitt and F.G. Worden (eds), *The Neurosciences: Third Study Program.* The MIT Press, Cambridge MA, and London, 1974.

Held, R. "Two Modes of Processing Spatially Distributed Visual Stimulation." F.O. Schmitt (ed), *The Neurosciences: Second Study Program.* The Rockefeller University Press, New York, 1970.

Jones, E.G. "The Anatomy of Extrageniculostriate Visual Mechanisms." F.O. Schmitt and F.G. Worden (eds), *The Neurosciences: Third Study Program.* The MIT Press, Cambridge MA, and London, 1974.

MacKay, D.M. "Perception and Brain Function." F.O. Schmitt (ed), *The Neurosciences: Second Study Program.* The Rockefeller University Press, New York, 1970.

Mountcastle, V.B. "The Problem of Sensing and the Neural Coding of Sensory Events." G.C. Quarton, T. Melnechuk, and F.O. Schmitt (eds), *The Neurosciences: A Study Program.* The Rockefeller University Press, New York, 1967.

Pollen, D.A. and J.H. Taylor. "The Striate Cortex and the Spatial Analysis of Visual Space." F.O. Schmitt and F.G. Worden (eds), *The Neurosciences: Third Study Program.* The MIT Press, Cambridge MA, and London, 1974.

Pribram, K.H. "How Is It That Sensing So Much We Can Do So Little?" F.O. Schmitt and F.G. Worden (eds), *The Neurosciences: Third Study Program.* The MIT Press, Cambridge MA, and London, 1974.

Stein, R.B. "The Role of Spike Trains in Transmitting and Distorting Sensory Signals." F.O. Schmitt (ed), *The Neurosciences: Second Study Program.* The Rockefeller University Press, New York, 1970.

Weiskrantz, L. "The Interaction between Occipital and Temporal Cortex in Vision: An Overview." F.O. Schmitt and F.G. Worden (eds), *The Neurosciences: Third Study Program.* The MIT Press, Cambridge MA, and London, 1974.

Werner, G. "Neural Information Processing with Stimulus Feature Extractors." F.O. Schmitt and F.G. Worden (eds), *The Neurosciences: Third Study Program.* The MIT Press, Cambridge MA, and London, 1974.

Wilson, J.P. "Psychoacoustical and Neurophysiological Aspects of Auditory Pattern Recognition." F.O. Schmitt and F.G. Worden (eds), *The Neurosciences: Third Study Program.* The MIT Press, Cambridge MA, and London, 1974.

Wright, M.J. and H. Ikeda. "The Processing of Spatial and Temporal Information in the Visual System." F.O. Schmitt and F.G. Worden (eds), *The Neurosciences: Third Study Program.* The MIT Press, Cambridge MA, and London 1974.

III. Motor Processes

Here is a good introductory article on the complex relations of the neurons of the cerebellum (which is more varied than that presented in the simplified model of the present text).

Llinas, R.R. "The Cortex of the Cerebellum." *Scientific American,* January, 1975.

These two excellent references from BYTE, The Small Systems Journal, contain material pertinent to many topics, but in particular they relate strongly to the problems of modeling the brain's motor functions.

Albus, J. "A Model of the Brain for Robot Control." BYTE, The Small Systems Journal. Volume 4, June thru September, 1979.

Powers, W. "The Nature of Robots." BYTE, The Small Systems Journal. Volume 4, June thru September, 1979.

This volume contains a number of articles on theoretical issues in motor organization of the brain at a more advanced level.

Stelmach, G.E. *Motor Control: Issues and Trends.* Academic Press, New York, 1976.

The following readings from the Neurosciences Study Programs are particularly recommended. The position on the organization of the motor system followed in the present volume is essentially that outlined in the Kornhuber article.

DeLong, M.R. "Motor Functions of the Basal Ganglia: Single-Unit Activity during Movement." F.O. Schmitt and F.G. Worden (eds), *The Neurosciences: Third Study Program.* The MIT Press, Cambridge MA, and London, 1974.

Evarts, E.V. "Sensorimoter Cortex Activity Associated with Movements Triggered by Visual as Compared to Somesthetic Inputs." F.O. Schmitt and F.G. Worden (eds), *The Neurosciences: Third Study Program.* The MIT Press, Cambridge MA, and London, 1974.

Ito, M. "The Control Mechanisms of Cerebellar Motor Systems." F.O. Schmitt and F.G. Worden (eds), *The Neurosciences: Third Study Program.* The MIT Press, Cambridge MA, and London, 1974.

Kornhuber, H.H. "Cerebral Cortex, Cerebellem, and Basal Ganglia: An Introduction to Their Motor Functions." F.O. Schmitt and F.G. Worden (eds), *The Neurosciences: Third Study Program.* The MIT Press, Cambridge MA, and London, 1974.

IV. Higher Functions

For an introduction to the functional, non-physiological aspects of human reasoning and problem solving, the relevant chapters of the following introductory psychology text are quite good.

Lindsay, P.H. and D.A. Norman. *Human Information Processing: An Introduction to Psychology,* Second Edition. Academic Press, New York, San Francisco, London, 1977.

A more advanced text in the psychology of cognition which covers the material in a series of chapters by various workers in the area:

Norman, D.A. and D.E. Rumelhart. *Explorations in Cognition.* W.H. Freeman & Co., San Francisco CA, 1975.

Here is a theoretically unified, although not universally accepted, approach to the psychology of human cognition which I personally enjoyed

Neisser, U. *Cognition and Reality: Principles and Implications of Cognitive Psychology.* W.H. Freeman & Co., San Francisco CA, 1976.

The following works all contain much of interest in regard to the physiology of human higher intellectual processes. The Luria reference is seminal and should not be missed. Most of these will require some background in basic brain function to read easily.

Bechtereva, N.P. *The Neurophysiological Aspects of Human Mental Activity.* Second Edition. Oxford University Press, New York, 1978.

Heilman, K.M. and E. Valenstein. *Clinical Neuropsychology.* Oxford University Press, New York, 1979.

Lenneberg, E.H. "Language and Brain: Developmental Aspects." *Neurosciences Research Program Bulletin.* Volume 12, Number 4, 1974.

Luria, A.R. *Higher Cortical Functions in Man.* Basic Books, Inc., New York, 1966.

The following article in the Neurosciences Study Program by Lenneberg contains some interesting thoughts on the physiology of language.

Lenneberg, E.H. "Brain Correlates of Language." F.O. Schmitt (ed), *The Neurosciences: Second Study Program.* The Rockefeller University Press, New York, 1970.

In addition to the references in this section, material relevant to "higher processes" will also be found in some of the references in the sections on hemispheric differences and lateralization and in the section on goal-directing functions, especially in those dealing with the frontal lobes.

V. Hemispheric Specialization

Some introduction to this complex topic at an elementary level may be found in the two Scientific American *articles below. They are particularly valuable for their excellent illustrations which clarify the cortical anatomy involved.*

Geschwind, N. "Specializations of the Human Brain." *Scientific American,* September 1979, pages 180 thru 199.

Lassen, N., D.H. Ingvar and E. Sinkhoj. "Patterns of Brain Function and Blood Flow." *Scientific American,* October 1978, pages 62 thru 71.

In the Neurosciences Study Program series, the entire group of articles in the third volume make an excellent set of complementary works. They are available reprinted together in paperback from the MIT Press as well.

Berlucchi, G. "Cerebral Dominance and Interhemispheric Communication in Normal Man." F.O. Schmitt and F.G. Worden (eds), *The Neurosciences: Third Study Program.* The MIT Press, Cambridge MA, and London, 1974.

Broadbent, D.E. "Division of Function and Integration of Behavior." F.O. Schmitt and F.G. Worden (eds), *The Neurosciences: Third Study Program.* The MIT Press, Cambridge MA, and London, 1974.

Cuenod, M. "Commissural Pathways in Interhemispheric Transfer of Visual Information in the Pigeon." F.O. Schmitt and F.G. Worden (eds), *The Neurosciences: Third Study Program.* The MIT Press, Cambridge MA, and London, 1974.

Darwin, C.J. "Ear Differences and Hemispheric Specialization." F.O. Schmitt and F.G. Worden (eds), *The Neurosciences: Third Study Program.* The MIT Press, Cambridge MA, and London, 1974.

Lieberman, A.M. "The Specialization of the Language Hemisphere." F.O. Schmitt and F.G. Worden (eds), *The Neurosciences: Third Study Pro-*

gram. The MIT Press, Cambridge MA, and London, 1974.

Milner, B. "Hemispheric Specialization: Scope and Limits." F.O. Schmitt and F.G. Worden (eds), *The Neurosciences: Third Study Program.* The MIT Press, Cambridge MA, and London, 1974.

Sperry, R.W. "Lateral Specialization in the Surgically Separated Hemispheres." F.O. Schmitt and F.G. Worden (eds), *The Neurosciences: Third Study Program.* The MIT Press, Cambridge MA, and London, 1974.

Teuber, H. "Why Two Brains?" F.O. Schmitt and F.G. Worden (eds), *The Neurosciences: Third Study Program.* The MIT Press, Cambridge MA, and London, 1974.

This is a collection of articles dealing with the curious syndromes which can develop with unilateral damage. These are an important source of information in attempting to understand hemispheric specialization.

Weinstein, E.A. and R.P. Friedland. *Hemi-Inattention and Hemisphere Specialization.* Raven Press, New York, 1977.

The following are good review articles. The Searleman reference is of particular interest in that it focuses on the question of just how much "left hemisphere function" is really seen in the right hemisphere.

Galaburda, A.M., M. LeMay, T.L. Kemper, and N. Geschwind. "Right-Left Asymmetries in the Brain." *Science,* Volume 199, February 24, 1978.

Searleman, A. "A Review of Right Hemisphere Linguistic Capabilities." *Psychological Bulletin,* Volume 84, Number 3, 1977, pages 503 thru 528.

VI. Goal-Directed Processes

The following book is outstanding in many respects. It is unequaled as a broad integration of a vast number of fields, including among others the neurophysiological aspects of its theme. It is unique in its insightful appreciation of the true role of motivational systems (values in his terminology) in directing the behavior of heuristic systems. It is all the more necessary on your reading list because the author is a computer scientist (among other things), and he gives examples of the implementation of goal-directed (value-driven) systems in working computer software that masters otherwise intractable problems. Highly recommended.

Pugh, G.E. *The Biological Origin of Human Values.* Basic Books, Inc., New York, 1977.

The next two works are good treatments of the subjects of motivation and reinforcement from the standpoint of the underlying physiology.

Rolls, E.T. *The Brain and Reward.* Pergamon Press, Inc., Elmsford NY, 1975.

Stellar, E. and J.D. Corbit. "Neural Control of Motivated Behavior." *Neurosciences Research Program Bulletin,* Volume 11, Number 4, 1973.

This monograph is interesting as an approach to the physiology of attention in its role as a goal-directing process.

Mountcastle, V.B. "The World Around Us: Neural Command Functions for Selective Attention." *Neurosciences Research Program Bulletin,* Volume 17, Supplement April, 1976.

In this same vein, the following book is an excellent source of readings on the frontal lobes in general (which I place in this section, but which, of course, also should be examined in relation to higher processes, sensory and perceptual functions, etc). The two chapters listed from it are particularly interesting in the present context.

Luria, A.R. and K. Pribram. *Psychophysiology of the Frontal Lobes.* Academic Press, New York, 1973.

Grueninger, W. and J. Grueninger. "The Primate Frontal Cortex and Allassostasis." A.R. Luria and K. Pribram (eds), *Psychophysiology of the Frontal Lobes.* Academic Press, New York, 1973.

Pribram, K.H. "The Primate Frontal Cortex-Executive of the Brain." A.R. Luria and K. Pribram (eds), *Psychophysiology of the Frontal Lobes.* Academic Press, New York, 1973.

From the Neurosciences Study Programs, the following articles may be read for their relation to goal-directing processes.

Scheibel, M.E. and A.B. Scheibel. "Anatomical Basis of Attention Mechanisms in Vertebrate Brains." G.C. Quarton, T. Melnechuk, and F.O. Schmitt (eds), *The Neurosciences: A Study Program.* The Rockefeller University Press, New York, 1967.

Zanchetti, A. "Subcortical ana Cortical Mechanisms in Arousal and Emotional Behavior." G.C. Quarton, T. Melnechuk, and F.O. Schmitt (eds), *The Neurosciences: A Study Program.* The Rockefeller University Press, New York, 1967.

VII. Memory

As I indicated in the text, the best work on memory processes is functional rather than physiological in nature. For a good introduction to the functional aspects of human verbal memory, I recommend the chapters on this topic from the Lindsay and Norman reference. The Rosenberg and Simon reference is a more advanced work in the functional area which is exceptionally interesting for its treatment of the problem of semantic level encoding and for its presentation of a computer simulation of the functional model presented.

Lindsay, R.H. and D.A. Norman. *Human Information Processing: An Introduction to Psychology.* Second Edition. Academic Press, New York, San Francisco, London, 1977.

Rosenberg, S. and H.A. Simon. "Modeling Semantic Memory: Effects of Presenting Semantic Information in Different Modalities." *Cognitive Psychology*, Volume 9, 1977, pages 293 thru 325.

Pribram, K.H., M. Nuwer, and R.J. Baron. "The Holographic Hypothesis of Memory Structure in Brain Function and Perception." D.H. Krantz, R.C. Atkinson, R.D. Luce, and P. Suppes (eds), *Comtemporary Developments in Mathematical Psychology VII.* W.N. Freeman & Co., San Francisco CA, 1974.

Westlake, P.R. "The Possibilities of Neural Holographic Processes Within the Brain." Kybernetik, Band VII, Heft 4, September, 1970.

The following references are review articles which will provide a good overview of the physiological material on memory. They are all reasonably advanced and should be read after reading the memory chapters in an introductory physiological text such as Carlson.

Chow, K.L. "Effects of Ablation."

G.C. Quarton, T. Melnechuk, and F.O. Schmitt (eds), *The Neurosciences: A Study Program.* The Rockefeller University Press, New York, 1967.

McGaugh, J.L. "Neurobiological Aspects of Memory." *Biological Foundations of Psychiatry.* Raven Press, New York, 1976.

Sperry, R.W. "Split-Brain Approach to Learning Problems." G.C. Quarton, T. Melnechuk, and F.O. Schmitt (eds), *The Neurosciences: A Study Program.* The Rockefeller University Press, New York, 1967.

VIII. Models of Brain Function

The following references contain a variety of approaches to modeling various aspects of brain function. I have grouped them here rather than under the functions to which they apply since they are all to greater or lesser degree speculative rather than descriptive of brain function. Although many represent attempts to model actual brain processes, others are simply attempts to mimic brain function, or to show that a particular function could be performed in a particular manner. Aside from setting off the works dealing with frequency-domain or holographic models together, I have made no attempt to classify this diverse collection of works (I doubt that I could). The reader is invited to browse amongst them where his fancy leads.

A. Frequency-Domain and Holographic Models

Grossberg, S. "Adaptive Pattern Classification and Universal Recoding: I. Parallel Development and Coding of Neural Feature Detectors." *Biological Cybernetics,* Volume 23, 1976, pages 121 thru 134.

Grossberg, S. "Adaptive Pattern Classification and Universal Recoding: II. Feedback, Expectation, Olfaction, Illusions." *Biological Cybernetics,* Volume 23, 1976, pages 187 thru 202.

B. Other Models of Brain Function

Albus, J. "A Model of the Brain for Robot Control." BYTE, The Small Systems Journal. Volume 4, June thru September, 1979.

Deutsch, S. *Models of the Nervous System.* John Wiley & Sons, Inc., 1967.

Powers, W. *Behavior: The Control of Perception.* Aldine Publishing Co., Chicago IL, 1979.

Powers, W. "The Nature of Robots." BYTE, The Small Systems Journal. Volume 4, June thru September, 1979.

Rakic, P. *Local Circuit Neurons.* The MIT Press, Cambridge MA, and London, 1976.

Szentagothai, J. and M.A. Arbib. *Conceptual Models of Neural Organization.* The MIT Press, Cambridge MA, 1972.

Trehub, A. "Neuronal Models for Cognitive Processes: Networks for Learning, Perception and Imagination." *Journal of Theoretical Biology.* Volume 65, 1977, pages 141 thru 169.

Willshaw, D.J. and C. Malsburg. "How Pattern and Neural Connections Can Be Set Up by Self-Organization." Proc. Royal Soc. Lond. B., 194, 1976, pages 431 thru 445.

INDEX

Text set in Garth Medium
by Byte Publications

Edited by Nicholas Bedworth

Design and Production Supervision
by Ellen Klempner

Copy Edited by Sheila S. Hayward

Figure and Table Illustrations by
Tech Art Associates

Printed and bound using 50# Finch Pub.
Web by Halliday Lithograph Corporation,
Arcata Company, North Quincy,
Massachusetts